BK 573.2 B787T
D1203538

THEORIES OF HUMAN EVOLUTION

Peter J. Bowler

Theories of Human Evolution

A Century of Debate, 1844–1944

The Johns Hopkins University Press
Baltimore and London

Printed in the United States of America

The Johns Hopkins University Press
701 West 40th Street
Baltimore, Maryland 21211
The Johns Hopkins Press Ltd., London

The paper used in this publication meets the minimum requirements of American National Standard for Information Sciences—Permanence of Paper for Printed Library Materials, ANSI Z39.48-1984.

Library of Congress Cataloging-in-Publication Data

Bowler, Peter J.
Theories of human evolution.

Bibliography: p.
Includes index.
1. Human evolution. I. Title.
GN281.B65 1986 573.2 86-3029
ISBN 0-8018-3258-6 (alk. paper)

For My Mother and Father

Contents

Tables and Figures

Tables

Figures

Preface

Some years ago I began to prepare the survey of the history of evolutionism that subsequently appeared as *Evolution: The History of an Idea.* It was intended partly as a textbook for university courses in the history of science, and one of its chief aims was to serve as a guide to the vast literature now available in the field. When preparing the sections on human evolution, I found to my surprise that in this particular area comparatively little work seemed to have been done. There were many studies that dealt with the reaction to the idea of human evolution in the original Darwinian debate, but only Loren Eiseley's *Darwin's Century* traced the issue into the late nineteenth century. There was certainly no general survey of early twentieth-century theories of human evolution available, and I was forced to cite books dealing with the discovery of fossil hominids and the Piltdown fraud. Although satisfactory for their own purposes, these books treated theories of human evolution merely as background to the fossil discoveries. I thus felt the need for a survey that would concentrate on the theoretical debates and dovetail more closely with the existing literature on the Darwinian revolution.

I had never worked in the history of anthropology or specialized in human evolution, although, like most students of the Darwinian revolution, I had a general interest in the religious and social implications of evolutionism. Still, I had a strong background in the late nineteenth- and early twentieth-century debates on the evolutionary mechanism, and I decided that it would be possible to use this as a different, but equally sound, foundation upon which to build the badly needed survey of human evolution. In the course of writing the book, I found that a number of anthropologists had already begun to tackle the issues involved, although none to my knowledge is preparing a comprehensive survey. A historian of evolution theory inevitably views the issues in a way rather different from that of a historian of anthropology, and the two approaches may generate a certain amount of tension. I hope the tension will be stimulating, however, and I have indicated in the Introduction how it may bear fruit. If this book helps to encourage a dialogue between the "Darwin industry" in the history of science and the anthropologists interested in the history of their own field, it will have served its purpose.

What I have provided here is only a skeleton outline of the theoretical debates and their most obvious implications. There are many topics left to explore, if only historians will tackle them. As in *The Eclipse of Darwinism,* for instance, I have scarcely mentioned the institutional framework within which the issues were debated. My excuse is that, as with the earlier study, an introductory survey cannot cover everything.

In case anyone is concerned about the dates given in the subtitle of this book, I should explain that 1844 was the year in which Robert Chambers's *Vestiges of the Natural History of Creation* sparked off the first major debate on the implications of evolutionism in the English-speaking world. The final date, 1944, is not meant to be taken too seriously, but does give a general idea of the limit I have placed on this study. It was the year in which George Gaylord Simpson's *Tempo and Mode in Evolution* showed that the Modern Synthesis of Darwinism and genetics could be applied to paleontology. Once this point was recognized in the field of human evolution, an entirely new framework of analysis emerged. Later publications are mentioned only to illustrate the extent of this revolution.

My thanks are due to the following for their advice and encouragement: Matt Cartmill, Donald Grayson, Michael Hammond, Misia Landau, and Frank Spencer. Special mention should also be made of Ian Langham, with whom I corresponded on the Piltdown affair, and whose tragic death is a major loss to the field. Mrs. Susan Ekin of the Science Library, Queen's University of Belfast, located some of the more obscure publications needed for my work through the interlibrary loan service. Travel grants from the Queen's University of Belfast enabled me to visit Cambridge University Library, and the manuscript was completed during leave of absence granted by the university in the Michaelmas term, 1984.

A Note on Surnames

Four British anatomists dealt with at length below have names that generate considerable confusion. They are Sir Wilfrid E. Le Gros Clark, Frederic Wood Jones, Sir Grafton Elliot Smith, and Sir Arthur Smith Woodward. All were widely known under what appear to be "double-barreled" surnames; thus: Le Gros Clark, Wood Jones, Elliot Smith, and Smith Woodward. The names are occasionally seen hyphenated, and in some bibliographies the double name is used as the surname. For Wood Jones, Elliot Smith, and Smith Woodward, it seems clear that the correct surname is the final name only, although the men themselves encouraged the use of the double name—for obvious reasons in the case of Smith and Jones. Le Gros Clark is rather more difficult. Many studies of human evolution still cite the surname as Le Gros Clark, as do some biographical surveys of modern scientists. Nevertheless, in the authoritative *Who Was Who* he is listed as Clark. In the list of members in the *Oxford University Calendar* (he was professor of anatomy at Oxford), he is entered as Le Gros Clark up to the 1950s but as Clark in the 1960s. I thus suspect that, as with the other three, the double name should not be cited as the true surname. In the text below I have used the double name for clarity whenever the character reappears after a substantial absence, but revert to the single name if it recurs constantly in the course of a paragraph or series of paragraphs. All four are cited under their final name in the bibliography, which is consistent with all library cataloging systems.

THEORIES OF HUMAN EVOLUTION

Introduction

The theory of evolution has always been controversial, but the controversies have centered on a number of different issues. By the early nineteenth century it was no longer fashionable to regard the Book of Genesis as a literal account of creation, although that position has enjoyed a revival in modern times. The paleontologists and naturalists of the nineteenth century at first continued to believe that supernatural agencies (if not actual miracles) were responsible for the production of new species in the course of the earth's long history. The idea of evolution threatened the traditional belief that God was somehow personally responsible for designing the succession of living forms. Some early versions of evolutionism tried to get around this problem by presenting the process as the unfolding of a preordained plan. Later in the century many neo-Lamarckian evolutionists argued that nature itself was a purposeful system, designed by its Creator to achieve certain goals. If God was not responsible for designing each species, he had at least created the self-designing forces of nature. More materialistic alternatives, including Darwinism, threatened even this comfortable halfway house. If evolution was a haphazard process driven by the natural selection of random variations in a struggle for existence, it was difficult to believe that a wise and benevolent God had designed it to work in this way. Many scientists—to say nothing of many ordinary people—found this vision too disturbing, and the more optimistic view of nature thus continued to flourish well into the twentieth century.

There could be little doubt that a comprehensive theory of evolution would have to account for the origin of the human race. This raised an entirely new set of problems. Traditionally, our mental and moral faculties were seen as the product of a distinct spiritual agency, the "soul," added to the physical body. Since animals were not supposed to have souls—at least in the Christian interpretation—any theory requiring an animal ancestry for mankind seemed to deny our spiritual status. To many it seemed obvious that such a move would threaten the foundations of morality and hence the stability of the social order. It was possible to argue that the human body had evolved naturally, but not the mind, although by the end of the century most thinkers saw this breach of the principle of continuity as too obviously an artificial compromise. If the fact of human evolution was to be accepted, the mind would also have to be seen as a product of nature. In this case, the image of evolution as a brutal struggle for existence did indeed seem to threaten morality. The only hope of preserving traditional values was, once again, to create a different view of nature. If evolution itself was purposeful, then the appearance of the human mind could be presented as the goal toward which the Creator had always intended the system to move.

The human race had already been included in the general theory of evolution described in J. B. Lamarck's *Zoological Philosophy* of 1809. Charles Lyell, who provided a critical account of Lamarck's theory in his *Principles of Geology,* was hostile largely because of his fears for the status of the human soul.[1] Although Lamarck probably had a little more influence in the early nineteenth century than historians used to think, the majority of naturalists remained hostile to evolutionism, and there was little incentive to explore the more unsettling implications of the idea. The situation changed dramatically in 1844, when Robert Chambers's anonymously published *Vestiges of the Natural History of Creation* discussed at length the philosophical consequences of treating the human species as the last step in a universal organic progression. There was an immediate uproar, with the question of the human soul being paramount to many critics.[2] By the time Darwin published the *Origin of Species* in 1859, no one could be in any doubt as to the implications of applying the theory of evolution to mankind.

Although the fear and disgust of the critics were real enough, Chambers's *Vestiges* actually did little to raise the level of discussion above that which had already been implicit in earlier materialistic accounts of the human mind, for instance, by phrenologists such as Franz Joseph Gall. Chambers simply assumed that evolution was progressive and that the development of organic complexity, particularly in the structure of the brain, would inevitably lead to the creation of a being capable of rational thought. In this sense, Darwin did introduce a new element into the debate, since his theory did not take progress for granted. It was thus necessary for him to ask how and under what circumstances mankind had evolved from an animal ancestor, and whether those circumstances had left a permanent legacy by influencing the structure of the mind. This questioning now opened up the possibility of a new kind of determinism based on the assumption that our instincts have been shaped by evolution and are beyond our control. We shall see, however, that this possibility was not consistently explored for some considerable time. Most of the late nineteenth- and early twentieth-century evolutionists shied away from so depressing a prospect. Instead they chose to retain the idea of progressive evolution, accepting that the human race had generalized social instincts, but insisting that these had been transformed by the inevitable growth in the power of rational thought.

Modern historians of science have produced a vast literature on the Darwinian revolution and its consequences. No survey of the topic is complete without a discussion of the issues raised by the question of *Man's Place in Nature,* the title of T. H. Huxley's classic book.[3] Darwin's views and the complementary or rival interpretations offered by Huxley, Lyell, and A. R. Wallace are explored at length. Yet most of the historians have limited their discussion to the participants in the original Darwinian debate. With the exception of Loren Eiseley's *Darwin's Century,* none of the standard works follows the topic of human evolution through into the twentieth century. What has become known as the "Darwin industry" has limited itself pretty closely to Darwin's own time. Perhaps this focus is understandable. There has been so much to do in building up an adequate picture of the original Darwinian debate that few historians of evolutionism have had the energy to extend their work into later decades. But the lack of interest in the subsequent development of ideas on human evolution may also stem from the tendency of our work on the implications of the Darwinian revolution to concentrate on a different type of issue. It is normal to talk about the evidence that was used to make plausible the fact of human evolution, and about the debates over the moral and religious consequences of giving mankind a place in nature. But in following up Darwin's own views on the actual process of human evolution, the discussion tends to concentrate on those who—like Lyell and Wallace—refused to accept the natural evolution of the mind. The substantial literature on "social Darwinism" also says surprisingly little about theories of human evolution.[4]

It is not the fault of the historians that their accounts say little about the process of human evolution, because there is simply very little on this topic in the writings of Darwin's followers. Darwin himself made a valiant effort to start a serious discussion of the issue in the *Descent of Man,* but his lead was not followed up until much later in the century. Of his supporters, both Lyell and Wallace decided that some form of supernatural intervention must be involved in the creation of the human mind. Huxley's *Man's Place in Nature* was an excellent survey of the evidence for the fact of human evolution, but said virtually nothing about how the process was supposed to have occurred. In Germany, Ernst Haeckel tackled the question of human origins rather more openly, but his views were somewhat confused. If one considers the excitement caused by the *implications* of human evolution, there is surprisingly little in the early Darwinian literature that serves as a lead into the more extensive debates on the *process* of human evolution that began toward the end of the nineteenth century.

The reason is that most of the issues that interested the participants in the Darwinian debates could be tackled without worrying too much about *how* the human race had evolved. The simple fact that it *had* evolved was enough to raise all the religious and moral problems that so concerned the critics of evolutionism. The Darwinists, of course, had to explain how the various human faculties had emerged from animal origins, but we shall see that writers such as Herbert Spencer and G. J. Romanes managed to do so in a surprisingly abstract way, without ever raising questions that could only be answered by a consideration of the actual course of human evolution. Those who discussed the social implications of evolutionism often used the mechanisms of biological evolution as analogies that could be applied directly to society. To oversimplify the situation, a social Darwinist needed to know only that all evolution worked by the survival of the fittest; human evolution was then only a special case of that general rule, and the precise details of how we had emerged from the apes did not matter. The same was true for those who based their social analogies on other mechanisms such as Lamarckism.

Historians of anthropology have devoted a good deal of space to the "evolutionism" of the late nineteenth century.[5] But the cultural evolutionism of E. B. Tylor and L. H. Morgan was a very different thing from the biological theory of the same name. It took for granted a series of cultural stages through which all human societies evolved, although at different speeds. At first there was little incentive to link these stages with steps in the biological evolution of mankind from the apes. Exponents of race theory came closer to establishing such a link, by presenting the "lower" races as earlier stages in the process by which the "white man" had ascended from apelike ancestors. What prevented these ideas from generating any serious discussion on the process of human evolution was the overwhelming faith in the inevitability of progress. If other races or cultures were seen merely as steps in the ascent of a fixed hierarchy of complexity, there was no point in inquiring about the forces that had shaped the direction of progress. The scale itself defined the direction, and was taken for granted by everyone. A simple assumption that more stimulating environments would push some races up the scale faster than others sufficed to answer all relevant questions about causation.

More intensive debates on the course and cause of human evolution began in the 1890s. In part they were a response to the growing quantity of fossil evidence available. Enough specimens of Neanderthal man had now been discovered to suggest that here was a genuine stage in our evolution from the apes, and the archaeologists were able to link the Neanderthals with the stone age culture known as Mousterian. Eugene Dubois's discovery in 1891–92 of *Pithecanthropus erectus* (Java man) seemed to reveal an even earlier stage in the process, and stimulated interest in the question of whether the human brain or the upright posture had evolved first. More important, though, was the collapse of the simple progressionist image of evolution that had dominated so much of mid-nineteenth-century thought. Anthropologists began to realize that neither cultures nor races could be ranked quite so obviously into a linear scale with White-Anglo-Saxon-Protestants at the top. Evolutionists somewhat belatedly realized that neither Darwinism nor any of the alternative theories was consistent with a linear progression toward a fixed goal. We shall see, in fact, that many of the new theories of human evolution succeeded only in preserving the progressionist viewpoint in a more sophisticated guise, but at least scientists recognized that the details of how the human species had evolved could be of considerable interest, and that the fossil evidence now opened the question up to scientific investigation.

Although academic historians of science have tended to ignore this explosion of interest in human evolution, there have been a number of popular accounts of the fossil discoveries by writers such as Ruth Moore, Herbert Wendt, and, more recently, John Reader.[6] The Piltdown fraud has, of course, generated a vast literature of its own, much of which has unfortunately been produced in the style of the detective thriller. The fascination of the fossil hominids is easy to understand—after all, they are the only tangible evidence of our origins. Yet it is obvious from the differing interpretations offered by scientists at the time that the fossils had meaning only to the extent that they could be fitted into theories of how human evolution occurred. Since the fossils were few, and consisted of fragments whose significance was often a matter of opinion, a wide range of debate over theoretical issues was still possible. Whatever the potential interest of the actual discoveries, it seems obvious that a comprehensive study of how understanding of human evolution has developed must focus on the theories, not on the fossils. The theoretical issues are far too important to be dismissed as mere background to the debates centered on the fossils.

More directly relevant to a study of theories of human evolution are the efforts made by physical anthropologists to uncover the history of their own discipline. Much of this material is published in the anthropological literature, and seems to have been largely ignored by the historians who contribute to the Darwin industry. C. Loring Brace's revival of the "Neanderthal phase of man" theory in 1964 contained a lengthy discussion of the theory's historical background, in which Brace tried to account for its unpopularity.[7] Brace has written a number of more recent historical articles, including a contribution to the excellent history of American physical anthropology edited by Frank Spencer.[8] As well as contributing to the same volume, Matt Cartmill has written a wide-ranging account of changing interpretations of the relationship between mankind and nature.[9] Perhaps the most controversial of these modern contributions are Misia Landau's efforts to explore the mythical character of theories of human evolution and the narrative structure of the stories they tell about our origins.[10]

Historians of science are often suspicious of historical studies written by working scientists. In some cases this suspicion is justified: the potted histories that appear as the introductory chapters of many science textbooks are notoriously unreliable because their only purpose is to make current theories seem the logical outcome of all earlier research. As T. S. Kuhn pointed out, the textbook and its historical introduction are means by which the current paradigm of scientific thinking seeks to legitimize itself.[11] The same tendency may sometimes be found in the anthropologists' writing on the history of theories of human evolution. Brace's account of the "hominid catastrophism" that had blocked acceptance of the Neanderthals as human ancestors sought to identify that theory with a fundamentally nonevolutionary viewpoint. Some years later Brace was, in fact, accused by Stephen F. Holtzman of indulging in what historians of anthropology call "presentism," the use of history to justify presently held positions.[12] Brace responded vigorously, although in chapter 4 I have tended to side with Holtzman. Nevertheless, I argue that historians of science are in no position to sneer at the work done by anthropologists, particularly in areas in which the historians themselves have refused to venture. We all have our preconceptions which are expressed in our attempts to reconstruct the past. A scientist's bias is at least obvious, and can be allowed for in a discussion of the historical issues. Writing as a historian of evolution theory who has recently penetrated into the area of human origins, I have found the work of the anthropologists extremely stimulating, and I hope this book will help to bridge the gap between their work and the Darwin industry.

The most interesting aspect of the anthropologists' approach is their concern to explore the narrative structure of the various accounts of human origins. Landau in particular has argued that such accounts generally specify a sequence of events by which our ancestors are thought to have been separated from the rest of the animal kingdom. Since the events are linked together in a narrative or storylike fashion, Landau and Cartmill see a parallel between the structure of paleoanthropological theories and that adopted by myths and folk tales. This structure enables the scientific theories to play the same cultural role as more traditional creation myths. The historian may now have to see early theories of human origins as—in Cartmill's humorous words—"the tale of Jack the Tree Shrew who climbed the arborescent beanstalk, shunned the temptations of specialization, and descended 60 million years later with the shining gifts of bipedality, toolcraft and intellect."[13] Perhaps our triumph over the apes parallels folk tales in which the clever younger brother outwits his more brutal siblings and takes over their birthright.

Historians may find this approach stimulating, but some scientists are disturbed by its potential implications. At least one anthropologist, Glynn Isaac, has wondered if Landau's work implied that modern theories too are little more than myths.[14] There are, in fact, a few professional biologists who argue that *all* attempts to reconstruct the details of the evolutionary process are so speculative that they lie outside the bounds of science. They insist that scientific biology must limit itself to the study of taxonomy, without inquiring into the origins of the forms to be classified. If this approach were applied to paleoanthropology, all theories of human origins would have to be dismissed as mere speculations, even if their structure did not have a mythlike character. Along with most other evolutionists, Isaac comforted himself by noting that the speculative nature of the theories may not be a problem so long as they can still be tested against an ever-growing body of evidence, and the more obviously false alternatives eliminated. The fossil evidence is still slender, leaving room for a good deal of inconclusive debate over the causes of human evolution. But it is fair to say that a number of alternatives have been *rejected* in the course of a century of research. Without getting into a debate on the philosophy of the scientific method, this situation would seem to imply that paleoanthropology is able to submit its hypotheses to critical testing. Those who would argue—on whatever grounds—that evolutionary hypotheses are mere speculations should remember that there is another group of anti-evolutionists who will welcome any excuse for dismissing natural accounts of human origins as no more "scientific" than the Book of Genesis.

The creationists aside, another group with an interest in stressing the speculative character of the early theories is composed of the sociologists of science. It is now frequently argued that all scientific "knowledge" is in fact an expression of the cultural values of the society that develops it.[15] Scientific theories represent ideas imposed on nature, not derived from it—ideas imposed by human beings trying to create an image of the world that will reflect the social environment in which they live. Early theories of human origins offer a fertile field of investigation for those who advocate this view of science. That so many different interpretations of the same few fossils were offered is a sign of the limited extent of empirical verification to which the field was subject. It is also clear that the theoretical positions were coordinated with their authors' social and religious opinions. In this area of science, what passed for knowledge was often mere opinion, at least until the fossil record began to improve in the late 1930s.

The sociologists of science are anxious to treat paleoanthropology as a genuine part of science, since if this area is amenable to social influences, then so, by implication, are all the others. By contrast, some historians of genetics and evolution theory may be tempted to doubt the scientific credentials of the early paleoanthropologists. The anatomists who interpreted the fossil hominids were out of touch with what have since emerged as the main currents in the development of modern evolutionism. It would be easy to dismiss them as bungling amateurs pontificating on matters they did not understand. Yet this would be going too far. If a fossil human bone were discovered, who else but an anatomist could assess its relationship to modern humanity? In the eyes of the public, the medical people who supplied the anatomical expertise were the only relevant scientific authorities. Nor were the anatomists' views on evolution so far removed from those still accepted by many naturalists at the beginning of the twentieth century. Non-Darwinian theories that are now seen as scientific dead ends were still popular at the time. Paleoanthropology was thus accepted as a science, albeit one existing at the very fringes of legitimacy. Almost all of the scientists involved admitted that their theories of human evolution went far beyond what the fossils could establish. Their excuse was, quite reasonably, that public interest demanded some evaluation of the slight evidence. The increasing number of fossils did eventually begin to limit the range of permissible speculation, along with the collapse of the anti-Darwinian theories of evolution.

To back up the claim that we should not dismiss the earlier theories as totally unscientific, I compared them with what most of us would regard as a thoroughly "mythical" account of human origins, the theosophist interpretation in Madame Blavatsky's *The Secret Doctrine* of 1885. Blavatsky was well read in the scientific literature of her time, and makes a surprisingly good effort to show that her theory of monsters, giants, and ancient civilizations is compatible with the evidence.[16] Yet the theosophist movement was still printing exactly the same text half a century later (the edition I consulted was published in 1925), by which time a considerable amount of additional evidence had been unearthed. It is this refusal to change the theory, or even to update the "sacred text" that marks the unscientific character of the movement, not the structure of the theory offered.

There is no automatic incompatibility between myth and science, or between myth and simple truth. As Cartmill points out, a story has a mythical quality if it appeals to us at a deep psychological level, evoking emotional responses that help us to identify how we feel about ourselves and the world we live in. Folk tales are myths whose origins are lost, and that appear to have no basis in fact, but there are real events that have taken place in the modern world that also have the ability to serve as symbols of something we care deeply about. Since theories of human evolution tell us about our own origins, they are bound to partake of this mythical character. Even if based on the best possible evidence, they would still tell us something about ourselves and our relationship to nature. If the evidence is loose enough to provide a good deal of leeway for rival speculations, it seems inevitable that our reactions to the theories will be determined as much by their emotional content as by their validity as science.

This last statement implies that there may be little difference between the anthropologists' appeal for us to recognize the mythical qualities of theories and the willingness of most historians to admit that scientific ideas intereact with the culture within which they are produced. There is a sense in which the Darwinian view of nature as a haphazard system driven by struggle and suffering has mythical qualities, as does the opposing view that the system is a harmonious one aimed at the production of a morally significant goal. Yet clearly Cartmill and Landau are trying to tell us something more. They talk of the mythical content of theories of human origins because they wish to draw attention to the narrative style of the theories. Darwinism expresses a particular view of how nature works as a general system, but a theory of human evolution often tells a story in which a series of particular events experienced by our ancestors is used to explain why we now differ from any of our primate relatives. The events produce effects that are governed by the general principles of biological evolution, but the end result is as much the product of the individual events as of the general principles. It is possible to characterize the human situation by stating the general principles and their implications, as the social Darwinist does, for instance, to justify the claim that social progress can only be achieved through struggle. Conversely, though, one could argue that human nature is governed by bloodthirsty instincts programmed into us by the hunting lifestyle of our early ancestors. This characteristic would depend on a particular event in the story of how we evolved, not on the Darwinian interpretation of the evolutionary mechanism. A similar view of society can thus be justified by appealing either to the general principles of evolutionism or to particular events in the human story, although, of course, there is no reason why both approaches may not be used together.

The Darwin industry is not intrinsically hostile to the claim that certain kinds of evolutionary explanation depend upon a narrative structure linking particular events, or to the belief that such narratives contribute to the general significance of the theory. A recent study by Gillian Beer compares the "plots" of Darwin's evolutionary narratives with those of Victorian novels.[17] Nevertheless, the majority of historians working on the impact of Darwinism have concentrated on the general principles of evolutionism. For Landau and Cartmill, by contrast, the narrative explanations of human origins by Darwin and later paleoanthropologists represent the more significant aspect of the attempt to define mankind's relationship to nature. Both levels of analysis may be valid, but the question we must decide is: which is the more appropriate at each stage in the history of evolutionism from Darwin's time through to the present?

I have already made clear my own feeling that, with the notable exception of Darwin himself, most early discussions of the implications of evolutionism were conducted in terms of general principles, rather than the specific details of human evolution. Consequently the appeal to the significance of narrative structures may strike many contributors to the Darwin industry as rather odd: it works for Darwin, but not for most of his contemporaries. Conversely, I think Cartmill is quite right to claim that the cultural significance of the images of the "killer ape" and "man the hunter" so popular in the 1960s and 1970s derives from a particular *story* about how we evolved. Modern reactions against the anthropology of aggression also make use of stories based on alternative interpretations of how our ancestors lived. The interesting period—which is also the main topic of this book—is the intermediate stage of the debate from about 1890 to the time of World War II. Here both levels of cultural significance can be seen at work, and our task is to evaluate their relative strengths.

Landau's work is the most detailed effort to explore the narrative structure of the theories produced in this intermediate period. Her analysis is presented as part of a general trend toward seeing the narrative style as the expression of a basic human need to tell stories. She quotes a number of authorities in fields such as linguistics, anthropology, philosophy, and history who have helped to found the field of "narratology," an enterprise based on the assumption that narrative is the primary mechanism of human understanding.[18] Central to the techniques employed by this school of thought is the recognition of certain underlying structural elements in any narrative explanation. The models for these structures are derived from the analysis of myths and folk tales, and Landau sets out explicitly to apply these models to theories of human evolution. The theories were the creation-myths of early twentieth-century civilization, and can be analyzed in the same terms as the myths of earlier cultures.

A possible objection to Landau's thesis is that the narrative style of explanation is by no means confined to the field of human origins. If one were to ask about the origin of the horse, or the origin of the mammals, one would expect to get the same kind of answer, in which a series of events directed a particular line of evolution toward the chosen goal. But perhaps this broader field of application is really an indication of the strength of the narrative approach. Many of the questions we find most interesting in the field of evolution theory can only be answered by the construction of hypothetical phylogenies and adaptive scenarios. The story of human evolution is merely the most emotionally loaded of this class of questions. The anatomists and paleontologists of the late nineteenth century had already begun to sketch in many such episodes in the history of life on the earth.[19] Their work created the framework within which the more sensitive topic of human origins could be tackled, once the anthropologists' commitment to a linear hierarchy of cultures and races had been broken down. From about 1890 onward, it was more widely recognized that human evolution must follow the same pattern as was found in the rest of the animal kingdom, and was thus more likely to be a branching rather than a linear process. With this recognition it became necessary to ask why the branches did not all run together in the same direction, thus creating the type of question that demanded a narrative answer.

The narrative element thus begins to figure more prominently in the theories of the early twentieth century, but it was still not—if I may use the term—the whole story. I shall argue that many of the theories proposed at this time still preserved some aspects of the old progressionism, although transmuted into a more sophisticated form. All too often the branches of primate evolution were portrayed as "experiments" by which nature tried, with varying degrees of success, to achieve its final goal of producing mankind. The advantages of becoming more intelligent were still taken for granted, and the problem was merely to explain why so many primates had been sidetracked away from their order's main evolutionary theme. In addition to this continuing faith in progress, prevailing ideas on the nature of the evolutionary mechanism still favored theories such as Lamarckism and orthogenesis, which emphasized the role of rigid phylogenetic trends and thus severely restricted the narrative element in the explanation of human origins. Although the new science of genetics repudiated these theories, anatomists and paleontologists continued to use the techniques that had been popular when the construction of hypothetical phylogenies had been at the forefront of biological research. The paleoanthropologists of the early twentieth century adopted the non-Darwinian concepts of evolution, and their explanations of human origins were thus shaped not only by the narrative possibilities of themes such as the acquisition of terrestriality, bipedalism, and toolmaking, but also by a belief in the existence of rigid, almost predetermined, evolutionary trends.

Landau's efforts to trace links between theories of human origins and folk tales provide us with valuable insights. Our prehuman ancestors were indeed portrayed frequently as "heroes" who had to undergo some kind of test before gaining their reward. The transition from the trees to the ground was often seen as an adventure undertaken by the boldest primates. Yet by concentrating on such key events in the story, Landau sometimes ignores the vigorous debates centered on evolutionary trends. Even when she discusses evolutionary process, she incorporates it into the narrative structure as the "donor" who gives the "hero" gifts such as bipedality. Yet some of the paleoanthropologists were not very interested in steps such as the acquisition of bipedality and the move out of the trees. They believed that many human characters have been produced by longstanding evolutionary trends, and some went so far as to argue that the similarities between humans and apes had been produced independently by parallel evolution in unrelated stocks. In such theories the trend itself explains the direction of human evolution, or at least creates a framework within which a narrative may be constructed to account for those human traits that are not produced directly by the trend.

My experiences as a historian of post-Darwinian evolutionism has no doubt encouraged me to emphasize the role of irreversible trends in the explanation of human origins. Anthropologists looking at the history of their own discipline may, perhaps, miss the significance of such trends because their thinking is conditioned by the very different situation created by modern evolution theory. Since the reemergence of Darwinism in the "Modern Synthesis" of the 1940s, theories based on predetermined trends and large-scale parallelism are no longer plausible. Evolution is seen as an open-ended or opportunistic process, and we can no longer assume that our large brains are the inevitable product of a longstanding trend. Darwinism forces us to explain all evolutionary developments by appealing to the adaptive benefits conferred by new characters in a particular environment. We must construct what are sometimes called "adaptive scenarios," which in the field of human origins will include those environmental or behavioral changes by which our ancestors became separated from those of the apes. As the term "scenario" implies, this kind of explanation is likely to have a narrative structure. Paleoanthropologists have always constructed these scenarios, but in a non-Darwinian climate of opinion, the significance of the adaptive changes was often more restricted than it appears to be in modern theories, because evolutionary trends were thought to be capable of generating certain products in any environment. The decision to stress the role of narrative throughout the history of paleoanthropology may thus be a more subtle form of presentism, in which the past is analyzed in terms dictated by our modern interests.

This question bears upon other issues discussed by philosophers of science concerned with evolutionism. David Hull has argued that the species concept should now be defined historically: the species is, in effect, an individual entity just like a human being, and hence can play a role in a historical narrative describing the course of evolution. This view of species is essentially a product of modern Darwinism. It is significant that in a recent survey of his position, Hull contrasts Darwin's opinions on the question with those of other naturalists (especially Lamarck and Lyell) for whom species and higher level taxa were eternal categories in the universal order—pigeonholes that will automatically be filled in the course of time.[20] Unlike Darwin, Lyell argued that if all the living mammals were wiped out, the reptiles would eventually reevolve a class identical to the one thus extinguished. In such a system of "repeatable" evolution, species would not be real individuals since they could be re-created over and over again. I suspect that Hull would also want to argue that historical narratives would play a less important role in the system. One might explain how a particular manifestation of the type appeared by means of a historical narrative, but the character of the type would not be explained by the narrative since it is already defined by the type's position in the universal pattern. When we find that many early twentieth-century paleoanthropologists still talked of evolution "striving" to create the human form, perhaps in a series of experiments, we are left wondering if they could possibly have seen the human species as an individual (in Hull's sense of the term). Humanity has become a predetermined goal of evolutionary progress, not a contingent product of a historical process that can be represented in narrative terms.

Landau and Cartmill are right to insist that, when a narrative structure exists within a theory, it generally corresponds to a "story" with deep emotional significance. The claim that our ancestors ventured onto the plains to explore a new environment conveys an optimistic message, just as the theory that they were bloodthirsty hunters evokes a sense of pessimism. Cartmill in particular insists that stories of human origins are told to emphasize how we came to differ from the rest of the animals, or how we became separated from nature. Yet the most controversial aspect of Darwinism was that it insisted on "man's place *in* nature." It undermined the traditional distinction between humans and animals (possession of a soul) and forced us to think of ourselves as products of natural processes. In such a scheme, the character ascribed to nature itself will also help to define its products. To explain the origin of humans by a Darwinian narrative is to accept that we are the accidental products of an undirected, and hence possibly amoral, historical process. Many post-Darwinian paleoanthropologists could not face up to this prospect, and therefore they preferred to assume that nature was inherently progressive and purposeful, with ourselves as the predetermined goal.

Theories of human origins had direct political and ideological implications. The most obvious area in which this factor can be seen in operation is in the link with opinions on race. The paleoanthropologists of the early twentieth century did little to shake off the earlier view that the white race stood at the head of the evolutionary scale of perfection. But their theories also had something to say about how the races could be expected to interact. Many believed that competition was inevitable, leading to the extermination of the inferior races. If this belief is called "social Darwinism," it must be remembered that the necessity of racial conflict was invoked by many who did not believe that natural selection was the mechanism by which new races or new species were created. As the theory of a linear sequence from ape to white man began to break down around 1900, the Neanderthals came to be dismissed as a degenerate race or even species of humanity, wiped out when modern races invaded their territory. Given the widespread appeal to the inevitability of racial conflict as a means of justifying modern empire-building, it is difficult not to see this view of the Neanderthals as an unconscious extension of the imperialist theme into paleoanthropology.

So far as I know, there is no general historical survey of theories of human evolution available, although there is a valuable collection of readings assembled by Theodore McCown and Kenneth Kennedy.[21] I have suggested some reasons that may account for the reluctance of historians engaged in the Darwin industry to enter into this field, and to the extent that my own experience has been shaped by participation in that industry, this book is a pioneering venture. My account almost certainly differs considerably from anything that would have been prepared by an anthropologist, and, as we have already seen, substantial differences of interpretation are possible. Nevertheless, I hope that anthropologists who are interested in the history of their own field will find this account stimulating, even if it does not tackle the issues in quite the way they themselves would do so. I also hope that historians of science, who have for so long neglected this aspect of evolutionism, will now be encouraged to look into it and begin a dialogue with the anthropologists.

A few words of explanation are needed here to guide the reader through the rest of the book. The first part consists of two chapters giving necessary background in the various fields that contributed to the debate on human evolution. Most historians of Darwinism are not familiar with the development of archaeology in the late nineteenth century, and may also be unaware of the parallel trends in physical and cultural anthropology. Conversely, archaeologists and anthropologists may have little knowledge of the history of post-Darwinian evolution theories. I thus felt that it was necessary to provide a brief introduction to the relevant fields, so that readers from a variety of different backgrounds would be able to prepare themselves to tackle the main study. Those who are already fully primed in any area may, of course, skip over the corresponding section. If they do read the sections on topics with which they are already familiar, I hope they will forgive any oversimplifications required by the need for brevity.

Part 2 is a fairly conventional look at the various attempts to construct a phylogeny for the human species, that is, to reconstruct its evolutionary history. It begins with the most obvious approach, which was to derive the human race from the great apes. The chapter on "Neanderthals and Presapiens" deals with an issue already familiar to historians of physical anthropology, an issue that shows how the theory of an ape origin became more complicated once the possibility of branching evolution was fully appreciated. The chapters on the tarsioid theory and polytypic theories outline unconventional views on human origins that stirred up a great deal of controversy in the early twentieth century. I have suggested that these heresies were regarded as particularly dangerous because they simply took the principles upon which everyone else was relying to extreme lengths.

Part 3 consists of several quite different cross sections through the complex of issues outlined in part 2. Here I have tried to uncover the differing interpretations of the causes that were thought to have shaped the path of human evolution. Was the human mind the product of a general evolutionary trend that necessarily increased brain size, or the result of a unique event such as the transition to upright walking or living on the open plains? It is here that the debate over the relative significance of narrative explanations and evolutionary mechanisms becomes relevant. I have argued that although the narrative element became more important in the early twentieth century, many paleoanthropologists continued to believe that primate evolution was governed by rigid trends. But were these trends functional or nonfunctional: was the increase in brain size the result of the inherited effects of use (implying a Lamarckian mode of evolution) or of an orthogenetic trend built into the germ plasm of the primates? We shall see that both alternatives were taken quite seriously. Finally, if there were general trends in human evolution, to what extent could they still be seen as an expression of the Creator's purpose? A few paleoanthropologists were still willing to argue quite openly that there was evidence of design in human evolution, and many others seem to have conceived of "nature" as a system striving to achieve a particular goal. At the same time, though, a harsher image of racial conflict fueled by bloodthirsty instincts laid the foundations of a much more pessimistic attitude toward human nature.

Part One Background

Chapter One The Evidence of Human Antiquity

The emergence of a serious debate on human evolution was conditioned by two prior developments. One of these was, obviously, the conversion of the scientific world to a belief in the general idea of evolution. Although the idea was widely debated in the years prior to the appearance of Darwin's *Origin of Species* in 1859, not until the 1860s did a significant body of scientists begin to take it seriously. Chapter 2 surveys the rise of evolutionism, noting that early views on the topic were frequently conditioned by non-Darwinian values which allowed evolution to be seen as orderly, progressive, and purposeful. But for the general idea of evolution to be applied to the human species, another scientific development formed an equally necessary foundation. This development was the widespread acceptance of the belief that the human species is immensely ancient, and had originally existed in a state that was both culturally and biologically "primitive" when compared with modern times. The emergence of this extended and progressionist interpretation of human history was dictated by changing attitudes toward the relics of ancient humanity. Because the earliest fossil human bones were so controversial, it was in fact the archaeological evidence for the existence of stone-age cultures that carried the burden of first establishing the true antiquity of the human species.

Archaeology

Acceptance of human antiquity coincided with the early phase of the debate over Darwin's theory. Important papers on the topic by Charles Lyell, Joseph Prestwich, and others appeared in the same year as the *Origin of Species*. This coincidence has generated a popular belief that Darwinism played a causal role in the formation of modern archaeology, but this idea has been dismissed by recent historians.[1] Evolutionism benefited from the widespread acceptance of human antiquity and the primitive status of early humans. If our species had evolved from some lower form, its culture would certainly have been primitive at first.

Furthermore, since Darwin assumed that evolution is an immensely slow process, he and his followers naturally expected the final stages of human evolution to be spread over a vast period of time. Archaeology thus contributed to the evolution debate, and may have gained some additional publicity thereby, but there seems little doubt that the establishment of human antiquity came about independently of the events leading to the publication of Darwin's theory. Some of the leading proponents of the antiquity of mankind, including Boyd Dawkins, Armand de Quatrefages, and even Lyell himself, were suspicious of evolution. The human form could be ancient, yet still have an independent origin. The two developments were in fact parallel expressions of the growing willingness of mid-nineteenth-century society to challenge the cultural values it had inherited.

The emergence of a belief in the vast extent of human antiquity has been described in detail by Donald Grayson, and need only be sketched in briefly here.[2] It is important to note, however, that archaeology took a great deal of its early inspiration from geology. The more recent geological epochs inevitably formed the timetable against which human antiquity was measured. In addition, the geologists' technique of using fossils to identify a temporal sequence of rock strata formed the natural model for archaeologists attempting to date a succession of deposits containing human remains and artifacts.

In the early nineteenth century, the French paleontologist Georges Cuvier and his followers established that a series of geological formations corresponding to successive periods in the earth's history could be charted by identifying the fossils contained in each stratum of rock.[3] The fossil record exhibited a sequence of strikingly different forms of living things that had inhabited the earth over a vast period of time. Cuvier founded the school of geological thought known as "catastrophism," based on the view that an abrupt change in the animal and plant populations occurred between one formation and the next. Catastrophic extinction and some mysterious process for the introduction of new species provided the "punctuation marks" in the history of life on the earth. The overall sequence was thought to represent a progressive development from the earliest periods, when only the simplest forms of life had existed, through successive epochs in which fish, reptiles, and finally mammals were introduced onto the earth's surface.

The extent of the discontinuities in the history of life was challenged by Charles Lyell's "uniformitarian" geology during the 1830s.[4] Yet Lyell agreed that there was a systematic change in the fossil population of successive geological epochs, and he himself helped to establish the currently accepted subdivision of the most recent or "Tertiary" rocks into a sequence of distinct formations. Later geologists extended his system to give the sequence outlined in table 1. It provides a breakdown of the Cenozoic era, more popularly known as the "Age of Mammals." In this era the mammals at last became the dominant form of vertebrate life, and it was in this era that the primates—the zoological order to which the human species belongs—evolved. It was widely assumed that the human form had appeared only during the latest, the Quaternary or Pleistocene, era.

In the mid-nineteenth century, Louis Agassiz popularized the theory that the Pleistocene had been affected by an ice age, a period in which much of Europe and North America had been covered by extensive glaciation. Later in the century scientists realized that there had been not one but a whole series of glaciations, separated by milder interglacial periods. In his *Great Ice Age* of 1877, James Geikie proposed a system of four major glaciations, followed by a less severe cold spell.[5] In the early twentieth century, A. Penck and E. Bruckner introduced what became the most popular terminology for the successive glaciations (see table 2). They argued that the human race first entered Europe in the Mindel-Riss interglacial, although this argument was disputed by other authorities.

There was also considerable debate over the absolute timing of the

Table 1. GEOLOGICAL SUBDIVISIONS OF THE CENOZOIC ERA (THE "AGE OF MAMMALS")

	Modern
	Quaternary or Pleistocene
Tertiary Series	Pliocene
	Miocene
	Oligocene
	Eocene
	Paleocene*

*The Paleocene was not generally distinguished from the Eocene during much of the period discussed here. A time chart showing the geological distribution of hominids according to modern evidence is given in the Epilogue.

Table 2. THE PENCK-BRUCKNER TERMINOLOGY FOR PLEISTOCENE GLACIATIONS

Glaciations	*Interglacial Periods*
WÜRM	
	Third Interglacial (Riss-Würm)
RISS	
	Second Interglacial (Mindel-Riss)
MINDEL	
	First Interglacial (Günz-Mindel)
GÜNZ	

Note: The names were derived from locations in Switzerland where deposits characteristic of each glaciation were found.

sequence of events in the geological record. In the late nineteenth century, geologists had been forced drastically to reduce their estimates of the earth's age because of the physicist Lord Kelvin's argument based on the rate of cooling of the planet's core.[6] Geikie allowed two hundred thousand years for the Pleistocene, whereas W. J. Sollas produced an estimate of four hundred thousand, based on rates of sedimentation. Arthur Keith accepted Sollas's figure, but also noted Penck's estimate of anything between half and one and a half million years.[7] Those who accepted the shorter period were working with a time scale for human evolution that seems highly compressed by modern standards. This situation may account for the willingness of a few authorities such as Keith to extend the antiquity of the human type back into the Pliocene. Only in the 1920s did the development of radioactive dating techniques by Arthur Holmes and others begin to drive back the limits of geological time, greatly extending the absolute time scale of human evolution.

The Pleistocene glaciations provided a framework for understanding the earth's recent past, but what evidence was there that human beings had existed at a time that (although geologically recent) was far beyond the antiquity of any known civilization? During the first half of the nineteenth century, it was widely accepted, on Cuvier's authority, that human fossils did not exist. The human species had appeared too late in the earth's history for its remains to be incorporated into even the most recent geological deposits. This opinion was challenged by Jacques Boucher des Perthes and others on the basis of stone tools found in deposits of what appeared to be considerable antiquity, since they also contained the bones of extinct animals.[8] Some of this early evidence was rejected because the stone tools were found in caves, where the deposits were frequently mixed up. But Boucher des Perthes had found stone hand axes in the clearly stratified gravel deposits of the Somme valley in northern France. Unfortunately, the first volume of his *Antiquités celtiques et antédiluviennes* of 1847 was written in the context of an extreme catastrophism that was already out of date. The stone tools were described as the products of a now extinct intelligent being which had been replaced by the human race. These odd opinions gave the critics all the excuse they needed to dismiss the discoveries themselves. Boucher des Perthes continued to work in obscurity until a dramatic improvement in his fortunes occurred in the late 1850s.

The revolution that led to the acceptance of human antiquity began in 1858 with the exploration of Brixham cave, near Torquay in Devon, by William Pengelly and Hugh Falconer.[9] They were trained geologists who could speak with authority on the association of flint tools and the bones of extinct animals in the cave. Their positive reports attracted more geologists and antiquaries to the cave, including Joseph Prestwich and Lyell himself. The years 1859–60 saw a flood of papers by eminent figures, guaranteeing the future acceptance of this type of discovery. Falconer went to see Boucher des Perthes and was convinced that his finds were also authentic. A campaign to convert the French scientific establishment led by the paleontologists Edouard Lartet and Albert Gaudry now began.

Among the most popular accounts of the new discoveries were Lyell's *Antiquity of Man* of 1863 and John Lubbock's *Prehistoric Times* of 1865. Whereas Lyell adopted a cautious attitude toward biological evolution, Lubbock openly sought to link the progress shown by the archaeological record with the evolutionary origins of the human species. To create his progressive scheme, Lubbock extended the "three age system" already devised by Scandinavian archaeologists. J.J.A. Worsaae and Christian Thomsen argued for a sequence of periods in which stone, bronze, and finally iron had been used as the chief materials for toolmaking.[10] Lubbock refined this scheme by dividing the stone age into two: the paleolithic (or old stone age, the period of chipped or flaked stone tools) and the neolithic (or new stone age, the period of polished stone).[11] Our ancestors had begun in the earliest paleolithic with the simplest forms of stone tools and had then progressed to more sophisticated use of the same material, before passing on to the discovery of metals.

Of the four stages in Lubbock's scheme, the paleolithic was clearly the most extensive. Already in the 1860s, efforts were being made to subdivide the paleolithic into a sequence that would more fully illustrate the development of human culture. These early efforts drew their inspiration from paleontology, since it was the link with the extinct animals that had first confirmed the antiquity of mankind. In 1861 Edouard Lartet surveyed the fossil mammals that were known in association with stone tools and outlined a sequence based on changes in the predominant animal remains found at different sites. Lartet's own work, based on extensive excavations in the Dordogne area of southern France, was made in conjunction with Henry Christy and later summed up in their *Reliquiae Aquitanicae* of 1875. The latest part of the paleolithic was called the reindeer period, of which the best known site was at La Madeleine.[12] Earlier sites showed fewer reindeer and were characterized by the remains of rhinoceros and mammoth. Earlier still came the age of the giant cave bear. It was widely believed that the tools found in the diluvial gravels of northern France by Boucher des Perthes corresponded to a yet earlier period with a warmer climate, during which early humans had lived in the open rather than in the caves explored in the south. Lartet himself later realized that some of his subdivisions could not be substantiated, and in the *Reliquiae Aquitanicae* he denied that the northern sites corresponded to an earlier period.[13]

Lartet's intention to date the sites by their associated animal remains would have provided a direct link to geological and climatic events such as the ice ages. Unfortunately this intention turned out to be impractical, since critics pointed out that the animals hunted by ancient humans would have differed from place to place according to local conditions. Cave sites and open sites might yield the remains of different animals, even when they were contemporary. The only alternative was to use the stone tools themselves as the basis for a system of relative dating, by identifying distinct toolmaking cultures and working out how they succeeded one another in the course of time. Since most archaeologists took their inspiration from geology, they tended to assume that different cultures could not have coexisted alongside one another. Like the sequence of geological formations identified by their fossil populations, there was a fixed sequence of toolmaking cultures through which humanity had advanced in the course of the paleolithic.

The leading figure in this movement was Gabriel de Mortillet. As a professor at the *Ecole d'Anthropologie* in Paris, de Mortillet assisted Lartet in arranging the prehistoric material for the Universal Exposition of

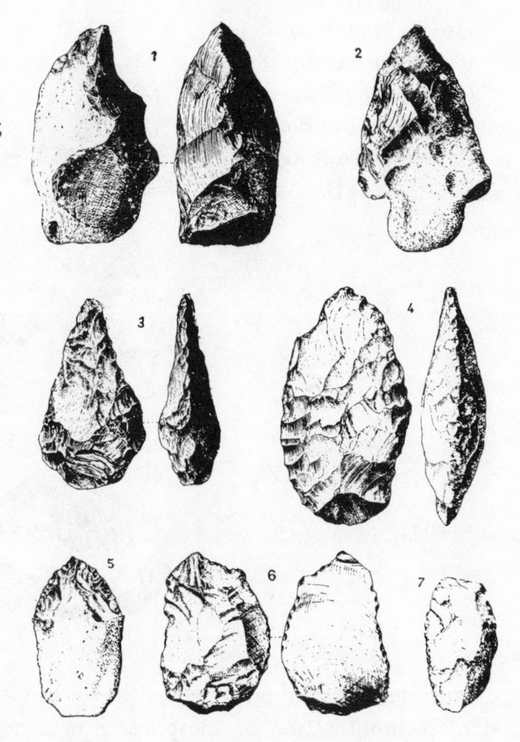

Figure 1. Chellean and Acheulian stone tools. Illustration 3 is the typical Chellean hand ax; 4 is the more rounded Acheulian style. From Marcellin Boule, *Les hommes fossiles* (Paris: Masson, 1921), p. 141.

1867. His own scheme was subsequently developed for the exhibitions in the National Museum in Paris. He abandoned Lartet's faunal links and divided the tools into cultures based on recognizable toolmaking techniques, each named after a prominent site where that style had been found. Since some sites had been occupied over long periods of time, excavation showed how the cultures had succeeded one another. De Mortillet was convinced that he could establish a sequence valid for the whole of European prehistory. In 1872 he proposed four periods of tool development in the paleolithic. Oldest was the Chellean, including the massive stone handaxes found by Boucher des Perthes (fig. 1). Here the tool was made by chipping off small flakes of flint to put a sharp edge on the original core. Following this period came the Mousterian, when the tools were made from the flakes, not from the core itself (fig. 2). Next was the Solutrean, a culture with beautifully shaped blades of chipped stone. The final development of the paleolithic was the Magdalenian, Lartet's reindeer age, when flint was used less and a high level of workmanship in bone was substituted. This scheme was outlined at length in de Mortillet's *Le Préhistorique* of 1883. With a number of subsequent

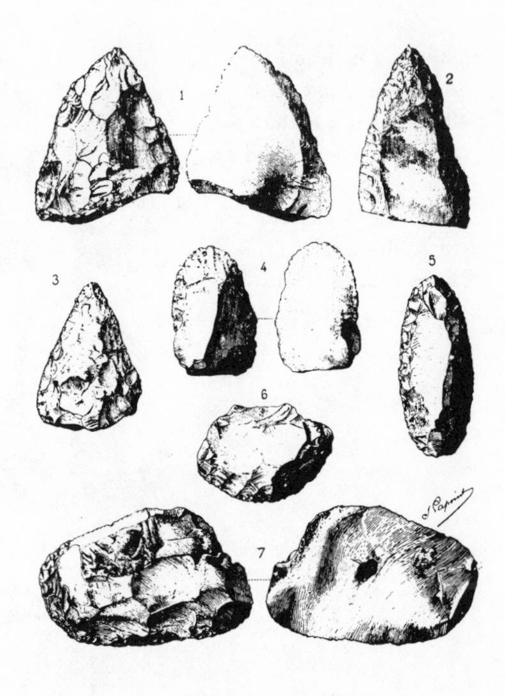

Figure 2. Mousterian stone tools from the cave at La Chapelle-aux-Saints. From Marcellin Boule, *Les hommes fossiles* (Paris: Masson, 1921), p. 187.

extensions and modifications, as shown in table 3, it became the standard technique used by most late nineteenth-century archaeologists.

The popularity of de Mortillet's scheme almost certainly derived from its analogy with the geologists' method of dating periods of rock formation by their fossils, but de Mortillet openly promoted an evolutionary interpretation of human development when this idea was still highly unpopular among French biologists. For some time after the Darwinian revolution in Britain, French paleontologists retained the traditional, discontinuous view of the fossil record, in which life advanced by the sudden or miraculous appearance of new types at the beginning of each geological period. Although de Mortillet admitted that the neolithic culture had entered Europe suddenly through the invasion of people with a higher civilization, he insisted that in the paleolithic each stage in the cultural development emerged naturally out of its predecessor by a process of gradual evolution.[14] The periods corresponded to stages through which culture must inevitably pass in the course of its upward march. He explained the relatively small number of Solutrean sites with the assumption that it was a very brief episode in the process, since to admit that it was only a purely local culture would have undermined the universality of the sequence.[15] De Mortillet's faith in the universality of progress was proclaimed in the conclusion of his account of the 1867

Table 3. GABRIEL DE MORTILLET'S SYSTEM OF ARCHAEOLOGICAL PERIODS, IN ITS FINAL FORM

	Period	*Site from Which Named*
	NEOLITHIC	
PALEOLITHIC	Azilian*	Mas d'Azil, Ariège
	Magdalenian	La Madeleine, Dordogne
	Solutrean	Solutré, Maçon
	Aurignacian	Aurignac, Haute-Garonne
	Mousterian	Le Moustier, Dordogne
	Acheulian	Saint Acheul, Amiens
	Chellean† (Abbevillian)	Chelles, Seine et Marne
	Prechellean, or eoliths	

*The Azilian was inserted by Edouard Piette and served to bridge the "hiatus" that de Mortillet believed to separate the palaeolithic and neolithic.

†De Mortillet's Chellean was later renamed the Abbevillian by Breuil. For classic expositions of this system, see M. C. Burkitt, *Prehistory;* G. G. MacCurdy, *Human Origins;* and H. F. Osborn, *Men of the Old Stone Age.*

Exposition; here he asserted that he had demonstrated the truth of the following propositions:

LOI DU PROGRÈS DE L'HUMANITÉ
LOI DU DÉVELOPPEMENT SIMILAIRE
HAUTE ANTIQUITÉ DE L'HOMME.[16]

From his materialist perspective, de Mortillet was convinced that cultural evolution was merely an extension of the later phases in the biological evolution of the human species. Although no human fossils were known from the Tertiary, he insisted that earlier forms of mankind must have been living at least during the later part of that era. These ancestors of modern mankind he named *Anthropopithecus* to indicate their intermediate status between apes and modern humans.[17] The Neanderthal race of the Mousterian period was a later stage in this development, still showing some signs of the ape ancestry.[18] He thus postulated a linear sequence of progress from ape to human, with the uniform development of culture taking place in the later phases of the process. Michael Hammond has shown that for de Mortillet this was not merely a philosophical position.[19] He was a prominent radical and socialist, who saw paleoanthropology as a weapon to be used in the fight for reform. If evolutionary and cultural progress was a universal law, then the triumph of the Left was assured.

As part of his campaign to establish the case for human ancestors in the Tertiary, de Mortillet supported those archaeologists who claimed to have found artificially worked flints in these more ancient deposits. He accepted the discoveries of the abbé Bourgeois, who had found what appeared to be stone tools as far back as the Oligocene.[20] Controversy raged around these "eoliths" throughout the later part of the century, and eventually the geologically most ancient discoveries were rejected. Marcellin Boule, for instance, showed how natural forces could fracture flints to give what might appear to be a primitive tool.[21] In the early twentieth century controversy also surrounded J. Reid Moir's discovery of apparently worked flints from deposits underlying the Red Crag formation near Ipswich in England. Moir had been inspired by the writings of T. H. Huxley, who had claimed that the ancestors of humanity might go back to great antiquity.[22] His discoveries were more plausible than those of Bourgeois because they came from deposits lying immediately below the Pleistocene, when, many authorities agreed, primitive humans might already have been in existence. Moir's later discovery of "rostro-carinate" or beak-shaped stones convinced the great zoologist E. Ray Lankester, and from this point onward the existence of primitive pre-Chellean industries was gradually accepted.[23]

From the beginning some doubted that de Mortillet's neat sequence of cultural periods corresponded to the real situation in paleolithic Europe. The Scandinavian originators of the three-age system had accepted that the stone and metal cultures had been brought into their countries by invading peoples and were not an "internal" development. Even de Mortillet had accepted an invasion to explain the introduction of the neolithic. Against de Mortillet, Boyd Dawkins argued that the toolmaking cultures corresponded to different tribal or racial groups that had inhabited Europe at the same time as one another—not a sequence of development through which all mankind must pass.[24] The Magdalenian "period" represented merely the culture of the ancestral Eskimos, who had migrated northward with the reindeer as the climate moderated. In the early twentieth century it became widely accepted that the Neanderthal race, with its Mousterian culture, had been wiped out when modern humans, makers of the superior Aurignacian tools, invaded Europe from the east (see chapter 4).

By 1912, the abbé Breuil, originally an exponent of de Mortillet's system, had begun to concede that the culture might represent independent developments rather than stages in each other's history.[25] In the 1920s Breuil and others began to distinguish three parallel culture groups in the lower paleolithic: the Clactonian, Levalloisian, and Tayacian. Louis Leakey's pioneering work in African prehistory led him to argue that the Clactonian culture had eventually developed into the Mousterian. Both this development and the Levalloisian industry were associated with the Neanderthal race.[26] The Acheulian had developed into the Aurignacian as a separate line of cultural evolution associated with modern humans. De Mortillet's simple progressive sequence had thus been replaced by a more sophisticated "branching" image of cultural development. This idea was linked to major changes in theories of human biological evolution, discussed in part 2. By the 1930s a considerable amount of fossil evidence had accumulated to illustrate the actual course of human evolution, and these discoveries form the basis of the paleontology section.

PALEONTOLOGY

Archaeology was crucial to the establishment of human antiquity and of a progressionist view of early human history because the stone tools gained wide acceptance by the scientific community much earlier than the handful of fossil remains available in the mid-nineteenth century. The tools indicated that some form of human hand had made them, even if the bones themselves were not preserved. A handful of ancient human remains were known, but these were so controversial that they could not be used to illustrate the earliest human form. Only gradually did an increasing number of fossils allow the first outline of the history of the human type to be reconstructed. Even in the early twentieth century, the evidence was so fragmentary that many different interpretations of human evolution were possible. The subsequent history of hominid paleontology is a complex affair, with the scientists often falling out among themselves in the full glare of publicity. The story of the important discoveries, and the controversies they endangered, has been told in a substantial body of secondary literature, and will only be summarized here.[27]

Almost all of those who actually discovered hominid fossils tended to exaggerate their importance by creating new taxonomic units for them. Thus "Peking man" was given generic rank by its discoverer, Davidson

Black, who named it *Sinanthropus pekinensis*. From the start, however, a few naturalists argued that the new form was merely another species within the same genus as "Java man," *Pithecanthropus erectus*. Nowadays the two are regarded as merely racial variants of the same species, and even the generic rank *Pithecanthropus* has been abandoned so that the species can be included in the human genus as *Homo erectus,* the forerunner of *Homo sapiens*. The early tendency for "taxonomic inflation" must be borne in mind whenever we discuss the fossils and the theories to which they were applied. For convenience this account will preserve the original names, as long as these were still in use into the 1930s.

Fossil apes were less controversial, but still of interest to those looking for the remote ancestry of the human type. As early as 1856, Edouard Lartet had announced the discovery of the remains of an ape from Pliocene deposits in France, to which he gave the name *Dryopithecus fontani*.[28] Further discoveries of a similar type were made later in the century, and *Dryopithecus* was frequently cited as the possible common ancestor of the modern ape and human stocks. That some of these fossil apes were found in Europe showed that early primates had been widely distributed over the Old World, thus making it difficult to pinpoint any particular location as the birthplace of the human form.

If one assumes that humans had evolved from apes, it was reasonable to expect that early paleolithic fossils would reveal some indication of their ape ancestry. With the exception of the controversial Neanderthal remains (discussed below), this expectation was not at first realized. Some of the first human remains to be discovered in the paleolithic revealed a surprisingly "modern" appearance and brain capacity. In 1868, excavations at the rock shelter of Cro-Magnon in the Dordogne uncovered a group of skeletons along with tools of the Aurignacian type. The bones, especially those of the "old man" of Cro-Magnon, were described at length by Paul Broca and Armand de Quatrefages, and were made the type specimen for the Cro-Magnon race by de Quatrefages and Hamy.[29] They revealed a tall, well-built race with fine features and a large cranial capacity—ancestors of whom anyone could be proud. Over the next few decades specimens identified with this type were found in several European locations, including the Grimaldi caves at Menton in the south of France.

Other late paleolithic races of *Homo sapiens* also came to light. In 1888 a skeleton was found at Chancelade, also in southern France, in association with Magdalenian implements. It was described as a racial type differing significantly from Cro-Magnon and bearing a strong resemblance to the modern Eskimos. This description seemed to uphold the theory that the people of this period had moved north with the reindeer at the end of the last cold spell. W. J. Sollas became a leading exponent of this view.[30] Later excavations at the Grotte des enfants cave at Grimaldi turned up new skeletons. R. Verneau saw these as another racial type with distinctly negroid characters.[31] The claim that negroid races had penetrated into Europe in the paleolithic naturally aroused a good deal of controversy.

Marcellin Boule argued that the Grimaldi negroids showed a modern form of humanity inhabiting Europe quite early in the paleolithic.[32] He went on to suggest that a fully human type may already have been in existence at the beginning of the Pleistocene. Arthur Keith took up a similar position in his *Antiquity of Man* of 1915, citing as evidence a modern-looking skeleton discovered in early Pleistocene deposits at Galley Hill, southern England, in 1888.[33] Most paleoanthropologists rejected this vast extension of the antiquity of *Homo sapiens,* dismissing the Galley Hill remains as a later burial. In 1932, however, Louis Leakey proclaimed a jawbone found at Kanam, near Lake Victoria in Kenya, as evidence of the very early appearance of the human form.[34] The Swanscombe skull fragment, unearthed in England in 1935, was also widely regarded as an Acheulian version of *Homo sapiens,* although the discoverer, Alvan T. Marston, treated it as an early Neanderthal type.[35]

If the very early specimens of *Homo sapiens* were of dubious provenance, what had the human type actually looked like in the early Pleistocene? A possible solution to this problem was provided by the eventual recognition of the distinctive Neanderthal race or species. The specimen from which this type was named had been found in the Neander valley (Neanderthal), near Düsseldorf, in 1857. A complete skeleton was found by workmen, but much was destroyed before it was rescued by a local schoolteacher. The remains were described by Professor D. Schaafhausen, who drew attention to their savage and brutal appearance, but did not suspect their true age. An English translation of Schaafhausen's paper contained additional notes by George Busk stressing the apelike features of the skull, especially its heavy supraorbital or brow ridges.[36] T. H. Huxley provided a detailed description of the skull in his *Man's Place in Nature* of 1863 (fig. 3).[37] Huxley accepted that it was the most simian human skull yet discovered, but pointed to the large cranial capacity and argued that the specimen fell within the possible range of variation for *Homo sapiens.* Already some, however, thought that the differences were so great that the individual must belong to a distinct

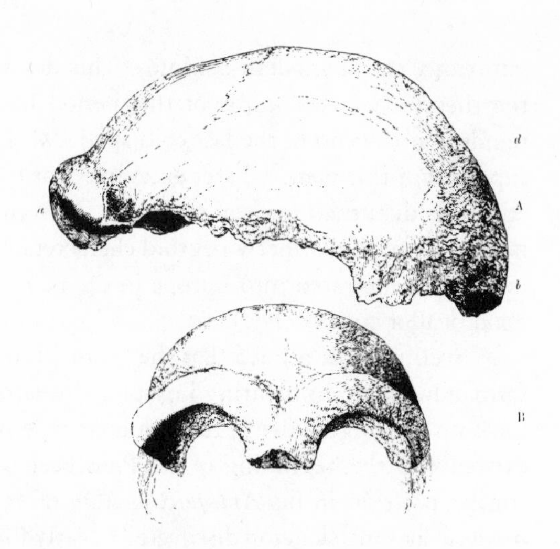

Figure 3. Side and front views of the original Neanderthal skull, showing the heavy brow ridges. From T. H. Huxley, *Evidence as to Man's Place in Nature* (1863), p. 139.

species of early humanity. The name *Homo neanderthalensis* was coined by William King in 1864.[38]

The Neanderthal remains became the subject of an intense controversy; many anthropologists refused to acknowledge the apelike characters as having any evolutionary significance. The noted pathologist Rudolph Virchow insisted that the simian features were merely the product of disease. The skeptics were finally confounded in 1886 by the discovery of several Neanderthal specimens at Spy in Belgium. The description by Julien Fraipont and Max Lohest not only linked these skeletons with the one from Neanderthal but also established a relative dating: the bones were found along with tools of the Mousterian culture.[39] A number of later discoveries confirmed this link. The most completely described specimen was that found at La Chapelle-aux-Saints in 1908 and sent to Marcellin Boule in Paris. Boule's monograph established the popular image of Neanderthal man, with his heavy features, low forehead, and apelike shambling gait.[40]

At the turn of the century, the Neanderthal type was often seen as a possible intermediate between modern humanity and its apelike forebears. The discovery of a yet more primitive form confirmed this interpretation for at least some paleoanthropologists, although controversy once again surrounded the find. The remains of "Java man" were unearthed by Eugene Dubois in 1891–92 and revealed a skull of much smaller cranial capacity than the Neanderthal type, yet still with the heavy brow ridges. A thighbone discovered later in the same deposit seemed to indicate that the creature had walked upright. Dubois, who had gone to Java in response to Ernst Haeckel's predicted link between humanity and the Asian apes, borrowed Haeckel's term *Pithecanthropus* to name the specimen. The full name was *Pithecanthropus erectus,* indicating the upright posture.

Although Haeckel welcomed the discovery, other paleoanthropologists were suspicious, and Dubois found himself embroiled in controversy.[41] Some claimed that the new form was merely an ape, whereas others saw it as fully human—a disagreement out of which Dubois was able to make a certain amount of capital.[42] Eventually he was forced to admit that the early Pleistocene date of the specimen indicated that *Pithecanthropus* was not our direct ancestor, but a side branch preserving the ancestral form into a later geological period. Even if only our uncle and not our grandfather, though, *Pithecanthropus* was "a venerable ape-man, representing a stage in our phylogeny."[43] Eventually Dubois became so frustrated by the critics that he retreated into isolation, refusing to let anyone else see the fossils. He now changed his position completely and insisted that *Pithecanthropus* was merely a giant gibbon, although still a gibbon that could have been transformed into a human by a sudden mutation in brain size.[44]

Despite the controversy, a significant body of paleoanthropologists, led by Gustav Schwalbe, accepted *Pithecanthropus* and the Neanderthals as stages in the evolution of the human form from the ape. Within a few years, however, the situation changed dramatically (see chapter 4). Boule's analysis of the La Chapelle-aux-Saints Neanderthal specimen was used to support his claim that the Neanderthal species was too apelike, and too late in the sequence of paleolithic development, to represent our direct ancestor. A more human type had already come into existence early in the Pleistocene and had eventually wiped out the backward Neanderthals. *Pithecanthropus* and the Neanderthals were both relegated to side branches of human evolution. The fossils themselves soon seemed to confirm this new image of human evolution as a branching tree. A jawbone discovered at Mauer near Heidelberg in 1907 was seen by many as being too massive for it to be linked with the *Pithecanthropus*-Neanderthal line. Far more critical, though, was the serious attention at first paid to what has now become known as one of the most celebrated cases of scientific fraud: the Piltdown remains.

The Piltdown story has been told often enough and needs only to be summarized here.[45] An amateur archaeologist, Charles Dawson, began to pick up fragments of a human skull in a gravel deposit at Piltdown in Sussex in 1912. A more careful search in conjunction with Arthur Smith

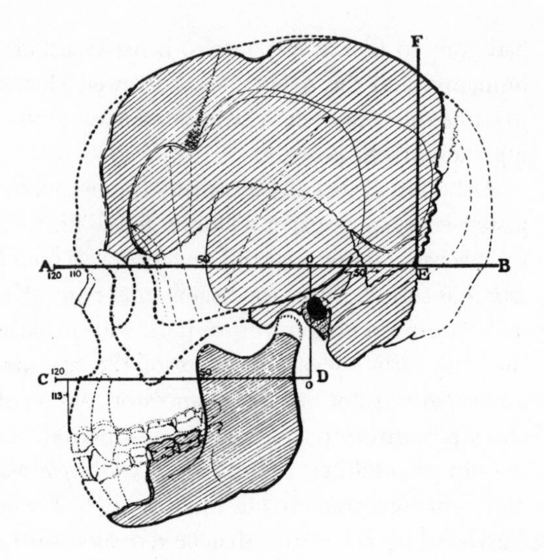

Figure 4. Arthur Keith's reconstruction of the facial profile of *Eoanthropus dawsoni*, "Piltdown man." Keith gave the skull a much larger capacity than other authorities would allow. From Keith, *The Antiquity of Man* (London: Williams & Norgate, 1915), p. 492.

Woodward of the British Museum revealed more pieces of skull and a jawbone with teeth. The skull, although thick, appeared to be of a modern type, without the characteristic brow ridges of the Neanderthal. The jaw was extremely apelike, but the teeth were worn in a manner typical of humans rather than apes. The remains were found along with primitive stone tools and animal bones indicating an early Pleistocene age. Woodward produced a reconstruction of the skull which gave it a cranial capacity of 1,070 cc, intermediate between human and ape, a figure accepted by the eminent brain specialist Grafton Elliot Smith.[46] Woodward named the composite creature *Eoanthropus dawsoni*.

British naturalists accepted the Piltdown remains as genuine—perhaps on the grounds of national prestige—and many argued that *Eoanthropus* represented the true ancestor of modern humanity. This position confirmed that *Pithecanthropus* and the Neanderthals were blind alleys in human evolution. Yet even within the British scientific community there were disagreements. Arthur Keith attempted to back up his claim for the extreme antiquity of modern humanity by reconstructing the skull with a much larger capacity (see fig. 4). He also objected to the apelike character of the face in Woodward's model. On this last point Keith was forced to give way when Teilhard de Chardin found a canine tooth at the Piltdown site, exactly fitting Woodward's prediction.[47] On the larger capacity of the brain case, however, Keith remained adamant throughout the rest of his career, and even tested his own skill by reconstructing a skull which had been deliberately broken to resemble the Piltdown fragments.[48] Keith also suspected that the Piltdown gravels might be of Pliocene age.

Outside Britain there was widespread skepticism over Woodward's hybrid *Eoanthropus*. In France, Boule was only too willing to accept the skull as an ancient form of mankind, but dismissed the jaw as that of an ape that had found its way into the same deposit.[49] In America, Gerrit S. Miller led a similar campaign to brand the jaw as merely that of a chimpanzee.[50] The remains lacked precisely those parts that would have allowed an unambiguous reconstruction of the skull and a clear link between the skull and the jaw. As Miller somewhat prophetically put it: "Deliberate malice would hardly have been more successful than the hazards of deposition in so breaking the fossils as to give free scope to individual judgment in fitting the parts together."[51]

The Piltdown remains gradually slipped into the background during the 1920s. They were simply too controversial for anyone to feel comfortable using them as important evidence for the course of human evolution. The first indication that they should be ignored altogether came from the problem of dating. When the Swanscombe skull was found in 1935, it revealed a comparatively modern human type from a gravel terrace of the same age as the apelike *Eoanthropus*. Marston, the discoverer of the Swanscombe skull, argued that the Piltdown jaw was that of an ape.[52] In the 1940s Kenneth P. Oakley developed a technique for estimating the age of fossils by their fluorine content. It revealed that Keith's Galley Hill man was not really ancient, but confirmed the antiquity of the Swanscombe find. Oakley demonstrated that the *Eoanthropus* remains were much more recent than the fossil animal bones found alongside.[53] His findings aroused the suspicions of Joseph Weiner and Wilfrid Le Gros Clark at Oxford, and additional tests soon showed that the remains were fraudulent: the skull was old by human but not by geological standards, and the jaw was recent. The teeth of an ape jaw had been artificially filed to give a human pattern of wearing. The remains had been stained to match the color of the animal fossils in the Piltdown gravel.[54]

The exposure of the fraud created a sensation, of course, and a good deal of embarrassment for scientists generally. Ever since, there has been a minor literary industry based on the suggestion of new suspects. Dawson was almost certainly involved in the deception, but most students of the affair believe that he must have had the support of a more competent scientist. Suggestions so far include Elliot Smith (whose views on the importance of the brain in evolution gained support from the discovery), W. J. Sollas (who held similar views, and was a bitter rival of Smith Woodward), Teilhard de Chardin (who found the canine tooth and later went on to develop a synthesis of evolutionism and religion), and most recently Sir Arthur Conan Doyle (who hated scientists because they laughed at his belief in the supernatural).[55] No one so far has suspected Keith or Woodward, both of whom rather pathetically defended *Eoanthropus* to the bitter end. I do not intend to play the detective here, but hope that a definitive solution to the mystery will eventually be forthcoming from the historians who are examining the archives of those associated with the original events.

One reason why the Piltdown remains sank into obscurity was additional discoveries of fossils that did not fit in with the pattern of human evolution implied by the human skull–ape jaw combination. The remains of "Rhodesian man" were found in a mine at Broken Hill, Rhodesia (now Zimbabwe), in 1921. There was general agreement on the Neanderthal affinities of the bones, but there were some differences from the classic European Neanderthals. Smith Woodward proposed a new species, *Homo rhodesiensis,* and the editor of the British Museum's report on the fossil tried unsuccessfully to make it a new genus.[56] In the end the Rhodesian type was accepted as a form of Neanderthal, indicating that this kind of humanity had also occupied Africa, perhaps acquiring distinct characters in that region.

Even more significant were the discoveries in the Far East which reemphasized the importance of the *Pithecanthropus* type. Inspired by the belief that Asia was the center of primate dispersal, the Canadian biologist Davidson Black went to China in the hope of discovering early hominids. At Chou Kou Tien, near Beijing (Peking), a fossil tooth was discovered in 1927, on the strength of which Black daringly announced a new genus, *Sinanthropus pekinensis*. More finds were made in 1928 and 1929, culminating with an almost complete skull. Black saw the resemblance between *Pithecanthropus* and *Sinanthropus,* but insisted that there were significant differences. He saw *Sinanthropus* as the direct ancestor of modern humans, whereas *Pithecanthropus* was a more specialized offshoot from the same stem.[57] The researchers also turned up evidence of fire and primitive stone tools at the site. These findings confirmed that *Sinanthropus* was an early human form, and "Peking man" became a scientific celebrity. Unfortunately, Black died shortly afterward, and the remains themselves were lost during the Japanese invasion of China.

In the 1930s additional *Pithecanthropus* remains were at last brought to light in Java by G.H.R. von Koenigswald. A more complete skull was assembled from the fragments recovered, showing that *Pithecanthropus* was indeed an early human form. In 1939 von Koenigswald joined with Franz Weidenreich, who was making a study of the *Sinanthropus* specimens, to proclaim the close similarity of the two forms.[58] Despite the problems created by the onset of war, Weidenreich was able to produce extensive monographs discussing the affinities of the two types.[59] He was now convinced that they were no more than geographical variants of the same stage in the evolution of the genus *Homo*. In the postwar years the two forms have been united into a single species, which, because of its status as a primitive, yet fully upright tool-user, has been named *Homo erectus*. This species is now accepted as the direct ancestor of *Homo sapiens*.

The reintegration of *Homo erectus* into the main line of human evolution also helped to bring about a new attitude toward the most exciting fossil discovery of the 1920s: the "Taungs baby." It was the skull of an immature hominid found in a lime quarry at Taungs, South Africa, in 1924 and described by Raymond Dart under the name *Australopithecus africanus*.[60] Dart argued that the creature represented an important stage in the development of modern humanity. The brain was scarcely larger than that of an ape, but the milk teeth were more human than apelike, and from the position of the foramen magnum (the hole through which the spinal cord passes) Dart inferred that it had already walked upright. The cave deposits at the Taungs quarry were difficult to date; Dart claimed a Pliocene age, but most authorities regarded them as early Pleistocene.

Arthur Keith dismissed Dart's find as irrelevant to human evolution, and it has been widely supposed that the misinterpretation promoted by the Piltdown fraud was responsible for delaying recognition of *Australopithecus*'s importance.[61] In fact, we shall see in chapter 7 that the situation was far more complex. Dart had not helped matters by publishing some deliberately provocative papers on neuroanatomy in London before taking up his position as Professor of Anatomy at Witwatersrand University in Johannesburg.[62] A few paleoanthropologists were, in fact, sympathetic to his claims, but pointed out that a mature specimen would be needed before a true estimate of *Australopithecus*'s affinities could be made.[63] If one ignored Dart's emphasis on the upright posture, it was fairly easy to fit *Australopithecus* into the scheme of human evolution commonly accepted during the 1920s. The most significant point against the Taungs skull was not its morphology but its location, since by this time the belief that humanity had evolved in central Asia was beginning to gain ground.

Dart's position was supported by the eminent, but eccentric, South African paleontologist Robert Broom.[64] In 1936 Broom began searching for adult specimens at Sterkfontein, and soon found a skull. Because it seemed geologically more recent than the Taungs fossil, he christened it *Australopithecus transvaalensis,* although he later gave it generic status by renaming it *Plesianthropus transvaalensis.*[65] In 1937 he found portions of a skull at a nearby site, but it was a more robustly built creature for which he created yet another new genus by naming it *Paranthropus robustus.* During the war years, Broom announced the discovery of an anklebone of *Paranthropus,* confirming that it had a fully upright posture.[66] He also began work on a substantial monograph covering all the South African specimens.[67]

Broom's discoveries helped convert the scientific community to acceptance of the Australopithecines as the earliest known homonids, a process that accelerated when Le Gros Clark threw his weight behind this interpretation.[68] With certain notable exceptions, principally that of Solly Zuckerman,[69] opinion now shifted to regarding the Australopithecines as an important step in the evolution of modern humans. It became clear that the upright posture had been achieved first, with the subsequent expansion of the brain giving rise to *Homo erectus* and finally *Homo sapiens.* For some time it was thought that *Australopithecus africanus* was the ancestor of modern humanity, with *A. robustus* (the modern name for Broom's *Paranthropus*) representing a side branch. More recent discoveries of early forms of *Homo* in Africa have thrown doubts on this conclusion, and it now seems that most of the known Australopithecine specimens must correspond to a parallel development that coexisted with the main line of human evolution. Don Johanson's "Lucy"—the much earlier *Australopithecus afarensis*—is now offered as a possible ancestor for both *Homo* and the later Australopithecines.[70]

Chapter Two The Framework of Debate

The archaeological and fossil evidence confirmed the antiquity of the human species but threw only indirect light on the process by which the species had evolved. Interest in the topic of human origins was so great, however, that the evolutionists were obliged to suggest hypotheses to account for this particular manifestation of their general philosophy. Inevitably, these suggestions were conditioned by preconceptions arising from religious and ideological positions. The origin of the human species was the most crucial area in which evolutionism threatened the traditional view of divine creation, and the scientists were forced to construct an alternative explanation that would satisfy their own feelings about human nature and nature itself. A tendency in archaeology to assume that cultural history was inherently progressive presented Western values as an inevitable goal toward which all development had aimed. This faith in progress necessarily shaped ideas about the biological origin of the human species. The progressionist viewpoint had direct applications to current affairs, since the "lower" races and cultures were all too easily seen as living fossils, frozen relics of past stages in the history of white Europeans. Such attitudes inevitably reinforced the prejudices upon which colonialism and imperialism were based.

Progressionism owed very little to the Darwinian theory of evolution by natural selection. In many respects, Darwin's mechanism challenged the most fundamental values of the Victorian era, by making natural development an essentially haphazard and undirected process. Few thinkers were prepared to accept the implications of applying such a system to the human species. Mankind might be incorporated into a scheme of natural development, but only if that scheme was itself progressive, so that our species could be seen as its inevitable goal. This teleological view of progress was not confined to the human sciences. In evolution theory itself, the late nineteenth century saw a massive rejection of Darwinian principles in favor of alternatives that implied that natural development was a more orderly and purposeful affair. Non-Darwinian mechanisms such as Lamarckism and orthogenesis were widely accepted by biologists, and formed a natural counterpart to the progressionist theories of the anthropologists and archaeologists. The purpose of this chapter is to emphasize the extent to which the late nineteenth century produced a framework for thinking about human origins that was distinctly non-Darwinian in character.

Evolution Theory

The phrase "Darwinian revolution" can easily obscure the complexity of the process by which evolutionism came to dominate scientific thought.[1] The early nineteenth century saw a number of developments that paved the way for the emergence of evolutionism, but these were not necessarily consistent with the approach that Darwin himself proposed. In one way or another, these attitudes continued to influence post-Darwinian evolutionism. The *Origin of Species* precipitated the conversion of the scientific community to evolutionism, but did not necessarily dictate the structure of the theory to be accepted.

At the very beginning of the nineteenth century the French naturalist J. B. Lamarck proposed a comprehensive theory of what we should call evolution in works such as his *Philosophie zoologique* of 1809.[2] Lamarck's influence on the development of evolutionism is difficult to assess, because later "neo-Lamarckians" tended to exploit a single aspect of his complex theory as an alternative to the Darwinian mechanism of natural selection. This aspect was the "inheritance of acquired characteristics," in which the individual organism acquires new characters as a result of the purposeful activity of its body responding to newly adopted habits. To use the classic illustration of the process, when the ancestors of modern giraffes began to eat the leaves of trees instead of grass, they stretched their necks to reach the trees and as a result their necks grew slightly longer. Lamarck assumed that such acquired characters could be inherited, so that the next generation of giraffes would be born with longer necks, and would continue the stretching process until after many generations the species would have achieved the present height.

In addition to his mechanism of adaptive evolution, Lamarck also believed that living forms gradually but inevitably progress toward higher levels of organization. His views were repudiated by authorities such as Cuvier and Lyell, but the element of progress was to intrude itself dramatically into the thinking of mid-nineteenth-century naturalists studying the fossil record. Once the fossil evidence for the successive introduction of the vertebrate classes had become available, naturalists were forced to account for such an apparently purposeful trend in the history of life. The most powerful method of approaching this problem stemmed from the idealist nature-philosophy now popular in Germany, in which the growth of the embryo was seen as a model for purposeful development on a universal scale.[3] In the work of Louis Agassiz, progress represented the unfolding of a divine plan of creation aimed at the production of the human form. Robert Chambers adapted this philosophy of development to give a transmutationist account of the origin of new forms in his anonymously published *Vestiges of the Natural History of Creation* of 1844.[4] Chambers saw the history of life progressing not through a series of miracles, but through a higher law of nature built into the universe by its Creator. The law caused each species to transmute itself from time to time into the next highest phase of the pattern. Here was a theory of linear progress, explicitly extended to include the human species as the high point of the process. Despite the suspicions of many scientists, it was in this form that the basic concept of evolution first became a talking point in Victorian society.

Darwin's theory came from an entirely different background. His first concern was not progression but adaptation to the environment. Natural theologians such as William Paley had interpreted adaptation as an indication of the Creator's benevolence. Darwin's study of biogeography on his voyage around the world on HMS *Beagle* convinced him that miraculous creation could not account for the way in which species are distributed around the globe. On his return to Britain he became interested in the artificial production of new characters in animals by breeders. This interest led him to the idea that a species may be systematically modified by selecting particular characters from the random variation that appears in any natural population. After reading Thomas Malthus's *Essay on the Principle of Population,* Darwin realized that the struggle for existence caused by the limitation of food supply would exert a similar selective force in nature.

The theory of natural selection thus depends upon the environment evaluating the individuals who make up a population according to the utility of the characters they are born with. Any individual with a character useful in the struggle for existence will breed more frequently than the average, whereas any with a harmful character will be eliminated. The population will thus adapt to any changes in its environment by a process that Herbert Spencer later called the "survival of the fittest." Darwin could not explain *why* individuals differ among themselves, but he held that it was obvious that variation did occur, and that it is not directed toward any useful goal. Selection alone exerts a guiding influence, constantly picking out those new characters which by chance turn out to be of some use in adapting to the environment.

Natural selection works solely to adapt a population to its local environment—it cannot give automatic progress toward a predetermined goal. It was clear to Darwin that evolution is not a linear process, but one of constant branching and divergence. Sometimes individuals from a single original species will migrate to a different area and become subject to new conditions. The group will then adapt to its own local environment in its own unique way. Eventually a single original species may produce a number of "daughter" species, each moving off along its own independent evolutionary path. Since geographical conditions undergo gradual change because of geological factors, evolution will be an essentially haphazard process, each species adapting to whatever conditions it encounters and splitting whenever necessary. Any population that cannot adapt quickly enough to a change in its environment will become extinct. The only kind of long-range trend permitted by this mechanism is continued specialization for a particularly successful way of life.

After many years of delay, Darwin published the *Origin of Species* in 1859. The book sparked off an immense controversy, but by the 1870s a substantial majority of scientists and interested laypersons had accepted the general idea of evolution.[5] Yet the world was not converted to "Darwinism," if we restrict that term to a belief in natural selection as the principal means of evolution. The situation varied enormously from country to country.[6] Only in Britain did a strong Darwinian camp emerge, led by T. H. Huxley and his scientific disciples. In Germany and America, a strong element of progressionism and Lamarckism was incorporated into many of the earliest evolutionary theories. France took much longer to convert, and here Lamarckism was the first prominent evolutionary mechanism. Even in Britain, Darwinism eventually, during the 1890s, declined in popularity in an episode that Julian Huxley was later to call the "eclipse of Darwinism."[7]

Because he realized that evolution is a highly irregular process, Darwin himself made few efforts to reconstruct the genealogy of particular modern forms. The fossil record gave a rough outline of the history of life, but was too fragmentary to be of any real use. Even Huxley at first refused to speculate about the evolutionary origins of modern forms, on the grounds that many types may have come into existence long before they first became preserved as fossils.[8] The late nineteenth century's fascination with the detailed reconstruction of the history of life was thus not a direct product of Darwinism.[9] The originator of this program was the German biologist Ernst Haeckel, who actually coined the term "phylogeny" to denote his hypothetical lines of descent. For Haeckel, evolution theory was part of a philosophy of cosmic progress. In his *Generelle Morphologie* of 1866 and his more popular *History of Creation* (translated in 1876), Haeckel synthesized Darwinism with a generous helping of Lamarck's inheritance of acquired characters and an almost idealist concept of life unfolding through a progressive sequence toward mankind. To flesh out his vision, he was eager to attempt a complete, if highly speculative, reconstruction of the history of life on the earth. He pioneered the use of treelike diagrams to illustrate how new forms have branched off from old ones in the course of evolution. Where fossil evidence was lacking, Haeckel used comparative anatomy to work out the most likely relationships, hence his willingness to see the lower forms of life as only slightly modified relics of earlier stages in the upward progress of the higher forms. Embryology also helped to fill in the gaps, and it was Haeckel who created the popular—and erroneous—image of embryonic growth recapitulating the evolutionary history of the species.

In the English-speaking world, the search for a more purposeful trend in evolution still had an explicitly religious basis. Even some of the biologists rejected Darwinism in favor of the belief that the evolutionary process has been directly supervised by the Creator. Chambers's belief in the gradual unfolding of the divine plan without miracles, once rejected as heretical, was now seen as the salvation of natural theology. Leading anatomists such as Richard Owen and St. George Jackson Mivart opted for this approach. To stress the inadequacy of natural selection, Mivart's *Genesis of Species* of 1871 drew attention to the apparently nonadaptive character of some evolutionary trends. Mivart believed that many aspects of evolution revealed regularities of development that were incompatible with the Darwinian mechanism and indicated a divinely imposed predisposition for life to evolve along certain fixed paths.[10]

One result of this continued interest in teleology was a growing respect for Lamarck's once-discredited mechanism of the inheritance of acquired characters. As the novelist Samuel Butler—a bitter critic of Darwin—pointed out, Lamarckism would allow one to believe that life was purposeful after all.[11] Instead of being designed by an external Creator, species have been given the power to design themselves by their own efforts: faced with a changed environment, animals will adopt purposeful new habits, and the inheritance of acquired characters will then produce appropriate changes in their physical structure. Butler's views were at first rejected by the biologists, but outside Britain Lamarckism had already become a potent force by the 1880s. In America there was a prominent neo-Lamarckian movement, and the paleontologist Edward Drinker Cope stressed a religious interpretation of the theory similar to that advocated by Butler. The Americans, however, came to Lamarckism in a unique way. Many were students of Agassiz, and they absorbed his idealist vision of development as the unfolding of a divine plan, mirrored in the purposeful growth of the embryo. They believed that evolution consisted of the steady addition of stages to the growth of the individuals making up the species. Eventually they realized that the inheritance of acquired characters would explain how such additions were made, as the species adapted itself to new habits. It was possible for a number of related forms to acquire parallel specializations adapting them to a shared way of life.

By picturing development as a process in which a number of parallel lines advanced in a common direction, the American school rejected the Darwinian image of haphazard, branching evolution. In some cases, they detected trends that seemed to go beyond the point of utility, driving a whole group inexorably to extinction. The term "orthogenesis" was introduced to denote this process of regular, nonadaptive evolution. A classic illustration was supposed to be the antlers of the so-called Irish elk, which were thought to have grown so large that they caused the species' extinction. Henry Fairfield Osborn, a pupil of Cope, continued to expound this mode of evolution well into the twentieth century. His work on the Titanotheres, a group of extinct mammals, seemed to show a series of parallel lines advancing through the same pattern of growth and overdevelopment. Osborn's work on parallel evolution was often cited as a model for theories of human evolution, although in this area the potentially harmful character of orthogenesis was conveniently forgotten.

In Darwin's theory of branching evolution, similarity of structure indicates community of descent: the modified descendants of a single ancestor still preserve an underlying similarity of structure inherited from that ancestor. But the Americans' theory supposed that a number of similar forms might actually belong to lines of development that had been separate for a long time, each independently acquiring the same new characters as the result of a shared tendency to vary in the same direction. To what extent, then, was similarity of structure a guide to evolutionary relationships? If several long-distinct lines could produce similar end products, phylogenies built on the assumption that similarities indicate close relationship would have to be reevaluated. If one applied this concept to human evolution, it might be argued that the similarities between humans and apes were acquired independently, so there was no close link between them. If one applied the same principle in a different way, it might be argued that the human races were the products of separate lines of evolution that had merely converged so closely together that they could now interbreed.

Since the question of the independent evolution of similar characters loomed large in discussions of human origins, it is necessary to have a clear understanding of the concepts involved. The two terms most frequently encountered in this area are "convergence" and "parallelism." Both are still in use today, but their meanings seem to have changed significantly since the reemergence of Darwinism in the 1940s. After a detailed survey of how the terms had been used up to 1946, Otto Haas and George Gaylord Simpson offered little more than a literal interpretation as the modern definitions: convergence occurs when two lines of evolution grow closer together, parallelism when they move in the same direction maintaining the same degree of separation.[12] Whatever the current usage, I suggest that in the early twentieth century these terms carried quite different implications for the evolutionary mechanism involved. Convergence almost always implied the coming together of two evolutionary lines *because they are subject to similar adaptive pressure.* If two forms adapt to essentially the same ecological niche, they will tend to solve the problems in the same way and acquire similar structures. Thus some of the Australian marsupials bear a striking resemblance to more familiar placental mammals—there is a marsupial "wolf" for instance. This resemblance is not the result of any inbuilt trends driving marsupials and placentals in the same direction, as witnessed by the kangaroos, for which there is no placental equivalent. But in some cases, similar adaptive problems have been solved in a similar way, and the result is convergence toward a superficially similar structure.

The term "parallelism," on the other hand, was almost always used to imply not an external (i.e., adaptive) but an internal (i.e., biological or genetic) pressure forcing two lines in the same direction. Even if exposed to different environments, the parallel lines would continue toward the same goal. The trend might originally have been acquired in response to an adaptive purpose, but once established it gained a momentum of its own that would continue whether or not the results were of any benefit, if necessary driving all the affected forms to extinction. In the early twentieth century, parallelism usually implied orthogenesis: evolution driven by internally programmed trends forcing variation along predetermined lines.

Orthogenesis, and hence parallelism in this original sense of the term, are inherently non-Darwinian concepts of evolution. When used to explain human origins, they implied that there was indeed a predetermined goal toward which all branches of human evolution were being driven. A Darwinian could, however, accept a limited amount of convergence, since here the driving force was still under the control of adaptation. In practice, though, a Darwinian could only accept the power of convergence to produce a superficial similarity of structure. Those anatomists and paleoanthropologists who thought that convergence could produce detailed similarities, enough to fool us into thinking that two distinct forms were closely related, were generally under the influence of a non-Darwinian theory of evolution. The long association between Lamarckism and the idea of guided or purposeful evolution encouraged a belief in the power of convergence to restructure distinct forms along fundamentally similar lines.

After 1900, the new experimental science of genetics began to undermine the foundations of both Lamarckism and orthogenesis. There seemed to be no evidence that the production of new genetic characters by mutation was shaped either by the needs of the organism or by an inbuilt tendency for the genes to mutate in a particular direction. Yet those biologists trained in more traditional fields, including anatomy and paleontology, paid little attention to these developments, with the result that Lamarckism and related concepts continued to be employed in these fields long after the experimentalists had repudiated them. The early twentieth century was a period of polarization within biology, with the older disciplines refusing to accept the right of the new genetics to dictate what they might include within their theories. It was always possible to argue that the Lamarckian effect worked so slowly that it could not be detected on the time scale of a laboratory investigation. Most of the scientists we shall deal with below came from the older professions. They were either paleontologists, such as Smith Woodward or Osborn, or anatomists from a medical background, such as Keith and Elliot Smith. They worked in the great museums or teaching hospitals, and had little contact with the brash new departments of genetics springing up in the universities. Only as biologists began to take stock after World War II was the Modern Synthesis of Darwinism and genetics able to sweep away the outmoded ideas and introduce the framework within which we still discuss the problem of human evolution today.

The failure of communication between the disciplines earlier in the century is evident from the lack of interest shown by paleoanthropologists in the first important initiative in evolution theory to come from the experimentalist camp—the "mutation theory" popularized by Hugo De Vries.[13] The mutations were much more dramatic than anything covered by the modern use of the term in genetics, since De Vries believed that new species could be produced instantaneously through the appearance of individuals with entirely new characters. Such a theory could certainly be applied to human evolution, and the experimentalists were aware of this. De Vries himself discussed and rejected the possibility that the human races might have been produced by mutations.[14] An American supporter suggested that the theory might vindicate the story of Adam and Eve.[15] Yet the majority of scientists concerned with human origins would have nothing to do with the idea. J. Arthur Thomson, a popular Darwinian writer, did note the possibility briefly, suggesting that "primitive man expressed a mutation, a sudden uplift, separating him by a leap from the animal."[16] Of the major figures, only Eugene Dubois appealed in his later career to the idea of a mutation producing a sudden increase in brain size.[17] By this time, however, Dubois was almost completely out of touch with contemporary thinking on the question.

The refusal of the anatomists and paleontologists to respond to this new idea is quite understandable, even supposing that they were aware of the experimentalists' literature. Mutations in this context could all too easily become an excuse for treating the appearance of the human mind as an unprecedented step into a new world, just as earlier thinkers such as Charles Lyell had appealed to a saltative origin for mankind to defend their belief in the human soul. The fact that mutations were produced by an unpredictable genetic mechanism would have undermined the assumption that human evolution could be analyzed by the techniques already developed by paleontologists for studying the course of life's history. The authorities might disagree over how the evolution of human characters had occurred, but they would have nothing to do with a theory that reduced the whole affair to a chapter of accidents produced by random genetic changes.

Meanwhile the geneticists had given up the claim that mutations create new species, accepting instead the modern view that mutations merely add genetic variability to the existing population. The emergence of the science of population genetics in the 1920s and 1930s reintroduced the concept of natural selection, since it was success in a given environment that seemed to determine the frequency each gene could maintain in the population. The Modern Synthesis was completed during the 1940s when naturalists such as Theodosius Dobzhansky, Ernst Mayr, and Julian Huxley showed how this revived Darwinism could be applied to a wide range of field studies.[18] George Gaylord Simpson demonstrated that the synthesis was compatible with paleontology, thereby undermining the evidence for Lamarckism and orthogenesis. The theories of directed evolution that had remained popular with paleontologists into the 1930s were now rapidly abandoned, forcing those concerned with human evolution to rethink their position accordingly. The inflated claims for specific or generic rank made on behalf of each hominid fossil were rendered implausible by the new Darwinism's emphasis on the variability of natural populations. Only now did the study of human evolution begin to take on a recognizably modern form.

Race, Culture, and Progress

Contemporary views on the nature of the evolutionary mechanism inevitably formed part of the foundation upon which a theory of human origins was constructed. Equally important, though, were the current opinions of anthropologists attempting to chart the physical and cultural development of the human race. A theory of human origins had to explain how "primitive" humans evolved from animal ancestors, and hence was influenced by prevailing ideas on the primitive state. It was widely believed that the "lower" human races were both physically and culturally primitive with respect to white Europeans. They were, in effect, the relics of ancestral stages through which the "higher" races and cultures had passed long ago. Anthropology thus dictated the goal toward which any account of human origins must be aimed. Furthermore, the progressionist image of cultural and racial development provided an obvious model for those seeking to extend their studies into an earlier period.

The predominant mode of thought in late-nineteenth-century anthropology was certainly progressionist. The "evolutionary" anthropology of E. B. Tylor and L. H. Morgan postulated a hierarchy of cultural stages through which all races were supposed to pass, and physical anthropologists were only too willing to depict the "lower" races as the bottom rungs of a ladder stretching from the ape to the white man. The concept of linear progress along a predetermined hierarchy was distinctly non-Darwinian in character, forming an anthropological counterpart to the theories of directed evolution still popular among the biologists. Lamarckism was often appealed to directly as a means of explaining how cultural and intellectual progress went hand in hand. Alternative models of racial and cultural development were available, and were more consistent with an image of branching evolution. But these alternatives were often subverted by the progressionist viewpoint in an area in which there was strong ideological pressure to justify the worldwide expansion of the white races.

Anthropology was "evolutionist" in the sense that it sought to describe the process by which modern societies had developed from some hypothetical primitive state. Opinions differed widely over what that primitive state had been. In the field of social anthropology, John Lubbock, John F. McLennan, and L. H. Morgan favored the view that the original state had been a "promiscuous horde," each offering his own interpretation of how the horde had acquired more complex social patterns.[19] They were opposed by Sir Henry Maine, whose *Ancient Law* of 1861 argued that modern legal systems had developed from patriarchal family groups. Maine later noted that his position was supported by Darwin, who realized that since apes live in family groups, it was likely that our own ancestors had followed the same pattern.[20] The popularity of the general concept of evolution can be measured by the extent to which late-nineteenth-century writers began to assume that human social behavior is conditioned by instincts inherited from our animal ancestors. Edward Westermark postulated an evolutionary origin for the instinct to avoid incest.[21] The leading exponent of this approach was William MacDougall, whose *Introduction to Social Psychology* of 1908 sought to explain all aspects of social life in terms of instincts built into the individual in the course of evolution.[22]

It was in the field of cultural anthropology that the most highly structured form of progressionism flourished. We have already seen how the archaeologists arranged their stone age cultures into a neat hierarchy of developmental stages. This approach was popularized by John Lubbock, who also exploited what became known as the "comparative method"—the view that modern primitives are relics of prehistoric stages in cultural evolution.[23] Archaeology was also an influence on the leading

British exponent of cultural evolutionism, Edward B. Tylor, who acknowledged the support of Henry Christy (Lartet's companion in the Dordogne excavations).[24] Tylor's purpose was to take the skeleton of technological progress revealed by archaeology and flesh it out by reconstructing the general culture of each stage in the process. He was particularly interested in the growth of religious beliefs, which he interpreted as efforts to rationalize certain confusing aspects of human experience such as dreams. He regarded modern primitives again as illustrations of the earlier stages of cultural development; by ranking the various levels of culture still seen in the world today, Tylor constructed what he assumed to be a universally valid scheme of progress.[25] All races had developed at different speeds along the same general path, as revealed by the essentially similar cultures and inventions acquired by savages all over the world. Although at first willing to acknowledge the possibility that ideas can be transmitted from one culture to another, Tylor became increasingly convinced that similarities among cultures must be the result of independent invention forced into parallel channels by the need to follow the same hierarchy of cultural stages.

Similar views were expressed in America by Lewis Henry Morgan, whose *Ancient Society* of 1877 defined the successive stages of savagery, barbarism, and civilization. The stages formed a "natural as well as necessary sequence of progress," and all except the earliest were still visible in the world today.[26] The growth of culture was guided by a "natural logic which formed an essential attribute of the brain itself" and which gave uniform results wherever it was applied.[27] The same approach to cultural evolution was adopted by Adolph Bastian in Germany. This system required a belief in what was sometimes called the "psychic unity" of all mankind: all races were subject to the same law of mental growth, although they may have ascended the hierarchy at different rates. To begin with, at least, there was no implication that the culturally more primitive races were intellectually inferior. J. W. Burrow has shown how, in the case of Tylor and the British anthropologists, cultural evolutionism arose not under the influence of Darwinism, but from a desire to preserve a faith in the worldwide unity of human nature inherited from the eighteenth-century Enlightenment.[28]

Whatever the origins of the progressionist interpretation of cultural history, it proved impossible for the anthropologists to resist the growing feeling that the culturally primitive races were also intellectually inferior. In his *Anthropology* of 1881, Tylor himself accepted that some races have shown themselves incapable of rising very far up the scale of civilization because of their smaller brain size.[29] Morgan too believed that intellectual capacity increased with the level of cultural development.[30] It has been argued that the link between cultural and intellectual development was founded on a kind of feedback loop sustained by Lamarckism.[31] An advanced culture stimulated the brain, which increased in size and so made possible a yet more rapid rate of cultural progress. Those races left behind in the ascent of the cultural scale are now left with smaller brains which prevent them from responding to civilization. The Lamarckian effect was appealed to explicitly by American anthropologists such as John Wesley Powell and W. J. McGee.[32]

The anthropologists did not necessarily believe that their theory offered a model for the evolution of the earliest human type from its animal ancestors. Tylor, for instance, explicitly denied any connection between cultural progress and the biological origins of the human race.[33] But for confirmed evolutionists such as Lubbock, the connection seemed inescapable. In the absence of fossil evidence for the emergence of humans from apes, archaeology showed that our ancestors were culturally and hence intellectually primitive. Cultural development was merely a continuation of the intellectual progress brought about by biological evolution. Lubbock was challenged on this point by the Duke of Argyll, a prominent opponent of Darwinism, who pointed out that technological progress was no guarantee of intellectual or moral development.[34] But the late nineteenth century was too heavily committed to progressionism for such an objection to be sustained. In France, Paul Broca even tried to show that there has been a significant increase in brain size among Europeans since the Middle Ages. A whole science of "craniometry" was founded to demonstrate the mental inferiority of the colored races.[35] This alleged inferiority was often linked to the supposedly apelike features of these races as a means of confirming their retarded evolutionary status. The neo-Lamarckian paleontologist Edward Drinker Cope wrote extensively on this theme, confirming the image of the lower races as intermediate steps in the ascent from the apes to the highest human type.[36] As late as 1910, careful studies were still being made to assign each race, living and extinct, to a position on a scale of brain size ranging from the ape to the most advanced human levels.[37]

Part 2 shows how this linear pattern of development was copied in some of the early theories of human origins. Conflicting tendencies in the social sciences, however, helped to create a more complex interpretation of cultural and racial development. These can be seen at work in the Victorian era's philosopher of progress Herbert Spencer. In his *Principles of Psychology,* originally published in 1855, Spencer adopted the Lamarckian view that habits eventually turned into biologically imprinted instincts. He also agreed that those races in which the brain had been used most actively, because of a more stimulating environment, acquired a higher level of intelligence.[38] He thus accepted the mutually stimulating interaction between cultural and biological evolution, a view expressed once again in his *Principles of Sociology* of 1876: "development of the higher intellectual faculties has gone on *pari passu* with social advance, alike as cause and consequence."[39] Yet Spencer was suspicious of the neat, linear pattern of cultural development established by Tylor and Morgan. He realized that if each race has developed in a different environment each may have acquired its own unique instincts.[40] Thus we cannot assume that modern primitives are the exact equivalents of earlier stages in the white race's development.[41] Progress only occurred when conditions were suitable, and each society was the product of a historical process shaped by the conditions through which it has passed. "Like other kinds of progress, social progress is not linear but divergent and re-divergent."[42]

Spencer could not escape the prevailing view that some races were superior to others, but he was prepared to insist that cultural and racial differences cannot be expressed in a simple linear hierarchy. In a system of branching development, differences are not necessarily a sign of superiority or inferiority. The implications of this point were increasingly recognized as the century drew to a close. In Germany, Friedrich Ratzel attacked the evolutionists' assumption of their own superiority and their tendency to regard cultural backwardness as a sign of racial inferiority.[43] Ratzel argued that the people of each area will have developed their own unique culture. When similarities are found among distant cultures, they should be attributed to borrowing rather than independent invention in accordance with a preordained law of mental development. In 1911 the British anthropologist W.H.R. Rivers was led to similar conclusions as a result of his failure to find a linear hierarchy of cultures in Oceanea.[44] The theory of cultural diffusion was eventually pushed to absurd lengths by Rivers's student, W. J. Perry, and by Grafton Elliot Smith (see chapter 9). Few anthropologists paid any attention to Smith's claim that all civilization has been exported from Egypt, or to his synthesis of diffusionism with a theory of human origins. By the 1920s, the social sciences in Britain and America had repudiated the racism and progressionism of the cultural evolutionists to such an extent that they were no longer interested in the biological origins of the human species. For Franz Boas and the students he trained in America, cultures were not constrained by biology, and each society was an independent development to be studied in its own terms, not by comparison with others.[45]

The anthropologists had now set up a branching, rather than a linear, model of cultural development, but they had dissociated it entirely from the question of biological evolution by assuming that human beings have acquired an infinitely plastic capacity for social interaction. Since Boas and his followers attacked the claim that some races are biologically inferior to others, their rejection of the linear model cannot have gone unnoticed by the physical anthropologists. There was, in any case, a parallel movement in physical anthropology which provided an alternative to the linear ranking of the races on a scale stretching up from the apes. Traditionally, the human races were held to be members of the same species, since all were descended from Adam and Eve. This "monogenist" position ran into difficulties, however, when it was realized that, to fit in with the biblical story, the separation of the races must have taken a comparatively short period of time. Thus James C. Prichard had been forced to assume that the human species is extremely malleable.[46] This assumption seemed to conflict with evidence suggesting that the races have remained stable throughout the period of recorded history. The alternative was to opt for "polygenism," the belief that the races are separate species with distinct origins and characters. Leading figures in this movement during the mid-nineteenth century were the Edinburgh anatomist John Knox and the Americans Samuel George Morton and Josiah C. Nott.[47]

The polygenists were convinced that the colored races were inferior to the white—indeed they wished to exaggerate the extent of the gulf between them—but they were not committed to a *linear* hierarchy of racial characters. Their position thus created a tension in the ranks of the anthropologists by encouraging a search for distinct racial types that were not always easy to fit into the progressionist model of cultural and racial evolution. Despite his emphasis on the scale of racial brain capacities, Paul Broca led the polygenist faction in France through his establishment of the *Société d'anthropologie de Paris* in 1859.[48] Broca accepted that the human species must be included along with the apes in the order of primates,[49] but he insisted that there was no simple hierarchy from ape to human.[50] He also rejected the Darwinian model of evolution, in which all related forms are derived from a common ancestor, arguing instead for many separate lines of evolutionary development.[51] The human races were, in fact, the products of separate lines, and Broca challenged the popular belief that they could interbreed freely.[52]

In Britain, James Hunt founded an Anthropological Society in 1863 to promote the polygenist view of race.[53] Significantly, most of the evolutionists, including Lubbock, Huxley, and Tylor, refused to join. Hunt rejected evolution precisely because it threatened to link together by common ancestry the racial types he wished to hold rigidly apart. Even those anthropologists who were not outright polygenists now began to search for distinct racial types, each of which was to be carefully described and classified. Within Europe itself, several different races were distinguished. One of the most important criteria used in this analysis was the "cranial index," a comparison of the width of a skull to its length. The Swiss anthropologist Anders Retzius emphasized the importance of this ratio, distinguishing the long-headed dolichocephalic types from the more round-headed brachycephalics. Such factors were difficult to accommodate into a linear scale, and more complex systems of classification were thus required.

By the turn of the century, a comprehensive system of racial classification had been established. Joseph Deniker's *Races of Man* of 1900 postulated six basic types, but with a multitude of subtypes to complicate the question. William Z. Ripley's *Races of Europe* distinguished three basic races in this area: the Nordic or Teutonic, the Mediterranean, and the Alpine.[54] Although not explicitly polygenist, these anthropologists saw little prospect of explaining how the races had become separated. Some historians of anthropology, including George Stocking and C. Loring Brace, have argued that this whole approach to the issue rested upon a pre-Darwinian viewpoint, with an almost Platonic concept of races as distinct natural categories.[55] The vast antiquity of the races was taken for granted, and used as an excuse to stop thinking about their evolutionary origin. Thus Broca's pupil, Paul Topinard, eventually backed away from polygenism to argue that the origin of the races was too remote for serious investigation. The modern races were mixtures of certain pure types formed in the distant past by a process no longer in operation.[56] Deniker too treated the races as pure types and dismissed the monogenist-polygenist debate as futile.[57]

Despite these evasions, however, the analysis of race was not without a potential to influence evolutionary thinking. Writers on the race question were anxious to construct a *historical* account of how the races came to be in their present locations. An early stimulus to this approach had come from the philologists, who had now traced both Sanskrit and most

European languages back to a hypothetical "Aryan" tongue spoken in the distant past. Since Sanskrit was the most ancient of the known languages, philologists such as Max Müller assumed that the original Aryan had been developed in Asia and carried into Europe by the invasion of a people who were naturally called Aryans.[58] By the 1890s this hypothesis had been strongly challenged, but interest in the concept of racial migrations did not decline.[59] Retzius had postulated that the original inhabitants of Europe had been brachycephalic, and had been subdued by the invading Aryan dolichocephalics. However, the discovery of the "old man" of Cro-Magnon, a skeleton dating from the late paleolithic, showed that a highly developed dolichocephalic race had been present in Europe long before the great migrations of the neolithic.[60] Ripley suggested that the Cro-Magnon race had evolved into the Nordic and Mediterranean types by adapting to the northern and southern parts of Europe. The Italian anthropologist Giuseppe Sergi believed that the ancestors of these dolichocephalic peoples had entered Europe from Africa. Later on, the brachycephalic Alpines had invaded from the east.[61] The population of the British Isles was thought to be composed of several racial groups, the earliest inhabitants having been pushed into marginal areas by later invaders.[62] Efforts were also made to seek out the remnants of paleolithic types displaced from Europe by the invasion of higher races, as in the popular belief that the Eskimos were the descendants of the Magdalenian reindeer hunters.

These ideas had strong political implications, of course. French anthropologists such as Broca were suspicious of Retzius's theory of an invading Aryan dolichocephalic race. Since the French are typically brachycephalic, the theory seemed to proclaim the superiority of the Aryan type more common in Germany. The French anthropologists themselves were quite willing to slander the racial character of their opponents in both the Franco-Prussian War of 1870 and World War I.[63] When the Aryan type was later identified with the locally evolved Nordic or Teutonic race, the way was clear for the creation of a whole pseudoscience of racial superiority which culminated in the mythology of the Nazis.[64] This was not a purely Germanic theme, however, since writers elsewhere in Europe and in America emphasized the Aryan or Anglo-Saxon's role in world history.[65]

The studies of race could form a direct model for theories of human origins. During the early twentieth century, the claim that superior races conquer new territory, displacing or exterminating the original inhabitants, was accepted as a major theme in human evolution. The insistence on the vast antiquity of racial types also had implications for human origins, since it implied that there were many parallel lines of evolution. It was possible to synthesize this position with the idea of progressive development by supposing that several lines of evolution have independently advanced toward the human form, at the same time acquiring distinctive characters suited to their area of origin. The basic principle of linear progressionism was thereby preserved, while admitting that the "lower" races have some distinctive features in addition to their preservation of a primitive evolutionary character.

The possibility of constructing a more sophisticated version of progressionism had been present from the beginning of the evolutionary debate. In 1864, Alfred Russel Wallace had tried to reconcile monogenism and polygenism by arguing that the races had separated into distinct evolutionary branches *before* participating in the final expansion of the brain to a fully human level.[66] In Germany, Ernst Haeckel openly maintained that the races were equivalent to distinct species, and traced the main types back to different species of ape-men (*Pithecanthropi*), each of which had independently achieved human status.[67] Haeckel retained the most important aspect of the linear scale by insisting that the woolly haired races, the Ulotrichi, preserved a greater proportion of apelike characters than the straight-haired Lissotrichi.[68] Yet his system then allowed for branching developments within the two types, coinciding with the advance to the human level. Pushing this technique to its limits, the radical German biologist Karl Vogt advocated a polygenist theory in which the human races were derived from separate ape ancestors.[69]

Evolution theory was thus flexible anough to accommodate both a goal-directed progress from the ape to the human form and the vast antiquity of distinct racial types. To do so, it exploited the concept of parallel evolution outlined above, in which distinct branches (each with its own minor distinguishing characters) can move independently through the same scale of development. Since this idea was still popular among paleontologists well into the twentieth century, it provided an obvious inspiration for paleoanthropologists seeking to emphasize racial differences while retaining the idea of progressive development toward the human form. Wallace, Haeckel, and Vogt thus anticipated the conceptual foundations of many later theories of human origins.

Part Two Phylogeny

Chapter Three Up from the Ape

Various approaches have been tried out in the construction of a phylogeny, or evolutionary tree, for the human race. We are dealing here with the course of human evolution, not its cause. The two topics are by no means unrelated, although a good deal of the literature on human phylogeny was written as though it were possible to put together a family tree for mankind by an abstract analysis of the data derived from comparative anatomy and paleontology. In fact, the different approaches to the question of how we are related to the rest of the animal kingdom do tend to presuppose different views on the nature of the evolutionary process, even though it may take a little detective work to uncover the links. Nevertheless, the question of phylogeny is, by itself, surprisingly complicated, and there is much to be said for trying to get a clear picture of the options before looking at theories of causation. It is generally assumed that evolution theory requires a belief in the descent of mankind from the apes. Certainly, a link with the apes is the most frequently explored avenue, but a minority of biologists have been prepared to consider views that exclude the apes altogether from human ancestry. Even when the ape link is accepted, there are many questions about the details of the relationship. Questions only increased once a number of fossil types had been discovered and had to be fitted into the picture.[1]

The popular image of human evolution, parodied in a host of cartoons, takes for granted our descent from an ape ancestor. The similarities between humans and apes had been widely commented on in the eighteenth century, and Linnaeus had classified both together in the order primates. The view that mankind might actually have arisen by transmutation from an ape was advanced by J. B. Lamarck in his *Philosophie zoologique* of 1809. Lamarck argued that mankind had probably arisen when a quadrumanous animal (an ape) had begun to walk upright and thereby gained supremacy over all the other animals. It was popularly supposed that he believed the ancestor of the human species to be the orangutan, although this supposition may have been an oversimplification of his position. Lamarck's was not a theory of divergent evolution; instead he tended to think in terms of parallel lines advancing independently along the chain of being. He thus needed only to suppose that the ancestral form of mankind had passed through an apelike stage, without postulating a link with the living apes. In fact, he noted that of the modern apes the orang of Angola (*Simia troglodytes*) is more advanced than the orang of the East (*S. satyrus*).[2] Despite his use of the term "orang"—Malay for "man" as in orangutan—the first-named species is the chimpanzee of Africa. Nevertheless, it was popularly believed that Lamarck had derived the human species from the orang, and this connection was seen as one of the most disturbing aspects of his theory.

Georges Cuvier used his immense prestige within French science in an effort to discredit Lamarck's transformist theory. He also attempted to render the link between mankind and the apes less plausible by adopting Blumenbach's classification based on two distinct orders, the Bimana and Quadrumana. The Bimana ("two handed") contained a single species, *Homo sapiens,* whereas the Quadrumana included all the apes, which were supposed to be "four handed" in the sense that their feet were adapted for climbing rather than walking and thus differed little in structure from their hands. Conservative naturalists thus challenged both the general idea of transmutation and the specific idea of human evolution from the apes. Nevertheless, Lamarck's theory was not without its supporters among the more radical French and British naturalists. We now know that the threat of the proposed link between humans and apes was sufficiently obvious in the 1830s to create extreme anxiety among those who found the religious implications of the link unacceptable. The geologist Charles Lyell and the anatomist Richard Owen were among those who were already shaping their scientific thinking in a manner designed to block any further development of the link.[3] Owen's later conflict with T. H. Huxley over the closeness of the relationship between mankind and the apes was thus a continuation of a policy he had already outlined much earlier in his career.

By the early 1840s, the threat of transmutationism seemed to have become less acute, at least in Britain, as the Lamarckians were discredited by the creation of a discontinuous interpretation of the progressive steps in the fossil record. The next phase of the debate opened in 1844 with the appearance of Robert Chambers's anonymously published *Vestiges of the Natural History of Creation.* Chambers accepted that evolution was the unfolding of a divine plan, but his claim that the human race was the product of the progression through the animal kingdom still emphasized the most disturbing implication of the theory. His book was popularly written, in an attempt to argue the case for transmutation directly to the general public, over the heads of the professional naturalists who had now consolidated against the idea. The scientists responded by criticizing the book severely, although the resulting debate certainly succeeded in promoting public awareness of evolutionism.

Curiously, however, Chambers did not revive the old fears of a direct link between mankind and the apes. Although he accepted the emergence of mankind from a lower form, his unique concept of how evolution proceeded led him to avoid the simple image of an ape ascending the scale to reach the human level. He renamed the primates the "Cheirotheria" (having grasping hands) and insisted that the apes, the Simiadae, were *not* the highest members of the order after mankind.[4] In his first edition he adopted a circular system of classification and claimed that the human form concentrated the qualities of all the members of the order. Later he abandoned this system, perhaps because it was difficult to reconcile with his generally linear view of progression, but he still made no effort to draw a direct link between mankind and the apes. In his chapter on the development of the human races, Chambers presented the "lower" races as immature versions of the white, but made no effort to characterize them as apelike.[5] The concept of increasing maturity fitted in with his idea that evolution works through the addition of stages to the growth of the individual. This view was shared later in the century by the American neo-Lamarckians, and was thus eventually linked to the hypothesis of an ape origin. Edward Drinker Cope in particular sought to identify some of the immature characters possessed by the lower races with their supposedly lower level of development from the ape form.[6] But Chambers himself did not make this link, and the opponents of his book seldom appealed to the image of an ape origin as a means of discrediting it. They concentrated instead on the general threat posed by a materialistic account of human faculties, and used the general discontinuity of the fossil record to undermine the whole idea of transmutation.

The popular belief that we evolved from the apes was thus largely a post-Darwinian phenomenon. It was Thomas Henry Huxley's detailed comparison of human and ape anatomy in his *Man's Place in Nature* of 1863 that established the case for a close relationship, which was sometimes called the "pithecometric thesis." Huxley implied that our relationship to the apes is so close that an evolutionary link is inescapable. Yet this thesis was not quite so straightforward as the cartoonists' image. The human race has certainly not evolved from the gorilla, although our ancestors may have passed through a stage resembling a less specialized gorilloid form. At best, the human race and the various apes shared a common ancestor, each branching in its own direction from the ancestral form. If the fossil record were to reveal the ancestor, it would probably turn out to be less specialized than any of the modern apes. It would still be recognizable as an ape, however, whereas the branch leading to mankind has acquired a whole range of new characters defining it as a separate family, the Hominidae.

It was difficult to predict exactly what the common ancestor might have looked like. Were the African or the Asian apes closer to the original form? Was it possible that the common ancestor was not a great ape at all, but a smaller creature more like a gibbon? E. Ray Lankester later suggested that the Piltdown discovery (a human skull with an apelike jaw) was responsible for shifting opinion away from the idea of a small ape toward a great ape ancestry.[7] Yet it is difficult to believe that the popular image of the Darwinian era was quite so far out of touch with the scientific viewpoint. Darwin and Huxley both linked the human race with the African apes, and Darwin supposed that our early ancestors had retained the great canine teeth of the larger apes. This supposition would imply that the common ancestor was an ape of substantial proportions, although it was most unlikely to have been as large as the modern gorilla. It was to fossil apes such as *Dryopithecus* that most authorities looked for evidence of the actual appearance of the more generalized apelike ancestor of mankind.

Darwin did not address the question of human evolution in the *Origin of Species,* confining himself merely to the hint that "light will be thrown on the origin of man and his history."[8] It was thus left for Huxley's *Man's Place in Nature* to present the first detailed discussion of the issue. This book originated in a dispute between Huxley and Owen on the extent of the differences between mankind and the apes. Owen still wished to preserve the conservative system of classification in which the two were separated into different orders. Blumenbach and Cuvier had concentrated on the structure of the feet and hands to create the two orders, Bimana and Quadrumana. Owen now switched his attention to the brain, arguing that three of the structures present in the human brain were totally lacking in that of an ape, the most important being the hippocampus minor. On this basis, Owen suggested that the human species should be placed in at least a suborder of its own. Huxley promptly challenged not only Owen's classification, but also the anatomical evidence upon which it was based.[9] Two years later, in 1863, *Man's Place in Nature* summed up Huxley's argument for such a close

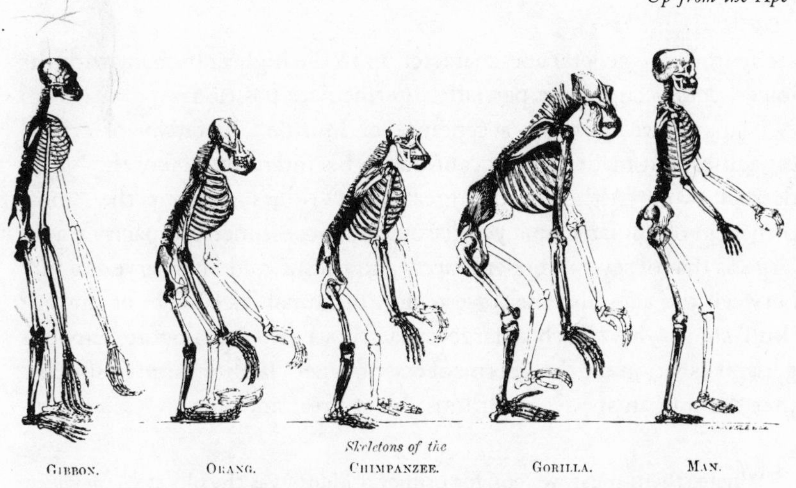

Figure 5. Comparison of the skeletons of a gibbon, orang, chimpanzee, gorilla, and human. The gibbon is drawn twice natural size. From T. H. Huxley, *Evidence as to Man's Place in Nature* (1863), frontispiece.

relationship between mankind and the apes that the two would have to be included within a single order, the primates.[10]

Besides insisting that careful dissection revealed a hippocampus minor in the ape brain, Huxley also discussed the evidence for the apes being quadrumanous. The gorilla's foot might look superficially like a hand, but a detailed study showed that the resemblance was only skin deep, and that in all essential respects it was a true foot.[11] Over and over again, Huxley insisted that at whatever organ we look the difference between man and the gorilla is less than that between the gorilla and the lower apes.[12] Huxley conceded that we cannot tell which ape is closest to the human form: he used the gorilla for his comparison of bodily anatomy and the chimpanzee for the brain (see fig. 5).[13] The impossibility of picking out a particular ape as the closest to the human form was stressed more clearly by Paul Broca, who showed that each ape species had its own points of similarity.[14] On the question of brain size, however, Huxley—like Broca—tended to adopt a linear model, borrowing his figures for cranial capacity from S. G. Morton. These figures implied that there was a greater difference between the largest and smallest human brains than between the smallest brain and that of the gorilla.[15] Huxley also noted the supposed ability of some of the "lower" human races to oppose the big toe like a thumb.[16] Without singling out a particular ape as the ancestor of mankind, Huxley was prepared to think in terms of a hierarchy from the general ape-character up to the highest human, with the lower races occupying a partially intermediate position.

It may have been his acceptance of Morton's hierarchy of cranial capacities that made Huxley cautious in his interpretation of the Neanderthal skull. Although the great brow ridges made it the "most pithecoid of human crania yet discovered," its estimated capacity was as large as that of several modern races, and thus it could not serve as a link between the apes and the lowest form of humanity.[17] That an ancient skull should have such a large capacity seemed to support Huxley's general position on the "permanence of type," leading him to suppose that the human species itself must be of great antiquity.

> Where, then, must we look for primeval Man? Was the oldest *Homo sapiens* pliocene or miocene or yet more ancient? In still older strata do the fossilized bones of an ape more anthropoid, or a Man more pithecoid, than any yet known await the researches of some unborn palaeontologist?
>
> Time will show. But, in the meanwhile, if any form of the doctrine of progressive development is correct, we must extend by long epochs the most liberal estimate that has yet been made of the antiquity of Man.[18]

Apart from this speculation about the timing of human evolution, Huxley offered no hypothesis on the factors that might have promoted the development. *Man's Place in Nature* played an important role because it established the link with the apes that most evolutionists took for granted, but the book made no significant contribution to the debate on the causes of human evolution.

Darwin, like many of his followers, drew on Huxley's study to illustrate the close link between mankind and the apes. In the sixth chapter of the *Descent of Man,* first published in 1871, Darwin listed a number of anatomical and embryological points confirming the affinity. He explicitly asked if the characters shared by humans and apes could have been acquired independently by "analogous variation" in two parallel lines of evolution, but rejected the possibility on the grounds of the sheer number of the resemblances.[19] The implication was that the similarities are a relic of the common ancestor from which the two lines have diverged. This ancestor would not, of course, have been a modern ape, but it would have had certain generalized ape characters. Darwin was always reluctant to draw up detailed predictions of the course of evolution, and the *Descent of Man* contains no diagrammatic representation of a phylogeny for the primates. In the Darwin papers at Cambridge University Library, however, there is a sketch of such a phylogeny, dated 21 April 1868.[20] It shows the human, ape, and baboon lines diverging from a generalized monkey ancestry. Significantly, Darwin made no effort to represent the line running through to humans as the central theme of the process. The human species was one primate among many, not the chief goal of the order's development. In the *Descent of Man,* Darwin ventured only the prediction that our apelike ancestors would have had prehensile feet, and that the males would have had great fighting canine teeth. He believed that these ancestors had probably lived in Africa, along with those of the gorilla and chimpanzee. He admitted, though, that it was useless to speculate too far on the location, since Lartet's *Dryopithecus* showed that apes had inhabited even Europe during Miocene times.[21]

Alfred Russel Wallace, the codiscoverer of natural selection, agreed with many of Darwin's points concerning human ancestry. Although his interest in spiritualism eventually led him to invoke a supernatural power to explain the evolution of certain human characters, Wallace aqcepted that this power had acted to transform a generalized ape in this unique direction. Darwin and Wallace thus came to disagree over the cause of human evolution, but not over its actual starting point in the apes. In his *Darwinism* of 1889, Wallace pointed to *Dryopithecus* as a large, gibbonlike ape with a more human type of dentition, and argued for a separation of the ape and human lines of evolution in the late Miocene.[22]

Darwin admitted that he would not have written the *Descent of Man* if Ernst Haeckel's *Natürliche Schöpfungsgeschichte* had appeared before he started work on his own book.[23] In fact, Haeckel's project was far more ambitious, as indicated by the title of its English translation, *The History of Creation.* It comprised a complete history of life on the earth, up to and including the appearance of mankind; only the later phases are of relevance here. Haeckel derived the apes from the Catarrhine or Old World monkeys via the Lemurs (which are called *Halbaffen* or semi-apes in German). He differed from Darwin and Huxley in preferring the Asian apes as the closest relatives of mankind. Along with the gibbon and the orang, this group gave rise to the hypothetical genus of *Pithecanthropi* or ape-men, several species of which independently evolved into the various human races.[24] The ape-men probably appeared first in an ancient continent now sunk beneath the Indian ocean, which had been given the name Lemuria by Philip Sclater.[25] Its disappearance would account for the lack of fossil intermediates. At least two species of *Pithecanthropi* had spread out into the world to become the ancestors of the straight-haired and the woolly-haired races of mankind, the latter retaining far more of the original apelike character. The overall effect was of a more or less linear development from the ape to the human stage of evolution, with the apes merely preserving an earlier form through which our ancestors once passed. In Arthur Keith's words, Haeckel saw the apes as merely "abortive attempts at man production."[26]

It was Haeckel's postulation of a link with the Asian apes that encouraged Eugene Dubois to go to the East Indies in the hope of finding a fossil intermediate. The result was the skullcap and thighbone popularly attributed to "Java man." Despite the fully erect posture indicated by the thighbone, Dubois presented his discovery as an intermediate between ape and man. He even borrowed Haeckel's terminology, adding only his own specific name in light of the upright posture: *Pithecanthropus erectus.* The skull capacity was intermediate between ape and human averages, and the skull also possessed apelike brow ridges.[27] The falling out of Dubois's opponents among themselves over whether *Pithecanthropus* was really an ape or a man illustrates the context within which the find was evaluated. In the genealogical tree that Dubois provided with some of his articles, the Miocene ape *Dryopithecus* was placed on a small side branch just before the main split leading to mankind and the modern apes. Another fossil ape, *Palaeopithecus* from the upper Miocene or lower Pliocene of the Siwalik hills in India, was placed close to the actual point of division.[28] *Pithecanthropus* itself, which Dubois believed to be upper Pliocene, came halfway along the line toward mankind. In his later career, Dubois rejected much of his early speculation and insisted that *Pithecanthropus* was only a giant gibbon (see fig. 6).

Despite the controversy surrounding *Pithecanthropus,* some authorities did accept it as a genuine intermediate between the ape and human forms. This position was adopted by another exponent of Haeckel's vision of human evolution, Gustav Schwalbe. Noting that both *Pithecanthropus* and the Neanderthal race had apelike skull characters, Schwalbe saw the possibility of arranging them into a sequence linking the ape and modern human forms. The Neanderthals were sufficiently human to be included in the genus *Homo,* but they represented an earlier species, which Schwalbe preferred to call *Homo primigenius.* In works such as his *Vorgeschichte des Menschen* of 1904, Schwalbe promoted the concept of a sequence from ape through *Pithecanthropus* and *Homo primigenius* to modern humanity: "Thus *Homo primigenius* must also be regarded as occupying a position in the gap existing between the highest apes and

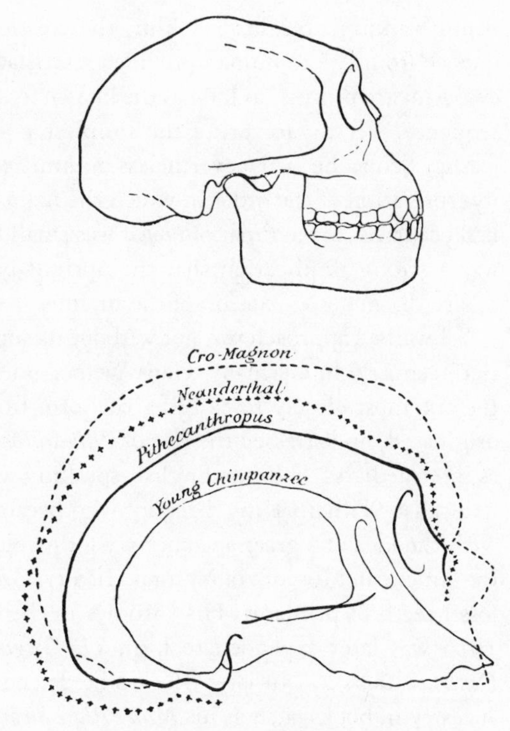

Figure 6. Restoration of the skull of *Pithecanthropus* and a comparison with the skull outlines of a Cro-Magnon, a Neanderthal, and a young chimpanzee. Note how the lower diagram implies a steady development from the ape to the human form. From Ernst Haeckel, *The Last Link* (London: A & C Black, 1898), p.25.

the lowest human races, *Pithecanthropus* standing in the lower part of it, and *H. primigenius* in the higher, near man."[29] Schwalbe was careful not to imply, however, that the morphological sequence necessarily represented the actual course of evolution, bearing in mind that both the Java and Neanderthal forms could be relics of earlier steps in the process preserved in isolation long after higher developments had taken place elsewhere.

Schwalbe's work upheld the case for a direct link between the great apes and mankind, and also established a particular sequence of intermediates by which the ape had been transformed. The known fossil hominids were accepted as representatives of genuine steps in the progress toward the final human form. In effect, Schwalbe followed Haeckel's program by setting up a more or less linear hierarchy of living or known fossil types. He did not admit the possibility of major side branches diverging away from the established line running from ape to human. Nor did he concern himself with the question of causation. Like Haeckel he did not see the necessity of trying to specify in detail the adaptive benefits that had encouraged a creature with an ape's brain to begin walking upright. For him, the creation of a morphological sequence from ape to human provided a satisfactory "explanation" of the evolutionary process, as long as the known fossils could be fitted into the sequence. Yet by accepting the simplest possible arrangement of the fossils, Schwalbe had nevertheless committed himself to a particular interpretation of the order in which the main steps in human evolution had occurred. Since *Pithecanthropus* was small brained yet bipedal, there was a strong implication that the upright posture had been achieved before the major expansion of the brain.

Schwalbe's approach was not without its supporters in the early twentieth century. In Germany, Hans Weinert identified the chimpanzee as the ape most closely resembling the form from which the human line originated, and stressed the role of *Pithecanthropus* and the Neanderthals as intermediates.[30] In the English-speaking world, it was the American naturalist William King Gregory who became most closely associated with the idea of a great ape ancestry for mankind. Early in his career at the American Museum of Natural History, Gregory began a study of the fossil teeth of primates. His "Studies in the Evolution of Primates" of 1916 was later expanded to form *The Origin and Evolution of Human Dentition* in 1922. In later decades he defended the idea of a great ape ancestry in books such as his *Man's Place among the Anthropoids* of 1934. Gregory believed that the gorilla provided the best modern equivalent of the kind of ape from which we evolved.[31] In a popular work, *Our Face from Fish to Man,* he presented an almost linear sequence of living forms to illustrate the upward course of vertebrate evolution, very much after the fashion of Haeckel. Yet in his technical work, Gregory was far more cautious. The comparison with the gorilla meant only that we had passed through a "gorilloid" stage, not that a creature identical to the modern gorilla was our ancestor. Even the known fossil intermediates were probably only relatives of our ancestors, not the ancestors themselves. *Pithecanthropus,* for instance, was an offshoot, not a direct ancestor, although it was the best available illustration of the earliest human stage.[32]

Gregory agreed with Arthur Keith that the gibbons represented the basic stock from which the line leading toward the larger primates (including the great apes and mankind) had evolved. On this theory, the upright posture had first been developed by the gibbons as an adaptation to brachation, or swinging through the trees suspended from the arms.[33] Brachation preadapted their heavier ape descendants for a partially upright gait on the ground, and minimized the extent of the final transformation to bipedalism in humans. The lower Oligocene *Propliopithecus* marked the point from which the modern gibbons and the larger apes had diverged. *Dryopithecus,* which Lartet had discovered in the European Miocene, was already a large ape, and Gregory speculated that it might indicate the point at which the ape and human stocks had separated. A new fossil from the Sewalik hills in India, *Sivapithecus,* marked the next stage on the human line, whereas *Pithecanthropus* showed a continuation through to bipedalism and an already enlarged brain.[34]

Schwalbe, Weinert, and Gregory assumed that the starting point for human evolution was a form similar to the modern great apes. In one way or another most early twentieth-century paleoanthropologists maintained a link with the apes. Indeed, a new line of evidence now became available to confirm the close evolutionary relationship between apes and humans. It stemmed from the biochemical investigations of George H. H. Nuttall, and demonstrated the close similarities between ape and human blood sera.[35] The anatomical links established by Huxley and others were thus reinforced. But evidence for a close relationship did not prove that the common ancestor had actually been a great ape similar to the gorilla or chimpanzee. Some paleoanthropologists now began to suspect that the ape and human lines of evolution had diverged from a common ancestor that was apelike in some respects, but which lacked the size and some of the distinguishing specializations of the living great apes. This approach replaced the almost linear progressionism of Schwalbe's scheme with an image of branching evolution in which apes and humans had diverged away from a common ancestor whose character was less apelike, and in some respects more human than any living ape. Such a position threw doubts upon Schwalbe's neat arrangement of the fossils, which rested on the assumption that the heavy brow ridges of the skull in *Pithecanthropus* and the Neanderthals were inherited from the great ape ancestor. If some early human types had developed apelike characters that had *not* been present in their ancestors, then the line leading toward modern humanity could have developed a large brain and a more human skull formation at a comparatively early stage in its evolution.

A number of factors helped to undermine the credibility of the great ape ancestry for mankind. One problem centered on the difficulty of imagining how the foot of the gorilla could have been transformed into the human foot, which is far more specialized for walking upright. The implication of Gregory's concept of a gorilloid stage was that the common ancestor, *Dryopithecus* perhaps, had already become substantially similar to a modern gorilla, with an upright posture that would facilitate the transition to bipedalism.[36] But Gerrit S. Miller, Jr., (better known as a leading critic of the hybrid "Piltdown man") argued that the transition to walking upright was far more critical.[37] A theory that required the direct transformation of a gorilla's foot into the corresponding human structure was almost as implausible as Wood Jones "tarsioid hypothesis" in which the apes were eliminated from human ancestry altogether. Miller favored a compromise in which the ancestral form had been a less specialized ape, the process of adaptation to upright walking being spread over a much longer period.

A similar threat to the idea of a great ape ancestry came from a study of the cranium. As Sir Arthur Smith Woodward pointed out in defense of his reconstruction of the Piltdown skull, the immature ape has a far more humanoid facial appearance than the adult.[38] On the basis of the still-popular theory that ontogeny recapitulates phylogeny, this observation would imply that the modern apes have evolved from an ancestor that was far more human in appearance. *Eoanthropus* would thus serve as a direct link between this less specialized common ancestor and modern humanity. The heavy brow ridges of *Pithecanthropus* and the Neanderthals showed that they had evolved along a quite separate line. Although affected by some humanizing trends, these forms had also been influenced by degenerative effects paralleling the main trend of ape evolution. They were thus only side branches of human evolution that had eventually been wiped out by a superior type.

From opposite ends of the skeleton, then, came evidence suggesting that the modern apes themselves had undergone a considerable degree of evolutionary specialization since the ape and human lines had diverged from a common ancestor. Far from being a shambling, heavy-browed type, the common ancestor might have had a more generalized appearance that was far more human than that of its ape descendants. Should it then be called an ape? This point was raised in the 1930s by the British anatomist Wilfrid Le Gros Clark, who feared that the term "ape" might be restricted to those creatures possessing the distinct, tree-dwelling specializations of the modern apes. With such a definition, the common ancestor of apes and humans would not itself have been an ape. But Clark himself argued against such a restricted use of the popular terminology.[39] The common ancestor could still be called an ape, if defined by reference to a particular grade of evolutionary development. Although lacking the specialized characters of the modern apes, the common ancestor would have had an equivalent evolutionary status in terms of its brain size. The modern apes had specialized while remaining within the "ape" grade, whereas our ancestors had advanced to a higher grade by acquiring a larger brain. Although admitting the divergent character of primate evolution, this approach nevertheless continued to focus attention on the expansion of the brain as the most important feature of the advance toward human status.

An overwhelming interest in the expansion of the brain was, in fact, the chief motivation that led many paleoanthropologists to reject Schwalbe's scheme of development through the *Pithecanthropus* and Neanderthal stages toward the modern human form and its implication that a fully upright posture had been achieved before the great expansion of the human brain had begun. Gregory implicitly accepted the priority of the upright posture, and was supported by another American, Dudley J. Morton. But Morton pointed out that their position led to a quite different problem of terminology. If the human line had become distinct through the adoption of the upright posture before the expansion of the brain, then it would be necessary to apply the generic term *Homo* to those pre-Pleistocene forms that had become human in their skeletal structure, but that had not yet developed the size of brain that was usually taken as the defining character of mankind.[40] This was precisely the opposite interpretation to that advanced by Le Gros Clark, and hardly anyone in the interwar years was prepared to accept Morton's position. If the brain was the most important of humanity's gifts, then the human line must surely have been marked by brain expansion since its initial divergence from the apes. It may be significant that Le Gros Clark dedicated his *Early Forerunners of Man* to Sir Grafton Elliot Smith, the leading exponent of the belief that the brain had led the way in human evolution.

The motives that led Elliot Smith and many others to stress the growth of the brain as the key to human evolution will be explored in chapter 7. Here we need only note the phylogenetic implications of the theory. The plausibility of such a reconstruction of human evolution was obviously linked to the rejection of a great ape ancestry for mankind. If the common ancestor of apes and humans had lacked the specialized characters of the modern great apes, then there was no need to treat *Pithecanthropus* and the Neanderthals as steps in a linear sequence from the ape to the human form. Once these extinct forms, with their heavy, apelike features, had been removed from the ancestry of *Homo sapiens,* the way was clear for the postulation of a new route from the generalized ape ancestor, in which the expansion of the brain preceded the acquisition of the upright posture. Not surprisingly, the supporters of this new phylogeny were inclined to take the Piltdown remains seriously, since

Eoanthropus did indeed seem to reveal a type in which the cranium had begun to approach the modern human form while the face still retained strong traces of an ape ancestry. Indeed, the real problem with Piltdown was that the jaw was *too* apelike even for the supporters of the "brain first" theory to accept. As a result, the Piltdown remains soon lost their position as the center of attention.

At the same time, the creation of a hypothetical ancestry for mankind in which the brain had developed first was bound to throw doubts upon Raymond Dart's interpretation of *Australopithecus,* after his discovery of the Taungs fossil in 1925. According to Dart, the skull had belonged to a creature that already walked upright, even though its brain was scarcely larger than an ape's. Such a creature would fit neatly into the gap between the ape and *Pithecanthropus* in the scheme proposed by Schwalbe and still being expounded by Gregory. We shall see that Gregory was, in fact, among the first to give a significant role to *Australopithecus* in his reconstruction of human phylogeny. Le Gros Clark also began to champion an Australopithecine stage in human evolution after the new discoveries of the late 1930s, but to do so he had to rethink his position completely.

The rejection of *Pithecanthropus* and the Neanderthals from human ancestry played a major role in conditioning early twentieth-century ideas on human evolution. The shape of human phylogeny was changed from a linear hierarchy to a multiple-branched tree, with many branches leading to totally different—and now extinct—forms of humanity. The desire to create a framework in which the development of the brain could be given first priority in human evolution certainly played a role in undermining the popularity of Schwalbe's scheme, but many other factors were involved.

Chapter Four Neanderthals and Presapiens

For those who advocated the descent of modern mankind from an ape resembling the gorilla or chimpanzee, the fossil relics of *Pithecanthropus* and the Neanderthals seemed ideal candidates for the "missing links" in the process. Further discoveries in the late nineteenth century had confirmed that the original Neanderthal remains were typical of an entire race of early humanity, whereas *Pithecanthropus* seemed to reveal an earlier an even more primitive form. Both had heavy brow ridges and generally apelike features. Although fully upright, *Pithecanthropus* had a brain capacity intermediate between that of the apes and modern humans. Some Neanderthals had large cranial capacities, but others were smaller, and by stressing the different structure of the brain one could argue that they had been less intelligent than modern humans. A more or less linear sequence thus led from a great ape ancestor through *Pithecanthropus* to the Neanderthals. The last stage in our evolution consisted of the final expansion of the brain, along with the loss of the Neanderthals' remaining apelike features.

This almost linear scheme of human evolution became popular in the first decade of the twentieth century, but then increasingly came under fire from a host of critics. Led by Marcellin Boule and Arthur Keith, a new generation of paleoanthropologists decided that the true process of human evolution was more complex, with *Pithecanthropus* and the Neanderthals representing side branches that led only to extinction. Modern humanity came from a separate line of evolution whose fossil remains were yet to be found, although various candidates were offered from time to time. They had already developed essentially modern characters while *Pithecanthropus* and the Neanderthals were stagnating in Southeast Asia and Europe. These early humans had finally invaded Europe and used their superior Aurignacian culture to exterminate the backward Neanderthals. Following the terminology of Henri Vallois, the postulation of this early form of true humanity is now known as the presapiens theory.[1] Throughout the interwar years, it was the most popular view of human evolution. From it came the classic image of the Neanderthals as degenerate, shambling ape-men, too brutal to be our own ancestors and fit only for extermination in the hands of the higher type.

The presapiens theory inevitably required an extension of the time scale of human evolution, and postulated that evolution itself is a slower process than earlier naturalists had believed. Given the prevailing belief that the Neanderthals formed a distinct species, the transition from *Homo neanderthalensis* to *H. sapiens* in the mid-Pleistocene represented quite a rapid transformation. The presapiens theory slowed the whole process down by supposing that more advanced forms of mankind existed earlier in the Pleistocene. Indeed, this tendency to extend the antiquity of mankind created a fragmentation within the supporters of the presapiens theory itself. Most were content to restrict the emergence of a recognizably human form to the Pleistocene, postulating a presapiens type only in the earlier part of that era as a development from a Pliocene ape. But a few extremists—the most notable of whom was Arthur Keith—wanted to go much further. They saw modern humanity emerging already in the Pliocene or even earlier, so that the human form had remained essentially static for hundreds of thousands of years. When fully developed, this tendency to extend the antiquity of mankind went beyond the presapiens theory altogether, as in F. Wood Jones's postulation of a line of human evolution stretching back into the early Tertiary, bypassing the apes altogether.

A few anthropologists continued to argue that the Neanderthals represented the transitional form between *Pithecanthropus* and modern *Homo sapiens,* a position now known as the "Neanderthal phase of man" theory, after Ales Hrdlička's 1927 Huxley Memorial Lecture. Others favored a compromise, the "pre-Neanderthal" theory, in which the classic European Neanderthals were regarded as a dead end, but earlier, less specialized Neanderthals were seen as the common ancestor of both the classic type and modern *Homo sapiens*. In the 1960s, C. Loring Brace led a move to reinstate the Neanderthal-phase theory, and the question is still unsettled today.[2] Many authorities still favor the view that the classic Neanderthals were displaced by modern humans, but their interpretation of the archaic form of *Homo sapiens* no longer fits the image created by the presapiens theory. Because *Homo sapiens* is now thought to have evolved from *Homo erectus* (*Pithecanthropus*), the intermediate stage might have certain Neanderthaloid characters, including heavy brow ridges, but not the whole complex of characters associated with the true Neanderthals. The once-popular belief that the Neanderthals represented a separate species, *Homo neanderthalensis,* has also been rejected, so that they can be included as a subspecies of *Homo sapiens.*

This area of confrontation between the linear and branching models of evolution has received more attention from historians of anthropology than the debate over the character of our ape ancestors. Brace included a substantial account of the origins of both his own and the opposing positions in his paper of 1964, and since has further commented on the issue.[3] Michael Hammond has made a detailed analysis of the social background to Boule's attack on the linear theory.[4] There are problems, however, with the interpretation offered by these studies. As a leading modern exponent of the Neanderthal-phase theory, Brace is perhaps not in the best position to take an objective view of the opposition. Inevitably, he sees the linear model as the most natural approach for an evolutionist to take, while he treats the presapiens theory, with its host of "dead end" side branches, as a highly artifical concept. At times, Brace suggests that the branching model is not really a theory of evolution at all, but an excuse to deny all evolutionary relationships among the known forms. Hammond adopts a similar viewpoint, treating de Mortillet's linear progressionism as the most obvious evolutionary model, and Boule's work as an almost nonevolutionary attack on it. For these writers, the crucial question is: why did so many anthropologists eventually abandon the search for an evolutionary link connecting the fossils, and adopt a model in which mankind originated from a totally hypothetical stock that was not represented in the fossil record?

Up to a point, one can sympathize with the assumptions that underlie this approach to the question. Certainly, the presapiens theory was used by its more extreme supporters as a means of escaping the shackles of the fossil record, leaving them free to postulate imaginary lines of hominid evolution that conformed with their own prejudices. It is true that Boule's teacher, Albert Gaudry, adopted a theory of multiple evolutionary branches as a means of justifying a "hands-off" approach to the study of the evolutionary mechanism. The linear scheme did indeed have the merit of being the simplest interpretation of the available evidence. Authorities as eminent as Haeckel portrayed linear progressionism as the correct interpretation of evolution. For the cultural anthropologist, furthermore, the very term "evolutionism" means the linear progresive sequence postulated by Tylor, Morgan, and others. Judged from this perspective, the Neanderthal-phase theory may well seem to be the most rational form of evolutionism, whereas the presapiens concept is an unwarranted complication that must be explained through the influence of nonscientific factors.

Unfortunately, the historian of evolutionary biology sees things in a very different light. Darwinism is essentially a theory of branching rather than linear evolution, which makes the presapiens theory seem intrinsically rather more plausible. One cannot help wondering if Brace's commitment to the Neanderthal-phase theory has led him to misjudge the issues surrounding the earlier debates. It must be remembered that even in anthropology the ideas of invasion and racial displacement had long been used as an alternative way of explaining European prehistory. When this point is added to the fact that all the available theories of biological evolution allowed or even encouraged a branching model of the process, the linear progressionism of the Neanderthal-phase theory no longer seems quite so straightforward.

Darwin's own refusal to speculate on genealogical trees was a necessary product of his belief in the imperfection of the fossil record, which in turn was a vital defense against the charge that the discontinuity of fossil sequences refuted evolution. The Darwinian theory promoted an image of haphazard, constantly branching evolution, with many lines ending in extinction. Gaudry's intentions may have been at variance with those of the Darwinians, but his concept of the actual "shape" of the evolutionary process was not incompatible with the selection theory. Even that great rival of Darwinism in the interpretation of the fossil record, the American school of neo-Lamarckism and orthogenesis, promoted a theory of *multiple* lines of parallel evolution. As the eminent British paleontologist D.M.S. Watson suggested in 1928, the ever-increasing variety of human fossils made it necessary to evaluate human evolution in terms of the principles established by the study of other animals.[5] It was most unlikely that a linear advance toward a single end product had occurred here, and nowhere else in the animal kingdom.

Watson emphasized that whenever evolutionists had committed themselves to a linear sequence of known fossils, further discoveries had always shown that the true course of evolution was far more complex. A classic example was the evolution of the modern horse from the small, four-toed Eocene *Eohippus,* a fossil discovered by O. C. Marsh. T. H. Huxley declared that the sequence from *Eohippus* through a series of fossil intermediates to the modern horse provided "demonstrative evidence of evolution,"[6] yet later discoveries showed that there was no straight line, only an irregularly branching process in which many of the known fossils were not on the direct line of descent. The supporters of the presapiens theory were simply assuming that the same thing happened in human evolution. Although the Neanderthals have now been partially reintegrated into our own ancestry, modern science still accepts the later Australopithecines as extinct side branches of the human family. If one judges by early twentieth-century or by modern standards, there is thus nothing intrinsically implausible about the concept of branching hominid evolution. Indeed, from the perspective of evolution theory, the real question becomes the very reverse of that asked by Brace and Hammond. What we should really be asking is: why did the grossly oversimplified image of a single line leading toward modern mankind become so popular around 1900, despite the fact that it was inconsistent with all known evolution theories?

The answer to this question depends on the uses to which the theories were put in the wider debate on the implications of evolutionism. We should not be surprised that scientists engaged in the campaign to uphold the basic idea of evolution should have had a strong interest in "missing links" and should thus have been tempted to regard the fossils, when they were eventually discovered, as components in the sequence leading toward modern forms. This temptation is clearly apparent in the study of horse evolution. Huxley saw the "public relations" value of appearing to explain the origin of such a well-known and highly specialized form via fossil intermediates. Marsh's discoveries of fossil horses were thus seized upon to construct a morphological sequence ending with the modern horse, and there was a natural temptation to assume that the sequence corresponded to the actual phylogeny. Subsequent discoveries were bound to show that this approach was an oversimplification, which Huxley could have predicted on the basis of his own views on the complex nature of the evolutionary process. The need for caution was easily forgotten, however, in the heat of debate, when there was a pressing need for positive evidence of continuous evolutionary sequences. Human evolution was still a controversial issue later in the nineteenth century, a situation that again ensured that the first relevant fossils were fitted into the simplest possible sequence from ape to human.

Outside the scientific debate, evolution theory was being used for all sorts of political and ideological purposes. The circumstances differed in each country, but in many cases the end result was a need to present evolution as an inevitably progressive force that justified the faith in continued human progress. In France, de Mortillet used archaeology to construct a progressive sequence emphasizing the demands for reform arising out of his socialist beliefs.[7] In Germany, Haeckel's campaign for reform took a very different approach but had an equal need for evolution as the guarantee of progress.[8] In Britain and America, the anthropological "evolutionism" of Tylor and Morgan also had political foundations. Tylor's scheme of linear development was designed to preserve a central feature of the Enlightenment's social philosophy, the principle of the uniform operation of the human mind in all circumstances.[9] By a curious paradox, Darwin's branching image of evolution also had its origins in utilitarian social philosophy, but his principal concern was adaptation, and he was thus led toward a mechanism that left evolution at the mercy of an ever-changing environment. Although Spencer tried to synthesize these two products of utilitarian philosophy in his less structured—and in some respects less optimistic—progressionism, the influence of Tylor, Morgan, and a host of Lamarckian anthropologists helped to ensure that cultural evolution was seen as a linear process. This view in turn dovetailed with the hierarchical image in which the "lower" races were seen as stagnant relics of earlier stages in the upward progress.[10] A whole host of influences thus turned late-nineteenth-century evolutionism into a form of linear progressionism, against which the logic of the Darwinian theory struggled in vain. It was thus almost inevitable that the first human fossil would be fitted into the simplest possible linear sequence of evolution.

The collapse of the linear view of human evolution in the early twentieth century was a necessary prerequisite for further development in the field, since it paved the way for a more sophisticated interpretation in tune with the broader principles of evolution theory. Yet in practice, the first efforts to create a theory of branching human evolution went too far in their willingness to dismiss the hominid fossils, especially the Neanderthals, as extinct side branches. This exaggeration can also be attributed to more than scientific overenthusiasm, since it too seems to have its roots in broader cultural trends. The early twentieth century saw a rapid loss of faith in the inevitability of progress, and the gap was soon filled by an alternative image of human development based on racial conflict. This theme had never been absent from nineteenth-century anthropology, but now it was elevated to become the principal analogy by which human evolution was to be understood. The extermination of the Neanderthals was merely a large-scale example of the process of racial extinction that had become an all too frequent consequence of Western imperialism—an illustration of the perils of racial stagnation from which lessons of national importance could be learned. The extension of the time scale of human evolution required by the presapiens theory also had implications for racial questions, since those who stressed the extent of the racial divisions within modern mankind demanded more time in which such a radical divergence of character could have occurred.

Origins of the Neanderthal-Phase Theory

Those anthropologists who established a racial hierarchy based on cranial capacities created the model for the Neanderthal-phase theory. They were quite prepared to take the brain size of the great apes as a starting point for their scale of measurement, fitting the "lower" races into an intermediate position between the apes and the white man. When coupled with the popular belief that the lower races had apelike skull features, this model led inevitably to the idea of an evolutionary scale of development. It was no longer fashionable to argue that the human form was the goal toward which all evolution was aimed, although this view was defended as late as 1884 by John Cleland, professor of anatomy at Glasgow.[11] Yet without using such explicitly teleological language, one could still imagine that the human brain was the product of an inevitable evolutionary trend. Once enough Neanderthal specimens became available to make clear their status as a distinct form of mankind, the new race could be fitted into its position on the scale of increasing cephalization at a point intermediate between the apes and the lowest modern races.

The most decisive step in the establishment of the Neanderthals as a distinct race was the discovery at Spy in 1886. As described by Julien Fraipont, these specimens gave clear evidence that the oldest form of mankind was a race inferior to any now living. Fraipont drew a link with fossil apes such as *Dryopithecus,* although he noted that they had a less simian jaw structure than the modern apes. He implied that two lines of evolution led from an earlier and even less specialized ape, one leading toward the modern apes and the other through the Neanderthals to modern mankind.[12] For many who read the reports of the Spy discovery, however, the Neanderthal race could be seen as a link with the great apes of today. In America, E. D. Cope, a leading exponent of the view that the lower races have apelike characters, followed Fraipont's description and emphasized the simian character of the Neanderthals.[13] He believed that the differences between the Neanderthals and the lowest living races were so great that the former would have to be designated as a distinct species, although there was an even greater gap between them and the apes.[14] Cope certainly saw the white race as having originated from the Neanderthals via a series of now extinct intermediates, thereby opting for what would later be called the Neanderthal-phase theory.[15]

Most authorities agreed that the Neanderthals were distinctly lower than any living race, but a few attempted to suggest that the Neanderthal type can still be seen in the world today. Huxley had pointed out the resemblance between the original Neanderthal skull and the Australian and Tasmanian aborigines, and some felt that these low races might represent a survival of the Neanderthal type, protected by their isolated location. If the various types were simply placed on a scale of increasing brain size, to interpret the significance of any gaps would be a matter of opinion.[16] As late as 1914, Richard Berry and A.W.D. Robinson still used such a linear scale, although they argued that the Tasmanians had developed far beyond the Neanderthals.[17] The last major proponent of a link between the living races and the Neanderthals was W. J. Sollas, professor of geology at Oxford. In his *Ancient Hunters* of 1911, Sollas tried to reconstruct the lives of our ancestors using comparisons with modern primitives. He championed the view that the Chancellade skull confirmed a link between the Magdalenian race and the Eskimos. The Tasmanians were supposed to have preserved much of the culture of the early Mousterian period.[18] Sollas even called the Australians "the Mousterians of the Antipodes."[19] Eventually Sollas fell into line with other authorities and accepted the distinct character of the Neanderthals, although popular writers on the race question found the concept of the "living Neanderthal" too useful a means of abuse to abandon so easily. In his *The Passing of the Great Race* of 1916, Madison Grant wrote that "ferocious gorilla-like living specimens of Palaeolithic man are found not infrequently on the west coast of Ireland and are easily recognized by the great upper lip, bridgeless nose, beetling brow with low growing hair and wild and savage aspect. The proportions of the skull which give rise to this large upper lip, the low forehead and the superorbital ridges are certainly Neanderthal characters."[20]

By contrast, many evolutionists saw the value of treating the Neanderthals as one of the links between the apes and the lowest form of modern mankind. Those who had postulated "missing links" were naturally anxious to see their hypothetical intermediates revealed in the fossil record. Not surprisingly, the leading proponents of linear progressionism in France and Germany hailed both *Pithecanthropus* and the Neanderthals as intermediates in the hierarchy linking the apes to modern mankind. Gabriel de Mortillet had postulated prehuman types in the

Tertiary to complement his acceptance of eoliths as the earliest stone tools. He repudiated Gaudry's suggestion that the eoliths might have been made by *Dryopithecus,* and argued for a being intermediate between apes and mankind, "animals of another genus, precursors of man on the chain of being, precursors to which I have given the name *Anthropopithecus.*"[21] The Neanderthals were presented as descendants of this hypothetical being, now human, but still retaining many apelike characters. De Mortillet insisted that the Neanderthals were gradually transformed into the higher races that replaced them.[22] In the 1903 edition of *Le Préhistorique,* Dubois's *Pithecanthropus* was linked into the sequence, intermediate between the apes and Neanderthals.[23] The continuity of the sequence was established by treating the Neanderthals as the lowest race of *Homo sapiens,* not as a distinct species.[24]

Curiously, Ernst Haeckel did not make a great deal of capital out of the apelike character of the original Neanderthal specimen, perhaps because he respected Huxley's interpretation of it. Haeckel was certainly ready to hail *Pithecanthropus* as a genuine intermediate between the apes and mankind, and it is significant that in his account of this new fossil he gave a diagram of skull forms showing both *Pithecanthropus* and the Neanderthals as intermediates.[25] Dubois also noted the similarity of the *Pithecanthropus* skull to the Neanderthals, except for its smaller capacity.[26] At the very end of his career, Haeckel lent his support to the interpretation of the Neanderthals offered in the detailed discussions of the Strasbourg anatomist Gustav Schwalbe.[27]

Schwalbe's position was a subtle one and has sometimes been misinterpreted by later writers. Because he insisted that the Neanderthals constituted a distinct species, *Homo primigenius,* it would be easy to assume that he denied their status as ancestors of *H. sapiens.* Ales Hrdlička, in fact, charged Schwalbe with originating the opposition to the Neanderthal-phase theory later taken up by Boule and Keith.[28] C. Loring Brace believes that Schwalbe began with a preference for treating the Neanderthals as intermediates, but eventually gave in to Boule's arguments and accepted them as a side branch.[29] In giving the Neanderthals the status of a distinct species, Schwalbe's original intention was certainly not to exclude them from human ancestry. He was merely stressing that the degree of morphological change from the Neanderthals to the modern form of mankind was so great that by normal taxonomic standards they would have to be assigned to separate species. At this time it was not unusual to argue that change within a continuous sequence eventually gave rise to a new species, without a speciation event involving the splitting of the original population into two, however controversial this procedure may be today.

Whatever his opinions on their actual relationship, Schwalbe accepted the sequence ape-*Pithecanthropus*-Neanderthal-modern mankind as a continuous morphological scale. But as early as 1906 he used a diagram to show two possible interpretations of the evolutionary relationships implied by this sequence.[30]

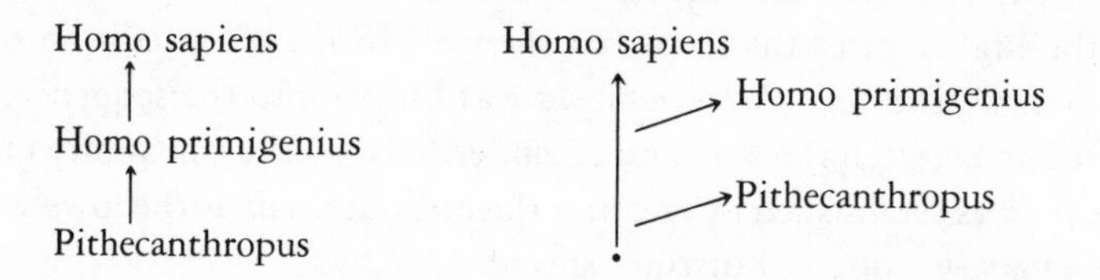

In suggesting that *Pithecanthropus* and *Homo primigenius* might lie off on side branches, Schwalbe seems to be allowing for branching evolution. Yet he declared that it does not really matter which of the two arrangements one prefers, since, in the second, *Pithecanthropus* and *Homo primigenius* can still represent stages in the development of *H. sapiens*.[31] He explained his meaning more fully in his response to Boule's work, in which he accepts more clearly that the second arrangement probably corresponds more closely to the true situation. Here he points out that since we do not know the actual transitional forms, we can still accept that the fossils assigned to *Homo primigenius* are reminiscent of the earlier form that is the true ancestor of modern mankind.[32]

In fact, Schwalbe had made only a limited capitulation to Boule's argument for the nonancestral status of the Neanderthals. He was working with the view that both *Pithecanthropus* and the Neanderthals (that is, their known fossil specimens) might be remnants of older phases in the evolution of modern mankind, preserved with little change outside the area in which further evolution was actually taking place. This is still linear, not branching evolution, but modified to include "branches" that merely preserve ancestral stages into later periods. It was already the standard explanation used by linear progressionists to account for the continued existence of "lower" races in the world today. The higher type was not thought to be literally descended from the known lower form, but from a closely related form that had represented the most advanced stage of evolution in an earlier period. Schwalbe did not believe that the Neanderthals were specialized offshoots that could never in principle have evolved into *Homo sapiens*. *Pithecanthropus* and the Neanderthals were both stages in the development of modern mankind, but Schwalbe was cautious enough to admit that the known fossils might not themselves correspond to our ancestors:

> Thus *Homo primigenius* must also be regarded as occupying a position in the gap existing between the highest apes and the lowest human races, *Pithecanthropus* standing in the lower part of it, and *Homo primigenius* in the higher, near man. In order to prevent misunderstanding, I should like to emphasize that in arranging this structural series—anthropoid apes, *Pithecanthropus, Homo primigenius, Homo sapiens*—I have no intention of establishing it as a direct genealogical series.[33]

In Britain, Schwalbe's concept of linear evolution was at first supported by W. J. Sollas and Arthur Keith, although both were suspicious of the claim that the Neanderthals differed enough from modern mankind to count as a distinct species. We have already noted how Sollas drew a comparison between the Neanderthals and the Australian aborigines in his popular book *Ancient Hunters.* In a more technical paper of 1908 he pointedly referred to the "Neanderthal race" (not species) and argued that they and the Australians "probably represent divergent branches of the same original stock."[34] This view seems to establish the Australians as a slightly modified form of an earlier stage in the evolution of the whole human race. Sollas fully accepted Schwalbe's morphological sequence: "the Neanderthals and *Pithecanthropus* skulls stand like the piers of a ruined bridge which once continuously connected the kingdom of man with the rest of the animal world."[35]

Arthur Keith published his first survey of human fossils in 1911, shortly after he had been appointed conservator of the Royal College of Surgeons. His *Ancient Types of Man* worked backward from the most recent remains and concluded with discussions of the Neanderthals and *Pithecanthropus.* Boule had already published his classic monograph on the Neanderthal skeleton from La Chapelle-aux-Saints, and although Keith refers to this study, as yet he did not accept Boule's conclusions on the status of the Neanderthals. He also repudiated Schwalbe's separation of the Neanderthals into a distinct species—there was no reason to suppose that they could not have interbred with modern humanity.[36] Despite its brutish appearance and stooped carriage (the latter was Boule's interpretation), this early form was "large brained, erect in posture, and in every sense of the biologist—a man."[37] Keith supposed "that the Neanderthal type is the precursor and ancestor of the modern type. The Neanderthal type represents an extinct stage in the evolution of man."[38] Already concerned about the antiquity of the modern races, Keith suggested that the Negroes and the whites were separate branches from the Neanderthal stock, each independently "shaping toward the modern form."[39] Like Schwalbe, however, Keith did not rule out the possibility that the European Neanderthals were left behind in the advance and were wiped out by higher races evolved elsewhere.[40] A cranium found at Gibraltar was presented as a transitional form between *Pithecanthropus* and the Neanderthals.[41] *Pithecanthropus* itself he regarded as so human that it must represent an early stage in our evolution, and he even suggested that it should be referred to as *Homo javanensis*.[42] Despite new evidence showing that the Trinil deposits were Pleistocene, Keith insisted that they might still turn out to be Pliocene. Even if only a primitive survival into Pleistocene times, though, *Pithecanthropus*'s status as a phase in human evolution was not invalidated.

In the first decade of the twentieth century, the belief that *Pithecanthropus* and the Neanderthals represented stages in the evolution of modern mankind from the apes had become fairly well established. It was widely conceded that the known fossil specimens might not represent our actual ancestors, but might belong to earlier stages surviving into later times. Yet this concession was seen as a qualification of, not an alternative to, the theory of linear development. Few went so far as to claim that the Neanderthals were still alive in the form of the "lowest" modern races, but this position had not yet been completely abandoned. At the same time, it was widely accepted that the emergence of modern humanity had been a relatively recent affair. At best, only a very low and apelike form of mankind (*Pithecanthropus*) had appeared by the end of the Pliocene, whereas the development through the Neanderthal to the modern stages had definitely occurred in the course of the Pleistocene. This belief did not seem unreasonable, given the knowledge that the Pleistocene had seen major climatic fluctuations that would have stimulated evolutionary progress. Sollas clearly expressed this view in his *Ancient Hunters:*

> Man, not only in the specific sense but also the broader generic sense (*Homo*) is a product of the Pleistocene epoch—the latest child of time, born and cradled amongst those great revolutions of climate which have again and again so profoundly disturbed the equilibrium of the organic world. Some thinkers deeply impressed with this reflection have gone so far as to suggest

> that these changes of the environment provided not only the opportunity but also the cause of his appearance.[43]

Seen in this light, the Neanderthals were not a degenerate offshoot, but a highly successful initial response of the human stock to the challenge of the ice ages.

The Presapiens Theory

By 1909, however, Marcellin Boule had already begun to attack the Neanderthal-phase theory. Originally trained as a geologist and an expert on stratigraphy, Boule had increasingly turned to work in the fields of archaeology and human paleontology. He became assistant to Albert Gaudry and in 1903 succeeded to the chair of paleontology at the *Musée d'histoire naturelle* in Paris. In 1914 he was appointed first director of the *Institut de paléontologie humaine.* Brace has argued that Boule's attitude toward evolution was largely negative, being derived from the essentially catastrophist tradition established by Cuvier early in the nineteenth century. Hammond and others have repudiated this suggestion, though, noting that Gaudry's image of evolution was a subtle one involving complex, branching trees, many details of which would be obscured by the imperfection of the fossil record.[44] By adopting this image, Boule may have made it less easy to investigate the causes of evolution, but he was promoting a cautious and sophisticated approach to the fossil evidence that was fully in tune with the thinking of most paleontologists. As Hammond emphasizes, he was determined to show that human paleontology fitted into the general pattern of evolution derived from the rest of the fossil record. It must also be noted, though, that Boule was heir to a more conservative ideological position, which repudiated de Mortillet's use of simple-minded progressionism as a justification for socialist policies.

Boule had worked on the excavations at Grimaldi in the south of France, where skeletons found in 1901 were attributed to a race resembling the modern Negroes. He believed that these could be dated to the same age as the Mousterian culture normally associated with the Neanderthals, although other authorities linked them to the Aurignacian. If a modern form of humanity had already existed at the same time as the European Neanderthals, then clearly those Neanderthals were not an-

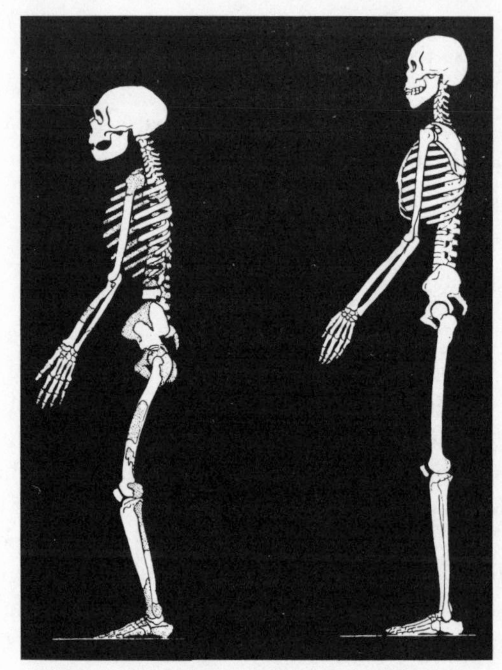

Figure 7. Marcellin Boule's comparison of the skeletons of the Neanderthal specimen from La Chapelle-aux-Saints and a modern Australian, showing the stooped posture attributed to the Neanderthal race. From Boule, *Les hommes fossiles* (Paris: Masson, 1921), p. 239.

cestral to the modern races. Boule waited for an opportunity to study a Neanderthal skeleton that was both relatively complete and clearly dated by association with the Mousterian culture. His chance came in 1908 with the discovery at La Chapelle-aux-Saints of the most complete skeleton of the Neanderthal type yet unearthed. It was sent to Boule in Paris, and in the following year his classic monograph appeared, arguing that here was a species so primitive that the modern humans of the Aurignacian could not have evolved from it.[45] The Neanderthals were thus a degenerate side branch of human evolution, eventually wiped out by a more fully evolved type coming into Europe from elsewhere.

In assessing Boule's position, one must understand the grounds upon which he dismissed the claim of a Neanderthal role in human ancestry. On a number of topics, he certainly chose to stress those characters that distanced his specimen from the modern type, and to ignore any factors that might have explained these differences in nonevolutionary terms. He emphasized those aspects of the skeleton that implied a stooping posture and, by ignoring the evidence for arthritis, implied that it was the characteristic posture of the race (see fig. 7). He stressed the prominent brow ridges and apelike appearance of the face. While admitting the comparatively large cranial capacity, he emphasized the different shape of the brain—lower at the front, where the higher intellectual faculties were supposed to be located—and suggested that the brain convolutions were simpler than in modern mankind. Yet by themselves, these points do not establish a position totally at variance with Schwalbe's interpretation. After all, Schwalbe himself accepted that the apelike characters of the Neanderthals were sufficient to justify putting them into a distinct species. Boule rejected Schwalbe's name, *Homo primigenius,* because he wanted to stress that the European Neanderthals were not the first form of humanity, but in many respects his characterization of what he preferred to call *Homo neanderthalensis* was similar to Schwalbe's.

If Boule identified the Neanderthal species by its retention of apelike characters, then, assuming the evolution of mankind from the apes, he could still believe that the Neanderthals preserved an earlier phase in the main line of evolution. This belief would contrast with Keith's later view which, as we shall see, stressed that in addition to their simian features, the Neanderthals also possessed specialized characters showing that their line of evolution had diverged away from that leading toward mankind. Boule certainly noted Nuttall's work on the similarity of the blood sera of mankind and the apes, which was generally taken as evidence of a close evolutionary link.[46] He also conceded that there might have been some "accidental" hybridization between the Neanderthals and the superior invading species.[47] While denying that any modern race was descended completely from the Neanderthals, he noted the resemblance to the Australian aborigines and admitted that this lowly form of modern mankind might have preserved ancestral characters shared with the Neanderthals.[48] He doubted Keith's suggestion that some Neanderthal characters might have a pathological origin, and later criticized Keith's argument that the "taurodont" structure of the Neanderthals' teeth was a distinct specialization.[49] In fact, Boule explicitly stated that the Neanderthals were not really a side branch, but "a belated form, a survival of ancestral prototypes."[50] In effect, Boule went beyond Schwalbe's position only in his desire to stress the reality of the Neanderthal extinction in Europe, and his willingness to extend the antiquity of the human form back into the early Pleistocene. It is not at all clear that his conception of the hypothetical intermediate between the apes and this early presapiens type would have been all that different from the late-surviving Neanderthals.

Schwalbe himself admitted the possibility that the European Neanderthals were a remnant of an earlier stage, but was not very interested in this interpretation since it did not destroy their value as a morphological link in the chain. Boule, in contrast, was anxious to stress that the sudden appearance of modern humanity in the Aurignacian must represent an invasion by a higher type that had evolved elsewhere. Schwalbe was a morphologist interested only in constructing hypothetical sequences, whereas Boule was a paleontologist strongly influenced by earlier debates in archaeology and anthropology. The difference between them was based on more than evolution theory, although certainly Boule was concerned to bring human paleontology into line with Gaudry's image of branching evolution.[51] Boule was a longstanding critic of de Mortillet's linear sequence of stone-age cultures, which he regarded as a gross oversimplification.[52] He was also well aware of the anthropologists' views on the complex history of racial types in Europe, in which invasions and at least partial exterminations were taken for granted.[53] These influences combined to encourage Boule's interest in a hypothetical line of presapiens evolution, even though at first he had no desire to challenge the theory of ape descent upon which even his own description of the Neanderthals was based.

The effects of Boule's commitment to a branching image of evolution can actually be seen more clearly in his views on *Pithecanthropus* and the origins of the hominid type. Because his position on the Neanderthals required that modern humanity had evolved much earlier in the Pleistocene than was normally supposed, *Pithecanthropus* could not be a direct ancestor. In his 1909 monograph, Boule anticipated Dubois's later position by dismissing *Pithecanthropus* as a gibbon which had independently acquired certain human traits. The primates had thrown out several rival lines of evolution, each trying to evolve into a higher type, and each acquiring some of the characters now associated with mankind.[54] In his book *Les hommes fossiles* of 1921 (translated as *Fossil Men* two years later), Boule went further and now insisted that mankind is not linked directly to the apes. Instead, two separate lines of evolution had started from the Old World monkeys, one leading to the apes and one to mankind.[55] In this later work, Boule now hinted that the "lower" characters of the Neanderthal species might best be described by comparing them to the monkeys rather than the great apes.[56] *Pithecanthropus* was still seen as a branch from the quite separate ape line, mimicking certain aspects of humanity.[57] By comparison, the dismissal of the Neanderthals as a "withered twig" on the human stem involved only a magnified version of the differentiation that has produced the racial hierarchy within *Homo sapiens*.[58]

Boule was now committed to the view that more advanced types than the Neanderthals would be found earlier in the fossil record. The Grimaldi Negroids were his first choice for this role, but he also saw the Piltdown discovery in the same light. Boule was suspicious of the ape jaw that Smith Woodward made part of *Eoanthropus;* for him the modern-looking cranium by itself indicated that the early Pleistocene already contained a type higher than the Neanderthals.[59] A new genus was certainly not justified for the fossil. The real *Eoanthropus* or dawn man would lie even farther back in time, and would be almost erect, with a smaller brain than any known human type.[60] Boule regarded the later discovery of *Sinanthropus* (Peking man) as merely a new species of *Pithecanthropus,* and was suspicious of the claim that it made use of tools and fire. The evidence for toolmaking showed only that the *Sinanthropus* bones were the remains of individuals killed and eaten by a more advanced form of mankind.[61] By now, however, Boule was prepared to concede that both *Pithecanthropus* and *Sinanthropus* were "prehominians," resembling the true ancestors of mankind who might eventually be discovered in the Pliocene or Miocene.[62]

By accepting the link between the Neanderthal species and the Mousterian culture, Boule was allowing for the coexistence of two separate toolmaking species on the earth. Later exponents of the presapiens approach cheerfully accepted the parallel evolution of distinct genera and even families up to a toolmaking level of intelligence—something that modern anthropologists find hard to swallow. Yet Boule's position required little more than an extension of the accepted view that geographical isolation could protect a culturally primitive race, at least for a time. Although the Neanderthals were a separate species, their culture resembled that of the lower human races, and hence there was a kind of parallelism in cultural evolution.[63] Here again, the implication is that the Neanderthals were merely an ancient form of mankind surviving into later times. Their extermination was comparable to the fate now being experienced by a number of technologically inferior races confronted by Western culture.

A more forceful case for eliminating the Neanderthals, or anything like them, from human ancestry was made by Arthur Keith. Keith's position at the Royal College of Surgeons gave him an unrivaled opportunity to study the anatomy of ancient human remains, and explains why he was so annoyed when the Piltdown discoveries went to the British Museum. Although in 1911 he still accepted the linear view of human evolution, in the following years his position underwent a transformation. His *Antiquity of Man* of 1915 emerged as a powerful, even an exaggerated, statement of the presapiens theory. Brace suggests that the "anti-evolutionism" of the French anthropologists was the major influence prompting this change of heart, but there is little evidence to support this.[64] Keith did make a visit to France in 1911, but he records that the museums were all shut for the summer, and it seems as if he toured the Dordogne sites without expert guidance.[65] He regarded Boule's work as merely confirmation of the trend to establish the Neanderthals as a distinct species which had been started by Schwalbe and continued by Adloff and Klaatsch.[66]

By Keith's own account, it was the archaeological evidence for the sudden disappearance of the Neanderthals that first made him suspect the prior existence of a presapiens type. In his *Antiquity of Man* he cited Hauser's discovery of a Neanderthal skeleton at Le Moustier in 1908, which appeared to show that the transition from the Neanderthals of the Mousterian to the Cro-Magnons of the Aurignacian was abrupt. This discovery "effected a revolution in our attitude toward the nature of Neanderthal man, and our conception of the antiquity of man of the modern type."[67] Keith's book also devotes a large section to the Galley Hill find of 1888, which he now began to suspect was a genuinely ancient form of *Homo sapiens,* and his autobiography refers specifically to his work on this specimen as a key step in his conversion.[68] But why did he now begin to take seriously a discovery whose antiquity was doubted by all other authorities? The answer almost certainly lies in his growing sense of the importance of racial distinctions in modern humanity and their role in the evolutionary process.

Like Boule, Keith saw his new position as a perfectly reasonable attack on the oversimplified concept of linear evolution:

> In our first youthful burst of Darwinism, we pictured our evolution as a simple procession of forms leading from ape to man. Each age, as it passed. transformed the men of the time one stage nearer to us—one more distant from the ape. The true picture is very different. We have to conceive an ancient world in which the family of mankind was broken up into narrow groups of genera, each genus being again divided into a number of species—much as we see in the monkey or ape world of today. Then out of that great welter of forms one species became the dominant form, and ultimately the sole surviving one—the species represented by the modern races of mankind.[69]

The comparison with the apes and monkeys was an excellent way of illustrating the artificiality of the old, linear viewpoint. For Keith, the modern races themselves represented branches of great antiquity, although not yet separated far enough to become species. He had already begun to emphasize this point in 1912, when he argued in a popular lecture that to create the other races from the primitive Australian aborigine would take the whole length of the Pleistocene.[70] If evolution worked this slowly, it was impossible to believe that the Neanderthals had been transformed into modern humanity in the course of the mid-Pleistocene. The true ancestors of *Homo sapiens* must already have achieved an essentially human appearance even before the brutish Neanderthals came to occupy Europe.

Keith was thus committed to the view that the Neanderthals must represent an entirely separate branch of hominid evolution. To substantiate this claim, he needed fossil evidence for the existence of more advanced human types which predated the Neanderthals of the Mousterian period in Europe. Much of the *Antiquity of Man* was devoted to the various discoveries of what appeared to be *Homo sapiens* in very ancient deposits, with Keith criticizing those who denied the true antiquity of these remains merely because of their "modern" character (fig. 8). Unfortunately, some of these early human types came from Europe, forcing Keith to postulate a temporary incursion of the lower Neanderthals into that continent during the Mousterian. The most important of the early human types was the Galley Hill specimen from England. Keith admitted that "our conception of the antiquity of man of the modern types—for undoubtedly the Galley Hill man is formed in the same mould as we are—turns on the authenticity of this discovery."[71] He did his best to refute the charge that the bones were a later burial intruded into an early deposit. A similar defense was offered for a number of other allegedly ancient human remains. If Keith's interpreation of this evidence was accepted, a modern type of humanity must already have come into existence in the Pliocene, thus completely upsetting the view that the Pleistocene ice ages had been the stimulus for the development of human intelligence.

Keith interpreted *Pithecanthropus* as a survival into the late Pliocene or early Pleistocene of the stage reached by our ancestors as early as the Miocene.[72] He no longer saw *Pithecanthropus* as the ancestor of the Neanderthals, since he believed that it already had a more upright posture than that attributed to the Neanderthals by Boule. Of even greater interest were the Piltdown remains. If possible, Keith would have pre-

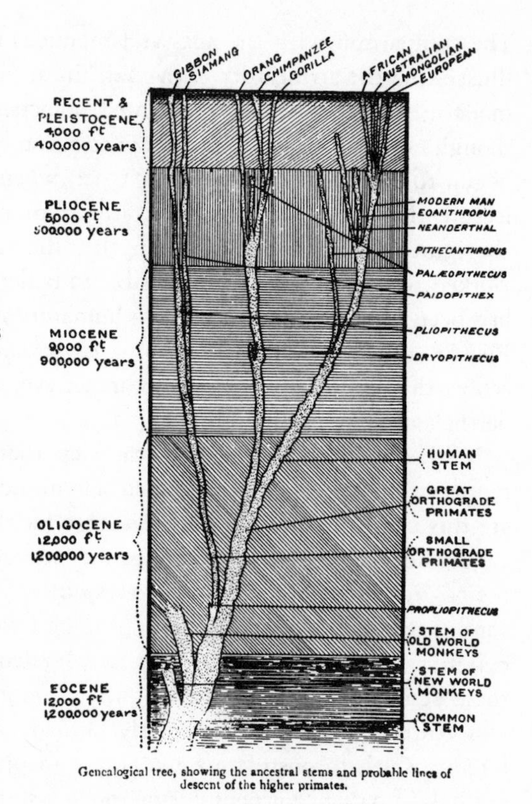

Figure 8. Arthur Keith's phylogony of the primates, showing a split between the ape and human lines in the late Oligocene and the emergence of modern mankind by the Pliocene. The durations of the geological periods given on the left-hand side of the diagram are far shorter than those accepted today, and Keith later reduced the estimates even more. Had he accepted the expanded time scale of earth history just starting to be indicated by measurements of radioactivity, his belief that human evolution has been a very slow process would not have required him to place the ape-human split in so early a period as the Oligocene. From Keith, *The Antiquity of Man* (London: Williams & Norgate, 1915), frontispiece.

ferred a Pliocene date for the Piltdown gravels, although most authorities saw them as early Pleistocene. The modern-looking skull of *Eoanthropus* certainly fitted Keith's idea of our Pliocene ancestor, but the apelike jaw was far too primitive. Keith might have followed Boule in rejecting the jaw, but as a patriotic Englishman he seems to have decided that he would have to work with the discovery as a unit. Nevertheless, he was determined to minimize the primitive aspects of *Eoanthropus* to fit it as closely as possible into his conception of the presapiens type. This view explains Keith's reluctance to accept the apelike character of the face given in Smith Woodward's early reconstruction, and his ongoing campaign to give *Eoanthropus* a larger cranial capacity than anyone else would allow. A large proportion of the *Antiquity of Man* was devoted to these issues. In the end, though, Keith could not make *Eoanthropus* the ancestor of modern mankind, since he was convinced that the apelike jaw pattern had already been transcended in the Pliocene. Nor could he see the modern-style cranium as a link with the Neanderthals. *Eoanthropus* thus had to be pushed out onto a side branch of its own, another experiment that had retained an apelike jaw, just as the Neanderthals had retained an apelike face.[73] This view at least had the merit of being consistent with the theory of multiple branches in evolution.

Keith was now firmly committed to the view that the Neanderthal species represented an entirely distinct branch of the human evolutionary tree, which had eventually been wiped out by the ancestors of modern humanity. It is significant that in his discussion of the Neanderthals' fate he referred to the work of Hermann Klaatsch (see chapter 6).[74] Klaatsch certainly promoted the idea that the Neanderthals were replaced by the invasion of a higher type, but he derived the Neanderthals and the modern black races from a gorilloid stock, and the modern whites from the orangutans. Keith was a leading critic of Klaatsch's polytypic theory, which he later dismissed as having had little influence.[75] Given Keith's own views on the extent of the racial divergence in modern humanity, though, one cannot help wondering if Klaatsch's theory—rather than Boule's—had triggered off Keith's willingness to dismiss the Neaderthals as an even more divergent side branch of human evolution.

Whereas Boule still retained a hierarchical interpretation of the stages through which the human type had advanced, Keith was convinced that the branches of human evolution had diverged in many different ways. He was not immune to the common prejudice that the black races were lower than the white, since as late as 1931 he claimed that no black race had ever developed a civilization.[76] Yet he was not a simple-minded racist. His autobiography notes that an early stay in Siam (Thailand) had made him aware of the extent of racial differences, but insists that he soon abandoned the urge to rank the races into superior and inferior types.[77] Keith realized that all the modern races are specialized in different ways starting from what he regarded as the primitive Australian prototype, with the blacks being in some respects *less* apelike than the whites.[78] His image of racial evolution was thus a branching rather than a linear one, with the racial differences being the raw material from which selection would produce those best fitted to expand into new areas.

Keith was already convinced that racial differences were the result of inherited variation in the glandular balance of the body, presumably the effect of long exposure to different environments.[79] It was this concept of divergent evolution that he applied to the Neanderthal question. He agreed with Boule that the Neanderthals had many apelike features, but suggested that these could be explained as the consequence of glandular balance, since they were duplicated today in individuals suffering from the disturbance of the pituitary gland known as acromegaly. Far more than Boule, Keith set out to argue that the Neanderthals possessed specializations in a direction opposed to that taken in the evolution of modern humanity. In at least one character—the lack of a nasal gutter—it was the modern form that was the more simian.[80] The teeth of the Neanderthals also showed an enlargement of the pulp cavity that Keith called "taurodontism." Adloff had noted it as a specialized feature, and Keith now accepted that it confirmed the distinct character of the Neanderthal evolutionary line.[81] The Neanderthal palate differed widely from the simian form, possibly an adaptation to a rough vegetable diet.[82] Whereas Boule emphasized the apelike character of the Neanderthals, Keith was prepared to recognize features that removed them farther from the apes than ourselves, in order to promote the view that their line of evolution branched off in a separate direction.

In effect, Keith turned the Neanderthal type into a "race" that had diverged so much earlier than the modern races that it could be classed as a separate species. Significantly, in speculating about the eventual fate of the Neanderthals, he drew an explicit comparison with the extermination of the "lower" races in the modern world. "What happened at the end of the Mousterian period we can only guess, but those who observe the fate of the aboriginal races of America and Australia will have no difficulty in accounting for the disappearance of *Homo neanderthalensis*. A more virile form extinguished him."[83] By 1916, Keith had begun to argue that the division of a species into competing racial and tribal groups was a vital part of the evolutionary mechanism, a view he was later to expand into a complete theory of human evolution.[84] The image of racial conflict seems to have fascinated Keith, and may thus have encouraged him to see the Neanderthals as the victims of a higher race's territorial ambitions.

The preface to the *Antiquity of Man* contains an addition written in 1915 noting that the outbreak of war had now changed many attitudes. Yet Keith's own views on racial competition may have been shaped by the wave of nationalist fervor that paved the way for war. Europeans had long taken it for granted that their superiority justified the domination of other races, but the rival nations now came to see themselves as competing for their share in world conquest. The prominent "social Darwinist" Karl Pearson openly applauded the displacement of the North American Indian and the Australian aborigine, but also warned of the threat posed to the British Empire by rival powers.[85] Imperial expansion and national rivalry were prominent themes in early twentieth-century European politics, both openly linked to the Darwinian notion of the "survival of the fittest." It would not be surprising, then, if Keith and the other supporters of the presapiens theory had borrowed this model of human relations for their explanation of the Neanderthals' sudden disappearance. Hammond has argued that Boule's work was inspired by a desire to refute de Mortillet's socialist interpretation of linear progressionism.[86] But this suggestion must be generalized if it is to serve an an explanation of the sudden conversion of many paleoanthropologists to the presapiens theory in the period around World War I. Boule's offensive against de Mortillet may, in fact, be only a manifestation of the broadening support for nationalist and imperialist policies in prewar Europe.

Keith's views on the evolutionary significance of race conflict had already been anticipated in a slightly different form by W. J. Sollas. In his *Ancient Hunters* of 1911, Sollas promoted the view that racial migration and displacement were important features of human evolution. Development did not take place continuously over the whole world; instead, higher types appeared from time to time and spread out from their original homes, dispersing or exterminating the inferior races in the surrounding areas.[87] Human evolution was thus a more complex process than had hitherto been assumed. Although attracted by the hierarchical model of cultural evolution, with modern primitives retaining the culture of earlier stages in human progress, Sollas soon perceived that his emphasis on racial conflict would fit neatly into the presapiens theory. In later editions of *Ancient Hunters* he reluctantly gave up the view that the Australians were relics of the Neanderthals and accepted that the Neanderthals were a distinct species not related to human ancestry.[88] The Piltdown discovery confirmed the branching character of human evolution, *Eoanthropus* now being presented as our direct ancestor.[89] The preface to the later editions records that the idea of racial displacement, considered heretical when first presented in 1911, had soon become the new orthodoxy.[90]

In the period around World War I, the majority of paleoanthropologists came to accept the extermination of the Neanderthals by a presapiens type. H. G. Wells's story "The Grisly Folk" of 1921 summed up the popular image of the Neanderthals as shambling ape-men, hunted to extinction by our own ancestors.[91] The harsher attitude toward national and racial interactions that became popular in the early twentieth century may well have helped to provide a climate of opinion receptive to this interpretation. Yet we cannot attribute the sudden rise of the presapiens theory to this factor alone, since some of the theory's advocates preferred not to dwell on the warlike implications of the Neanderthals' disappearance. Several additional factors can, in fact, be cited.

First, the archaeological evidence now began to indicate a comparatively sudden replacement of the Neanderthals by a modern type in Europe. The fossil record also became rapidly more complex at this time, inevitably throwing doubt on the linear scheme of human evolution proposed by Schwalbe. The discovery of the Heidelberg jaw in 1907 was the first step in this process. Although of the same age as *Pithecanthropus* (early Pleistocene), this jaw was so massive that few authorities thought it could belong to the same type. Piltdown, of course, added yet another branch to the early Pleistocene tree, and seemed to confirm the view that the Neanderthals were not ancestral to modern humanity. In this respect, the Piltdown forgery did help to deflect theorizing about human origins up a blind alley, although it was by no means the only factor involved. The discovery of Rhodesian man in 1921 added to the confusion, since most anthropologists saw its relationship to the Neanderthal type, but could not accept that it belonged to the same species. So yet another branch had to be added to the human family tree.

As we have seen, Boule and Keith both insisted that a branching model of human evolution was more realistic than a linear one. The fossil discoveries of the early twentieth century created a climate of opinion in which this perfectly reasonable point could be accepted—and then exaggerated out of all proportion. There were enough fossils to indicate that branching had occurred, but not enough to determine how and when. Scientists were thus free to use their own judgment, and the result was a plethora of mutually contradictory phylogenies which was to bring the whole field into disrepute. Later critics have alleged that the presapiens theory was used as an excuse to invent totally hypothetical lines of human ancestry, ignoring the known fossil record. The root cause of the desire to place all the known fossils on side branches was almost certainly the anthropologists' exaggerated claims as to the novelty of their discoveries. When every new fossil was described as a new species, if not a new genus, there was no room for them all to be accommodated on a single line of descent. The critics quite correctly accuse the early twentieth-century paleoanthropologists of exaggerating the diversity of the human family tree, but this exaggeration was hardly a simple product of the decision to treat the Neanderthals as a form that had become suddenly extinct.

Those who did not share Keith's enthusiasm for the significance of racial conflict had other reasons for challenging the concept of a Neanderthal ancestry for modern humanity. The most obvious motive was a desire to uphold the increasingly popular view that the expansion of the brain had led the way in human evolution (see chapter 7). Grafton Elliot Smith was the leading exponent of this position, and welcomed the Piltdown discovery as evidence for the early emergence of a modern brain structure. If not quite our true ancestor, *Eoanthropus* was closely related to the early Pleistocene type from which we had arisen.[92] As the codiscoverer of the Piltdown remains, Arthur Smith Woodward naturally presented his reconstruction of *Eoanthropus* as the ancestor of modern humanity, and this reconstruction necessitated the expulsion of the Neanderthals to a degenerate side branch.[93] Most authorities now came to regard *Pithecanthropus* as yet another deadend, an example of a Pliocene grade of evolution that had been preserved into Pleistocene times. An alternative that preserved something of Schwalbe's scheme was to treat the Neanderthals as descendants of *Pithecanthropus,* with *Eoanthropus* as the true ancestor of modern humanity, as suggested in a popularly written book by H.G.F. Spurrell.[94]

Spurrell's suggestion pushed the separation between the *Pithecanthropus* and presapiens lines well back into the Pliocene, but raised problems for the orthodox view that the Pleistocene glaciations had played a major role in stimulating the evolution of modern humanity. Boule and Keith both had adopted what most of their contemporaries regarded as highly exaggerated opinions on the antiquity of the presapiens type. Both, in fact, believed that an essentially modern human form had already come into existence by the end of the Pliocene. Keith admitted that his position was dismissed as an "amusing heresy" by almost everyone else in the field.[95] Sollas, Elliot Smith, and Smith Woodward all vigorously repudiated Keith's efforts to increase the capacity of the Piltdown skull, because they were convinced that an early Pleistocene type must have a brain significantly smaller than the modern average (although its general organization might be similar). The extension of human antiquity needed to be only enough to sidestep the Neanderthals within a time scale still defined by the Pleistocene. Talk of a fully human type in the Pliocene was nonsense: the Galley Hill find was a later burial, and *Eoanthropus* must show significant evidence of its still lowly status, whatever its promise for the future.

One paleontologist who did accept something resembling Keith's position was the American Henry Fairfield Osborn. His *Men of the Old Stone Age* was written to stress the more advanced character of Paleolithic culture. To bring out the more human aspects of our Paleolithic ancestors, Osborn opted for an early separation of the Neanderthal and presapiens stocks.[96] Even *Eoanthropus* was not a true ancestor but a side branch lying between the two main groups. Apart from placing *Pithecanthropus* on the Neanderthal stem, Osborn's position agreed closely with Keith's. Significantly, Osborn later went on to extend the antiquity of the human line even farther back in time—so far, in fact, that the apes were eliminated from human ancestry altogether (see chapter 5).

Keith eventually began to back away from his claim for the extreme antiquity of modern humanity, but Osborn's theory shows that the temptation to extend the origin of the human type as far back as possible could influence other workers in the field. Support for Keith's original position eventually emerged from the work of Louis Leakey in East Africa, and Leakey went on to become a leading exponent of the early and independent evolution of the genus *Homo*. Leakey was born in Africa, and after graduating at Cambridge went back to Kenya with the intention of applying the archaeological techniques developed in Europe to uncover the as yet unknown prehistory of the area.[97] His 1926 East African Archaeological Expedition convinced him that Africa was the source of the Chellean-Acheulean culture that eventually spread to Europe. In 1931 he began to uncover the remains of an early stone culture in what is now called Olduvai Gorge. He suggested that this "Oldowan" culture was the origin of the Chellean tradition.[98] He also began to find the bones of stone-age races that seemed as distinct from one another as are the races of today.[99] In 1932 he discovered the Kanam jaw, which seemed to indicate the existence of a form closely resembling *Homo sapiens* in the early Pleistocene, thereby confirming the position already staked out by Keith.

Leakey's views were shaped by both the archaeological and anthropological evidence coming out of Africa. The antiquity of racial types led him to support Keith's belief that racial evolution was a slow process and hence that *Homo sapiens* must have been in existence for a long period.[100] The sheer complexity of cultural evolution in both Africa and Europe showed that a linear image of cultural progress was inappropriate. Leakey now believed that cultures were spread by migrating peoples, who could move in and out of a territory, occupy the same area as other culture groups, and sometimes interact with other peoples to give new cultural developments. It then seemed obvious that the cultures must be identified with different racial groups and, if necessary, with distinct species of early mankind. Leakey pointed to two great cultural traditions, the Oldowan-Chellean-Acheulean-Aurignacian sequence, and the Clactonian-Levalloisian-Mousterian complex. All the evidence suggested that the Mousterian, and hence by implication the others of that group, were the products of the Neanderthals. Leakey argued that the other sequence must belong to modern mankind, which inevitably pushed our ancestry much farther back in time.[101]

In 1913 a German archaeologist, Hans Reck, had found modern-looking human remains in what he took to be an extremely ancient East African deposit. Reck accompanied Leakey's third expedition in the hope of vindicating his claims, but despite his initial optimism, he soon discovered that the remains must have been a later burial.[102] On the same expedition, however, the discovery of the Kanam jaw and the more recent Kanjera remains seemed to confirm Leakey's view that early *Homo*, if not *Homo sapiens*, had originated the Oldowan culture in the early Pleistocene. On the strength of the Kanam jaw's apparent lack of Neanderthal characters, Leakey reconstructed in his imagination a fairly modern form of mankind, *Homo kanamensis*. He believed that other controversial examples of early *Homo* should also be taken more seriously.

Leakey's views on cultural and biological evolution were summed up in his popular *Adam's Ancestors* of 1934. Here he made it clear that the makers of the different cultures were far more distinct than mere races: they were separate species, if not separate genera. Like Keith, Leakey pointed to the existence of four genera within the modern apes, and asked why we should expect human evolution to be any less liable to branching (fig. 9).[103] The Neanderthals could be distinguished from modern mankind by a series of absolutely characteristic anatomical features, each of which showed a closer link to the apes.[104] In his conclusion, Leakey insisted that the Neanderthals constituted a distinct genus which could be subdivided into its own species. The European Neanderthals became *Palaeoanthropus europaeus*, while Rhodesian man was *P. rhodesiensis*.[105] Leakey agreed with Solly Zuckerman's suggestion that *Pithecanthropus*, *Sinanthropus*, and the Neanderthals should be linked into

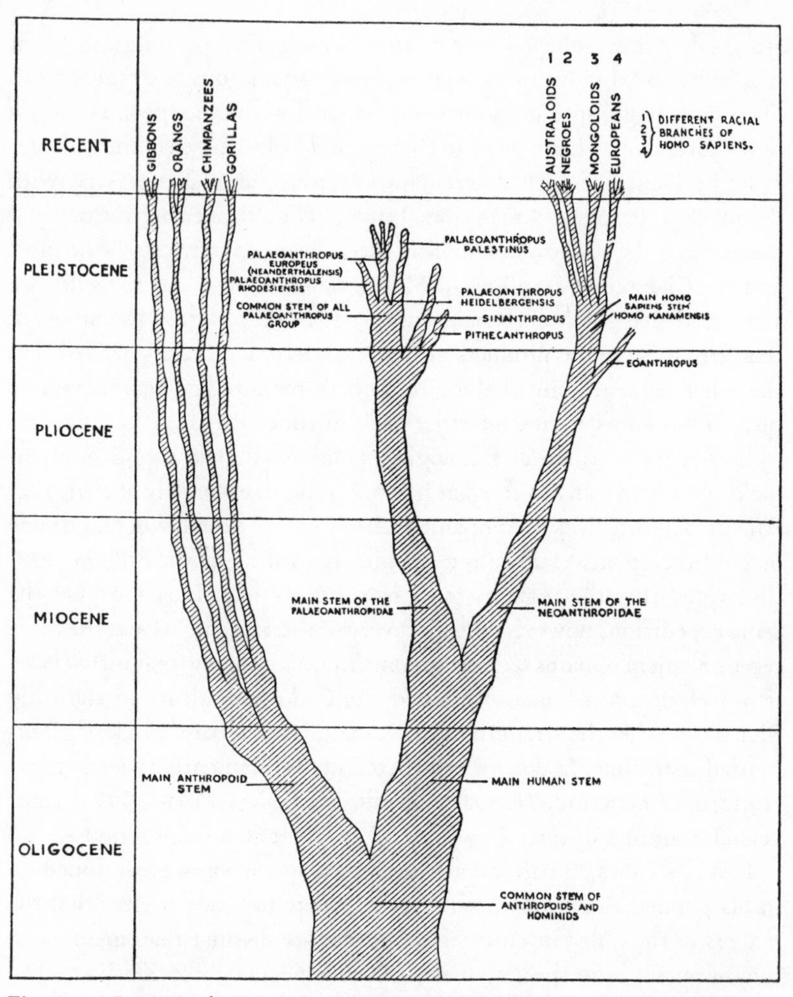

Figure 9. Louis Leakey's phylogent of the primates. Like Keith, he places the main ape-human split in the Oligocene, but then adds a second major branching in the human stem between the Palaeoanthropidae and Neoanthropidae in the Miocene. From Leakey, *Adam's Ancestors* (London: Methuen, 1924), p. 227.

a subfamily, the Palaeoanthropidae, which separated in the Miocene from the human stem, the Neoanthropidae.[106] Piltdown man was regarded as a primitive side branch of the Kanam stock, and thus quite close to the ancestry of modern mankind.[107]

In the postwar years, Leakey became widely known for his later fossil discoveries in East Africa, and for his refusal to admit that the Australopithecines were ancestral to modern humanity. He continued to defend the view that the extinct hominids must all be assigned to distinct genera, and retained the separation into two subfamilies. He regarded the Australopithecines as merely another branch of the Palaeoanthropidae, paralleling those leading toward *Pithecanthropus* and the Neanderthals.[108] Although this system of classification was rejected by most other workers in the field, the discovery of very early members of the genus *Homo* did confirm that the later Australopithecines were not ancestral to modern mankind.

By modern standards, Leakey's view that different species or even genera could independently develop enough intelligence to begin toolmaking seems rather suspect. It requires a level of parallel evolution that is inconsistent with modern Darwinism and a good deal of faith in the willingness of early mankind to tolerate rivals occupying the same territory. Admittedly, the generic distinctions of the 1930s are not the same as those of today. Yet it is clear that Leakey did indeed believe he was dealing with distinct species—this was not just a case of using a specific name to designate a mere race. Leakey treated the specific and generic distinctions among fossil hominids as fully equivalent to those accepted among living animals. For him, all such distinctions were based on anatomical differences and had little to do with genetic sterility between the populations. He accepted that the different cultures could interact with one another, and even that the various forms of early mankind could interbreed successfully.[109] As late as 1953 he insisted that animals from two distinct subfamilies could produce fertile hybrids.[110] Leakey seems to have believed that cultural barriers prevented the various species of early hominids from merging together, a point that fits in with Keith's interest in the tribal instinct as a factor that can preserve the integrity of biological groups. Unlike Keith, however, Leakey had no interest in seeing intergroup conflict as the means by which the inferior types are eliminated. He did not dwell on the problem of *how* the Neanderthals disappeared, and it may be that his exaggeration of the differences between them and ourselves was a way of avoiding the implications normally associated with the presapiens theory. If the Neanderthals were so remote, their disappearance might be due to natural causes, rather than extermination by a higher type. Leakey was thus free to promote a more harmonious image of racial interactions, strongly at variance with the imperialist message normally derived from the presapiens theory.

Leakey's system of classification was no more than an extreme version of the position commonly accepted during the 1930s. In the postwar years, such a position was still defended by Leakey himself and by other writers whose opinions had been formed in earlier decades. Teihard de Chardin—who had been active at the Piltdown site—was convinced that there were several parallel branches of human evolution, and used this belief to support his broader thesis that the whole development of life on the earth has been drawn toward a spiritual goal. His *Le phenomène humain* (*The Phenomenon of Man*) was eventually published in 1955 and attracted wide popularity even at this late date. In 1948 the once-prominent geneticist R. Ruggles Gates still argued that the sterility criterion for distinguishing species was worthless, and that the human races were the end products of independent lines of development.[111] Teilhard and Gates used the concept of parallel development for very different purposes, but their work reveals how difficult it was for thinkers from a variety of backgrounds to throw off the belief in directed evolution. Nevertheless, during the postwar years a majority of scientists at last began to turn away from the idea, as the revived Darwinism of the Modern Synthesis destroyed the plausibility of the mechanisms upon which parallelism had been based.

In the 1930s the majority of paleoanthropologists established their species on the basis of morphological differences, and they all exaggerated the significance of what we now take to be merely local variations within a single species. Given the resulting multiplication of species and genera among fossil hominids, an inflated view of the number of the evolutionary side branches was inevitable. Once the Neanderthals were dismissed as a distinct species, the possibility that fully modern forms of humanity could have existed early in the Pleistocene was opened up. Many anthropologists seemed only too willing to throw off our relationship to the brutish Neanderthals, and to allow extra time for the formation of major racial divisions within our own species. The reformation of hominid taxonomy that took place in the 1950s reintegrated the Neanderthals into *Homo sapiens* and thus allowed them to be perceived as a variant of our own primitive ancestors. But if the presapiens theory has thus collapsed, the predictions of Keith and Leakey have been fulfilled at a different level. The evolution of mankind has indeed turned out to be a branching process, with the later Australopithecines as side branches leading to extinction. We no longer expect to find modern forms of humanity in the Pliocene, but the discovery of early members of the genus *Homo* has confirmed that many fossil hominids were not our ancestors.

Defense of the Neanderthal-Phase Theory

Some time before Leakey began to develop his version of the presapiens theory, Keith had backed away from the extreme position staked out in the first edition of his *Antiquity of Man*. In the second edition of 1925 he conceded that the evidence did not favor his original conclusions.[112] It was still possible to make out a case for the antiquity of Galley Hill man, but in the absence of any additional confirmation "it becomes easier to doubt this evidence than to believe that human evolution ever becomes stationary."[113] Keith now accepted that Rhodesian man, which had some Neanderthal and some modern characters, was close to the common ancestor of both types.[114] Following the discovery of Neanderthal and modern remains alongside one another at Mount Carmel in Palestine, Keith and McCown argued for the evolution of the Cro Magnon race in the East, leaving Europe as a backwater in which the more primitive Neanderthals were preserved until wiped out by an invasion of the higher type.[115] Without abandoning the concept of the Neanderthal extinction, Keith thus adopted the view that our own ancestors had passed through a phase distinguished by at least some Neanderthal characters. He thus adopted what Vallois called the pre-Neanderthal theory.

This intermediate position must be born in mind when evaluating the status of the Neanderthal-phase-of-man theory in the interwar years. There were some efforts to defend linear evolutionism against the branching approach of the presapiens theory, although Brace argues that they were largely ineffective.[116] But if it was possible to continue arguing for the extinction of the European Neanderthals yet still accept a partially Neanderthaloid phase in the ancestry of modern mankind, the gulf between the opposing positions was in fact diminishing in a way that foreshadowed the rehabilitation of the Neanderthals in modern times. W. K. Gregory, although he did not specialize in the study of the later stages of human evolution, admitted that there was something to be said for the idea that the Neanderthals were survivors of an earlier phase in our development.[117] The excitement generated by the discovery of Peking man also helped to boost a viewpoint favorable to the Neanderthal-phase theory, since *Sinanthropus*—like *Pithecanthropus*—had heavy brow ridges that seemed to anticipate the Neanderthal type. Davidson Black certainly regarded *Sinanthropus* as ancestral to modern mankind, although he preserved a branching concept of evolution by suggesting that *Pithecanthropus* was a degenerate offshoot from the main stem.[118] Once what is now known as *Homo erectus* was accepted as our early Pleistocene ancestor, it was difficult to avoid the conclusion that something like a Neanderthal phase had intervened, even if the European Neanderthals had been wiped out by a form that had evolved more rapidly elsewhere.

Some continued to defend the old form of the Neanderthal-phase theory, including the belief that the European Neanderthals could have been involved in the transition to modern mankind. In America, Ales Hrdlička was the most prominent exponent of this position. In some respects he was an unlikely candidate for such a role, since he was a physical anthropologist who shared Keith's opinions on racial diversity. Yet this view did not prevent him from seeing the modern races as comparatively recent branches which had diverged after the Neanderthal phase had been passed. It seems probable that Hrdlička was reluctant to accept the claim that the European Neanderthals had been driven to extinction because it did not fit in with his views on racial migration. In 1926 he included a defense of the Neanderthal-phase theory in an article challenging the common opinion that mankind had evolved in central Asia. According to Hrdlička, the evidence favored a more westerly origin, and there was no valid reason for abandoning the old belief that the human type first appeared in Europe.[119] By thus reversing the direction of migration, he was forced to reject the extinction of the European Neanderthals and consider them as our potential ancestors.

Hrdlička had supported this conclusion as early as 1913,[120] although his most famous defense of it came in his Huxley Memorial Lecture of 1927 entitled "The Neanderthal Phase of Man." Here he cited Keith and Boule as leading exponents of the presapiens theory, but suggested that the movement could be traced back to Schwalbe.[121] This suggestion involves a misunderstanding of Schwalbe's position—which Brace attributes to Hrdlička's dislike of all things German—although Schwalbe did not rule out the possibility that the Neanderthals might be survivors of an earlier phase. Hrdlička argued that the Neanderthals should be defined as the makers of the Mousterian culture, and insisted that this culture was linked by transitions to those lying above and below it.[122] If the development of culture was continuous, then so might be biological evolution. There was also a wide variation in the characters of the Neanderthal specimens from different locations, some being far closer to the modern type than others. Opponents of the Neanderthal-phase theory assumed that there was no consistent trend involved, but Hrdlička pointed out that the lack of any precise dating of the sites left a reasonable possibility that there might have been a steady transition from the Neanderthal to the modern type. The presapiens theory implied an invasion of Europe at the height of a glacial period, which seemed most unlikely, especially since no trace of the hypothetical presapiens type had been found anywhere else in the world.[123] In fact, the rigors of the glacial climate had stimulated evolution and produced the rapid development of the modern form from its Neanderthal ancestor.[124] The presapiens theory implied "a long double line of human evolution, either in near-by or the same territories; a sudden extinction of one of the lines; and evolutionary sluggishness or pause in the other," all of which amounted to "an outright polygeny—which is undemonstrable and improbable."[125] In the diagram illustrating his position, Hrdlička showed not a single line of evolution but a "bushy" one, with a multitude of small side branches corresponding to the racial variants thrown off at all phases of human development.

A version of the Neanderthal-phase theory had already been defended in the 1924 Huxley Memorial Lecture by the French anthropologist R. Verneau. Verneau had suggested that some of the Grimaldi specimens had a negroid appearance, and he now revived the old theory that the black race was an early form of mankind which retained a higher proportion of characters left over from our ape ancestry. The Neanderthals were an even earlier and more apelike phase in the process, although a remnant of this phase could still be seen in the Australian aborigines.[126]

The most consistent defense of the Neanderthal-phase theory came from German anatomists who had been influenced by Schwalbe. Hans Weinert emerged as an outspoken supporter of the view that modern mankind had evolved from a chimpanzeelike ancestor via phases corresponding to *Pithecanthropus* and the Neanderthals. In 1927, Weinert persuaded Dubois to throw open his collection of fossils to further investigation, revealing more *Pithecanthropus* specimens from the original discovery. Weinert also worked on a new reconstruction of the *Pithecanthropus* skull, with the intention of revealing its status as an ape-human intermediate. He conceded, however, that the known *Pithecanthropus* remains were probably too late to represent the actual ancestors of the later phases.[127] Although Weinert is best known for his work on *Pithecanthropus,* his position almost inevitably required him to treat the Neanderthals as the next stage of human evolution. Significantly, he shared the belief that the ice ages had been an important stimulus to the last episodes in the process. His views were summed up in his survey *Ursprung der Menschheit* of 1932.

The renewed interest in *Pithecanthropus* sparked off by the discovery of Peking man played a prominent role in the career of another German supporter of the Neanderthal-phase theory, Franz Weidenreich. Originally a lecturer at Frankfurt, Weidenreich left Germany during the Nazi period to work first in China and then in America. As early as 1928 he developed a form of the Neanderthal-phase theory, citing Hrdlička's views as similar to his own.[128] Starting from Schwalbe's interpretation of *Homo primigenius* (the Neanderthal species), he argued that Rhodesian man was the earliest form of the species, which then migrated from Africa to Europe and underwent further development. He thought the black races of today were derived from the African version of the Neanderthal species.[129] In his later work, Weidenreich generalized this thesis by supposing that the Neanderthal phase developed into the modern one across the whole world, meanwhile preserving the racial distinctions already apparent in the Neanderthals.

Weidenreich's work on the new *Pithecanthropus* specimens uncovered by von Koenigswald, and on the *Sinanthropus* remains, convinced him that these were variants of the earliest human form, which had evolved into modern mankind via an intermediate Neanderthal phase. There were no side branches in human evolution after the *Pithecanthropus/Sinanthropus* stage, since the European Neanderthals had developed into the modern type along with other representatives of the Neanderthal phase over the whole world. Rhodesian man illustrated the lower end of the Neanderthals' morphological range, and the Mount Carmel specimens formed the upper end, where this phase blended into the lowest forms of modern humanity.[130] In Weidenreich's scheme, the Palaeoanthropinae and Neoanthropinae were the chief phases of a single line of development, not two branches leading toward separate end products.[131]

The most original part of Weidenreich's thesis was his claim the *Pithecanthropus* and *Sinanthropus* were racial variants of the same form, and that such geographical distinctions have been preserved intact while mankind has evolved into its modern form via the Neanderthal phase. He identified Solo man as the Far Eastern variety of the Neanderthals. He summed up his overall thesis: "(1) We now have evidence of the direct transformation of the Neanderthaloid form into *Homo sapiens;* (2) The Neanderthalians need not be considered as having become extinct without having left any descendants behind; (3) Racial or regional differentiations are recognizable within the Neanderthalians themselves and can be traced from these to races of modern mankind."[132] Even in more recent

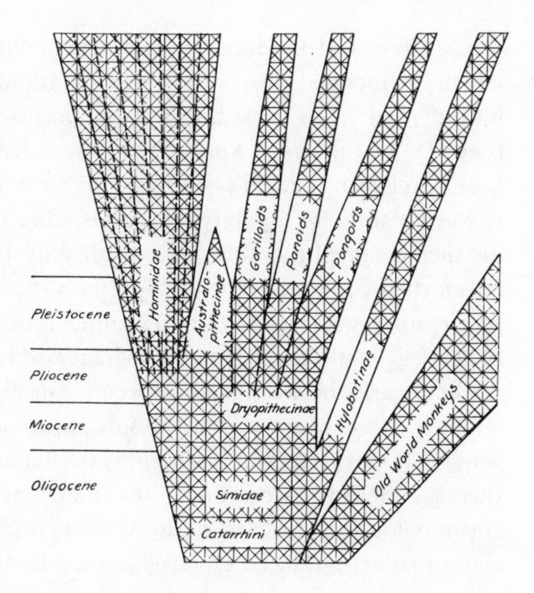

Figure 10. Franz Weidenreich's phylogeny of the primates. Note the absence of branching in the hominid stem. The horizontal lines represent stages through which the vertical lines, representing the racial types, advance in parallel. From Weidenreich, *Apes, Giants, and Man* (1946), p. 24.

times, there had been a universal tendency for long-headed racial types to convert themselves to the round-headed form.[133] In the diagrams given in a number of Weidenreich's later works, he shows a series of racial lines moving in parallel through a sequence of phases from the ape to the modern human form (see fig. 10). The Australopithecines were typical of the earliest phase of the process, although Weidenreich had to admit that the later members of this group formed a short side branch no longer connected to the multiple lines leading toward modern mankind.[134]

It would be easy to see in this view an almost orthogenetic theory in which some mysterious compulsion urged separate lines to evolve together in the same direction toward a predetermined goal. Weidenreich's approach to evolution has certainly a degree of linearity that betrays its origins in Schwalbe's morphological sequences and does not mesh easily with modern Darwinism. Yet Weidenreich himself believed that the general direction in the trend of human evolution was determined by the growing size of the brain, which had adaptive benefits in all circumstances. The various lines of development were in contact with one another, and some interbreeding took place at all stages, enough to ensure parallel evolution, but not enough to blend the racial characters together.

In the postwar years, the American anthropologist Carlton Coon tried to show that Weidenreich's interpretation of the antiquity of racial types was consistent with the Modern Synthesis of Darwinism and genetics. In

1939, Coon had produced a replacement for Ripley's *Races of Europe* which maintained that nonsapiens hominid species, including the Neanderthals, had contributed to the formation of the modern racial types.[135] His *Origin of Races* of 1963 was dedicated to the memory of Weidenreich and offered a defense of his views on race. Coon argued that it was possible for a polytypic species, that is, a species consisting of distinct geographical subspecies with only limited interbreeding between them, to preserve its subspecific variants intact while the whole species undergoes a transition to a higher form. The new genes responsible for a general adaptation such as increased brain size can "leak" across the boundaries between the subspecies, causing rapid evolution in each, while the local characters of the subspecies do not diffuse in the same way.[136] If this statement is a valid extension of Weidenreich's position, then parallelism in the non-Darwinian sense is not required. Many anthropologists, however, regard Coon's position as merely an excuse to claim that certain racial types achieved fully human status more slowly than others.

In the immediate postwar years, Weidenreich's position on the Neanderthal phase of development was almost as controversial as his theory of racial origins. Few believed that the classic Neanderthals of western Europe could have evolved rapidly enough to explain the sudden appearance of the Aurignacian type. Yet the general idea of a Neanderthaloid phase through which our ancestors must have passed gained plausibility as the reinterpretation of the Australopithecines made the sequence *Australopithecus–Homo erectus–Homo sapiens* seem an increasingly likely candidate for the general outline of human evolution. This view reopened the possibility that something resembling the Neanderthal type might fit in neatly as the intermediate between *Homo erectus* and modern *Homo sapiens*.

The evidence also began to suggest that earlier workers had exaggerated the gulf between the Neanderthals and the modern races. In 1957, William L. Straus and A.J.E. Cave showed how Boule's image of the slouching Neanderthal had been influenced by his refusal to acknowledge that the specimen he was studying had been crippled by arthritis.[137] If the Neanderthals had walked erect, they were much more appropriate as a possible ancestor of modern humanity, and they soon came to be regarded as merely an early subspecies of *Homo sapiens*. In 1964, C. Loring Brace launched a deliberate attack on the view that the European Neanderthals had been wiped out, arguing for a transition from the Neanderthal to the modern phase in all parts of the world.[138]

He revived and extended the arguments of Hrdlička, Weinert, and Weidenreich, claiming that they had been consistently ignored over the previous decades. Brace's historical analysis may have been influenced by his desire to present the presapiens theory as an essentially nonevolutionary posture, but the effect of his article on anthropologists was considerable. The initial reaction—included with the original article—was mixed. but in recent years many experts have come to accept that something like a pre-Neanderthal type represents the oldest form of *Homo sapiens.* Many still believe that the European Neanderthals represent a specialized form of our species that disappeared suddenly, although it is possible that when more modern types invaded Europe there was absorption as well as extermination. The archaic form of *Homo sapiens* was not a classic Neanderthal, but it may well have possessed certain Neanderthal-like features, such as the heavy brow ridges, inherited from *Homo erectus.* The presapiens theory is thus refuted: all human remains from the early Pleistocene are now attributed to *Homo erectus,* which is thought to have spread from Africa and then evolved into archaic *Homo sapiens* in several parts of the world. We still have a theory of branching evolution farther down the scale, though, since the later Australopithecines now occupy the side branch once assigned to the Neanderthal "species" in the first efforts to break down the linear progressionism of the late nineteenth century.

Chapter Five The Tarsioid Theory

Huxley's demonstration of our close affinity to the great apes ensured that most subsequent speculations on human evolution would start from the assumption that we had evolved from an apelike ancestor. Most evolutionists accepted that this hypothetical ancestor would have been recognizably a great ape, even if not quite so specialized as the modern gorilla or chimpanzee. But once they realized that the modern apes had changed to some extent from the common ancestral form, it became possible to wonder about the extent of their additional specialization. Was it conceivable that the apes had changed much more than we had at first imagined, so that the common ancestor would not have resembled the modern apes closely at all? Could the common ancestor have had a more human appearance, and the apes have been formed by a process of degeneration at the same time that the human form was finally perfected? Taking this line of thought further still, could the similarities between the apes and mankind be due entirely to parallel evolution, so that the common ancestor lay much farther down the family tree of the mammals? In this case the line of human evolution would never have passed through anything resembling an ape stage. In the extreme form advocated by Frederic Wood Jones, this theory entailed an entirely separate line of human ancestry striking out from an Eocene form resembling the modern tarsier. This "tarsioid theory" was always something of a heresy, but it was widely discussed, and in the late 1920s a slightly less extreme position was advocated by an authority as eminent as Henry Fairfield Osborn.

Although pushed to what most anthropologists considered a ridiculous extreme, the tarsioid theory was based on principles that were merely extensions of well-established scientific ideas. Darwin himself had speculated that the similarities between the apes and mankind might be due to parallel evolution, but had backed away from the idea because he found it inconceivable that such extensive resemblances could be acquired independently.[1] But just how great was the power of parallel or convergent evolution to generate morphological similarities in unrelated forms? Anyone who accepted a theory of evolution in which such powers were exaggerated would naturally be tempted by the possibility of the independent evolution of the apes and the human stock. Such theories were still quite popular among anti-Darwinian naturalists in the early twentieth century. In addition, the recapitulation theory could be used to suggest that the ancestors of the modern apes had been more human in appearance, since the young ape of today lacks the heavy, specialized features of the adult form. This point was made by Smith Woodward in his efforts to show that the Piltdown skull, with its strikingly human character, was a likely ancestor for modern mankind.[2] In such a view, the human form was in some respects a very primitive one, whereas the apes had acquired their more brutal form through additional specialization. Woodward's great rival, Arthur Keith, extended the antiquity of modern mankind back to the Pliocene or even earlier, invoking a separate line of evolution for which there was at best only dubious fossil evidence. Keith and Woodward still accepted that the common ancestor of mankind and the modern apes would be recognizably an ape, even if an unspecialized one. But if the logic of their positions were extended by someone with a greater faith in the power of parallel evolution, then it might become possible to eliminate the apes altogether from human ancestry.

Why should any scientist have been so anxious to postulate an entirely separate line of human evolution? One answer to this question lies in the open hostility expressed in some quarters against the late-nineteenth-century form of Darwinism. Strictly speaking, Darwinism meant the theory of evolution by natural selection, but thanks to Huxley's work many people felt that the theory of ape descent was an integral part of a more broadly defined form of Darwinism. Those who were hostile to the selection theory and its apparently materialistic implications thus felt obliged to deny the close link between mankind and the apes. Anti-Darwinian evolutionary mechanisms such as Lamarckism and orthogenesis provided just the foundation needed for a theory in which the apes and mankind were separate but parallel developments. Wood Jones adopted a critical attitude to Darwinism and had a strong leaning toward Lamarckism. He believed that the inheritance of acquired characters was a more powerful agent of adaptive evolution than natural selection, so that in his view two unrelated forms that adopted a similar lifestyle would soon develop close anatomical similarities. Osborn was a leading exponent of orthogenesis, a theory in which multiple lines of evolution advanced in parallel under the influence of an internal (or purely biological) driving force.

Significantly, Osborn did not extend his theory to mankind until late in his career, and the motivation for his change of heart must almost certainly lie outside the field of science itself. It can hardly have been a coincidence that Osborn began to exclude the apes from human ancestry shortly after he had become involved in the defense of the general theory of evolution against the attacks led by fundamentalist religious leaders that had begun with the "monkey trial" of John Thomas Scopes in 1925. Elimination of the "ape stigma" from human descent was certainly a bonus in the drive to defend evolutionism against the charge that it reduced us all to animals. Wood Jones had strong religious beliefs throughout his career, and his first attack on the ape theory was actually published by the Society for the Promotion of Christian Knowledge. For him, the tarsioid theory helped to minimize the brutalization of human nature that he believed to be a product of the integration of mankind into the animal kingdom. To the extent that the idea of an ape ancestry had become a symbol of this lowered status, a link back to a more presentable creature such as the tarsier could be a valuable weapon in the fight to create a more humane version of the general theory of evolution.

A hypothesis that was sometimes confused with the tarsioid theory was Edward Drinker Cope's emphasis on the role of a lemurlike stage in human evolution. Cope had studied the fossil lemuroids of the Eocene, and postulated that the line leading toward the apes and mankind had started from such an ancestry. Although some thought that Cope wanted to trace the ancestry of mankind back to the lemurs without an intermediate ape stage, he insisted that this was not his purpose. He did, however, wish to eliminate the monkeys from human ancestry, arguing that they and the anthropoid apes had diverged from a common ancestry in the lemurs.[3] In fact, Cope accepted the apelike Neanderthals as a stage in the evolution of modern mankind from a great-ape ancestor.

The real originator of the attempt to exclude the apes from human evolution was the leading opponent of Darwinism, the anatomist St. George Jackson Mivart. His *Genesis of Species* was widely regarded as the most powerful attack on the selection theory. It is significant that one of Mivart's chief objections was based on the supposed regularity of evolution, which he believed to give clear evidence of trends imposed by a higher Power.[4] In his *Man and Apes* of 1873, Mivart launched a direct attack on the thesis of Huxley's *Man's Place in Nature.* Claiming that the evolution of mankind from the apes was now being accepted uncritically, he set out to undermine the anatomical evidence for the ape link. His policy here was to question the extent of our resemblance to the gorilla, arguing that there were also many parallels with the structure of the other apes, the monkeys, and even the lemurs. Huxley would not have denied this argument, and was well aware that mankind had not evolved directly from the gorillas, but Mivart suggested that the complexity of these relationships made it impossible to construct any evolutionary hierarchy leading up to mankind. The relationships formed a network, not a ladder.[5]

Insisting that he was not opposed to the general idea of evolution, Mivart nevertheless argued that it was impossible to use anatomical resemblances to work out genealogical relationships. Some similarities might be inherited from a common ancestry, but others might be acquired independently by separate, parallel lines of development. If we cannot tell which is which, we cannot work out the course of evolution from this kind of evidence.[6] Mivart referred to his more general attack on Darwinism to substantiate the claim that similar characters could be acquired by what were obviously quite separate lines of evolution.[7] He was prepared to consider the possibility of a more complex link between man and the animals, but warned that if we postulate a separate origin for the human soul, we might as well accept that the body too was a distinct creation.[8] Few later naturalists would accept this last point, which reveals the theological origins of Mivart's opinions. Yet his analysis of the complications introduced by the possibility of parallel evolution was to become the foundation for many of the heretical ideas on the origin of mankind.

Significant opposition to the theory of ape ancestry did not build up until the very end of the century. In 1897 Robert Munro, a strong advocate of the belief that the development of an erect posture was the key breakthrough in human evolution, suggested that the great apes have degenerated to a more quadrupedal form of locomotion than that of the common ancestor.[9] A similar thesis was advanced on anatomical grounds by the German anthropologist Hermann Klaatsch. In a popular account of human origins published in 1902, Klaatsch developed the unorthodox idea that our species had evolved in the isolated and protected environment of Australia.[10] In opposition to the claim that the gibbon represented an important stage in human evolution, he argued that the arm of the gibbon could never have been transformed into the human form, so the ancestors of the gibbon must have been more human in their structure than their ape descendants.[11] Mankind and the apes were separate developments from a half-upright climbing form with equally proportioned arms and legs. Furthermore, the line leading from this common ancestor toward mankind was the shortest, because our structure has preserved many primitive traits.[12] *Pithecanthropus* was rejected as being not at all the sort of missing link we should expect to find.[13] Haeckel took the trouble to reject Klaatsch's views in one of his last publications, arguing that since no one believed we had evolved directly from the living apes, it was unnecessary to eliminate the apes altogether from our ancestry.[14] Subsequently, Klaatsch dropped his opposition to the ape theory, but went on to develop an even more controversial view of polytypic evolution, to which we shall return in the next chapter.

Another line of argument against the belief in a great-ape ancestry for mankind was based on the theory that ontogeny recapitulates phylogeny. When Smith Woodward invoked this theory to defend his claim that *Eoanthropus* was the true human ancestor rather than a form with a more apelike skull, he was continuing an already established trend. In 1897 the German anthropologist J. Ranke appealed to the pattern of individual growth to suggest a reinterpretation of the origin of racial skull characters.[15] J. Kollmann took this theme further in 1905. The most controversial aspect of Kollmann's theory was his claim that the modern races had evolved from a smaller, pygmylike ancestor, a view that Schwalbe went out of his way to refute because it entailed the removal of the great apes from the human family tree.[16] Kollmann was also forced to dismiss *Pithecanthropus* as merely an ape. He insisted that the true ancestor, although apelike in some respects, was a smaller creature with a much more human skull form than the modern great apes. He appealed to the more rounded skulls of young apes, via the recapitulation theory, to justify this hypothesis.[17] Franz Weidenreich later grumbled that Kollmann and other supporters of recapitulation simply ignored all the fossils in order to allow more scope for their speculations.[18] Nevertheless, it is easy to see why Smith Woodward should have linked this position, with its implication that a more rounded skull had preceded the apelike Neanderthals, with the Piltdown discovery.

The recapitulation theory also figured prominently in a critique of the *Pithecanthropus*-Neanderthal sequence of human evolution by the Canadian Charles Hill-Tout in 1921. Like Woodward, Hill-Tout stressed that the modern skull form of *Eoanthropus* made it impossible to believe that the more apelike *Pithecanthropus* and Neanderthals were the ancestors of mankind.[19] *Pithecanthropus* was merely an ape, whose partly human appearance could be explained by the fact that it had not yet acquired the full ape specialization.[20] The common ancestor of the apes and mankind had been more human than simian in appearance, a point confirmed by the rounded skulls of young apes.[21] Hill-Tout insisted that the modern human form was a very primitive one, that we have changed far less than any of the other primates and have preserved the ancestral form of the order.[22] Although these supporters of the recapitulation theory did not go so far as to deny the link with the apes altogether, they were convinced that the common ancestor bore little resemblance to the highly specialized modern apes. Hill-Tout also echoed Klaatsch's suggestion that the human form preserves a very primitive primate character. In many ways this position foreshadowed that adopted by the more extreme critics of the theory of ape descent. If the hypothetical primitive form could be identified with another, less apelike creature such as the tarsier, then the apes would be dismissed altogether as a side branch having no direct connection with human evolution.

In addition to using anatomical and embryological arguments to eliminate the great apes from human ancestry, some anthropologists tended to look for a way of stretching the process of human evolution over a longer period. We have already seen how Keith expanded the presapiens theory into a case for the Pliocene origin of the genus *Homo,* although he never abandoned the ape link completely. In 1907, P. Adloff—to whom Keith gave credit for first recognizing the taurodont specialization of the Neanderthal teeth—published a study of ape and human jaw structures. He referred to Klaatsch's theory, but extended its logic to postulate multiple parallel lines of evolution stretching back to the most primitive form of the primates.[23] In effect, Adloff set up a hierarchical sequence of stages through which a number of separate lines moved independently, some having passed farther along the scale than others. While accepting *Pithecanthropus* and the Neanderthals as offshoots from the human stem, he saw this stem as quite distinct from those leading toward the great apes and the gibbon. Parallel evolution on a massive scale accounted for the similarities between mankind and the apes, and each of the lines had independently passed through a lemuroid and a monkey phase in the course of its ascent. In 1913 the Italian anthropologist G. Sergi published a theory with even more parallel lines of development stretching far back into the past. Referring to Osborn's views on parallel evolution in the mammals, Sergi argued that the various modern races of mankind were separately evolved from animal ancestors.[24] He saw each of the ape species as the product of a separate line of evolution, and the apes as a whole as a collection of parallel developments having no direct contact with the human group.[25]

Perhaps the closest anticipation of Wood Jones's position came from the Dutch zoologist A.A.W. Hubrecht in 1897. Hubrecht was particularly interested in the spectral tarsier, a curious creature from the forests of the East Indies which many believed to be a primitive form of primate. He made a detailed study of the reproductive process of the tarsier, and then linked his conclusions with a comparative study of its other characters to argue for a reconstruction of the classification of the primates. The lemurs were normally regarded as primitive primates, remnants of the ancestral type from which the monkeys and later the apes had evolved. The tarsier was treated as merely an anomalous form of lemur. Hubrecht maintained that the lemurs were not ancestral to the true primates, but he removed the tarsier from its association with the lemurs so that it could stand as a better representative of the ancestral type.[26] The ancestors of mankind and the apes had remained distinct from the monkeys and the tarsiers throughout the Tertiary, each being derived from a small, but erect and large-brained Mesozoic insectivore.[27] Citing Huxley's 1863 speculations on the extreme antiquity of mankind, Hubrecht also argued against Dubois's interpretation of *Pithecanthropus* and the whole idea of a close relationship between mankind and the apes.[28] The modern lemurs, monkeys, and apes, instead of being remnants of ancestral stages through which the line of human evolution had passed, were now seen as independent groups evolving in parallel from a much more remote ancestral mammalian type.

In fact Hubrecht went even further to postulate a radical thesis on the origin of the mammals themselves. He supported Osborn's view that the monotremes, marsupials, and placentals are not stages in a single evolutionary process, but independent developments from a much earlier and more primitive common ancestor.[29] He then challenged the traditional view of the origin of the mammals by arguing that the placentals might have evolved directly from a Paleozoic amphibian, rather than from a reptile.[30] (It may be added that in 1907 the German anatomist Robert Wiedersheim hinted that a distinct primate line of evolution might be traced back to the Paleozoic.[31]) Hubrecht concluded his analysis with a general argument for the value of embryological studies and an appeal for more effort to be put into the search for fossil evidence that might substantiate his hypotheses. This appeal reveals the weakness of this whole approach: postulating lines of evolution for which there was no fossil evidence magnified the imperfection of the fossil record, making paleontology seem virtually useless. Robert Broom's work on the mammal-like reptiles of South Africa soon exposed the weakness of Hubrecht's speculations on the origin of the mammals. Similarly, the major argument used against Wood Jones's position was that it systematically ignored all the fossils that were seen to throw light on the conventional picture of human evolution.

Frederic Wood Jones trained as an anatomist in London, and in his early work acknowledged his debts to Keith and Elliot Smith. His reputation was based on a book, *Arboreal Man,* derived from a series of lectures given to the Royal College of Surgeons in 1915–16. It was a polemical work in some respects, but it did not challenge the whole traditional interpretation of human evolution. At this point, Jones made no effort to deny the link between the apes and mankind; indeed he claimed that nothing could be added to the "brilliant generalizations" of Huxley or to the "careful analysis" of primate structures by Keith.[32] He attacked those who claimed that mankind had originated from a quadrupedal form, and that the acquisition of an erect posture was a crucial, and quite recent, stage in our evolution. Jones insisted that our mobile forelimb and hand are not recent developments, but very primitive structures retained from the earliest mammals, perhaps even from the generalized limbs of the therapsid reptiles from which the mammalian class had sprung.[33] Mankind was not a "high" form in the sense that it stood at the head of some absolute hierarchy of progression. Our forelimbs are extremely primitive, more primitive even than those of the other primates.[34] The foot was the truly distinctive feature of mankind, since unlike the hand it represented an advanced evolutionary specialization.[35] It was the arboreal lifestyle of our early mammalian ancestors that had preserved the flexibility of the forelimb upon which our modern dexterity is based. Any lengthy period of terrestrial life would have produced quadrupedal specializations blocking the path of evolution toward the human hand.

By adopting this position, Jones could ally himself with Keith's view that the brachiating locomotion of the gibbons had preadapted the lower part of the body for upright walking and with Elliot Smith's claim that arboreal life stimulated the growth of the brain.[36] He thus gave no indication of a desire to challenge the new orthodoxy on human evolution. Yet already his emphasis on the primitive character of some human features was pointing him in the direction of a theory in which the human stock became uniquely differentiated at an early stage of evolution, perhaps "in that dawn period when the Mammals themselves were evolved from some possible Theromorph ancestor."[37]

Jones soon went on the challenge the theory of ape ancestry more openly in a pamphlet published by the Society for the Promotion of Christian Knowledge in 1918 under the title *The Origin of Man* (it was subsequently reprinted as an article in a general work on evolution and its implications).[38] He now expanded his claim that mankind retained primitive characters into a general attack on the linear, hierarchical view of evolution. The vertebrates had not evolved from the highest invertebrate, nor the mammals from the highest reptile.[39] In each case, a lower, less specialized form had made the breakthrough to a new level of organization—a well-known evolutionary generalization often attributed to E. D. Cope. Jones suggested that the popularity of the belief that mankind had evolved from the apes was due to the influence of Huxley and Haeckel, who had promoted an uncritical acceptance of the oversimplified linear view of development.[40] The warnings of Owen and Mivart based on the *differences* between ape and human anatomy had been ignored, because the Darwinians were unwilling to admit the extent to which similar habits could produce a convergence toward similar structures in unrelated lines of evolution.[41] Jones insisted that the lemurs were not closely related to the apes and should not be included among the primates, since their superficial similarities were another product of convergence.[42] The general view that the tarsier was merely a specialized lemur had blinded many paleontologists to the significance of the Eocene fossil tarsioids: in fact they were the primitive ancestors of the primates (see fig. 11).

Jones itemized a series of anatomical differences which, he claimed, were enough to prove that the human stem had evolved separately from that of the apes.[43] Mankind lacked the specializations of the chimpanzee and gorilla, retaining instead many primitive features of our tarsioid ancestry. Those features in which we do resemble the apes must therefore be due to convergent evolution in the two lines. Even the human brain was a development of the primitive mammalian structure, not a continuation of the line of development shown by the other primates.[44] The human stock had thus remained distinct from that of the apes from the base of the primate stem, and the missing link was not an ape-human intermediate, but a small creature similar to the modern tarsier.[45] Jones was thus forced to repudiate the conventional ape-man image of Piltdown and the Neanderthals. In conclusion, he clearly stated what he took to be the wider consequences of his position.

> Man is no new-begot child of the ape, bred of a struggle for existence upon brutish lines—nor should the belief that such is his origins, oft dinned into

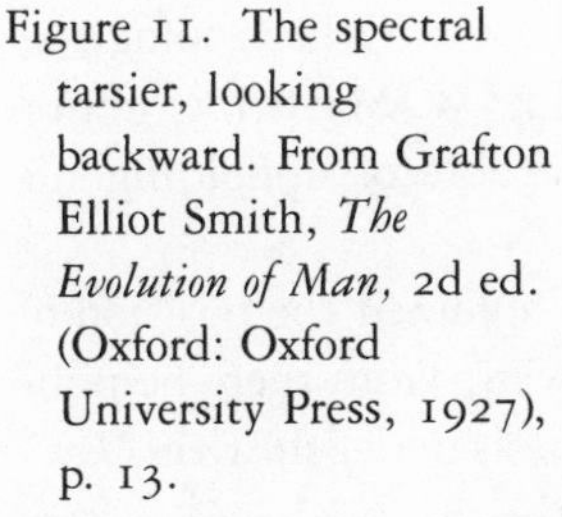

Figure 11. The spectral tarsier, looking backward. From Grafton Elliot Smith, *The Evolution of Man,* 2d ed. (Oxford: Oxford University Press, 1927), p. 13.

> his ears by scientists, influence his conduct. Were he to regard himself as an extremely ancient type, distinguished chiefly by the qualities of his mind, and to look on the existing Primates as the failures of his line, as misguided and brutish collaterals, rather than as his ancestors, I think it would be something gained for the ethical outlook of Homo—and also it would be consistent with present knowledge.[46]

Jones's thesis was debated at a meeting of the Zoological Society of London in 1919, at which Elliot Smith emerged as the leading critic. He objected to the "fashionable and severely overworked doctrine of convergence," without denying that some convergence could be traced between the various branches of the primates.[47] He also insisted that the lemurs *were* primitive primates, complaining that to exclude them from the order because they were not tarsiers was as illogical as excluding the dogs from the Carnivora because they were not bears.[48] Smith believed that the monkeys evolved from the Eocene tarsioids, and from them in turn evolved the anthropoid apes and mankind. The modern *Tarsius* had developed an overspecialized reliance on its visual powers and had escaped extinction only by adopting nocturnal habits. Other zoologists, including J. P. Hill, P. Chalmers Mitchel, and E. W. MacBride, joined in rejecting Jones's arguments as totally incapable of upholding his unorthodox view of human evolution.

Jones himself left Britain and subsequently obtained the position of Professor of Anatomy at the University of Hawaii. From there he published in 1929 a more extensive defense of his thesis under the title *Man's Place among the Mammals.* He complained that the theory of ape ancestry was a dogma that had been imposed on Darwinism despite Darwin's own deep suspicion of the linear image of development.[49] Again he emphasized the ability of convergence to produce similar structures when different evolutionary branches adopted similar habits.[50] Paleontology showed how often separate lines had evolved in parallel toward a similar goal, although many of the examples cited were popularly attributed to nonadaptive orthogenesis.[51] Jones now criticized W. K. Gregory as a leading exponent of the view that mankind had recently evolved from the apes. If Gregory wrote of our "gorilloid" ancestors, he could hardly complain if ordinary people thought he meant that we evolved from the modern gorilla.[52] He noted the humanlike character of the tarsier's reproductive system as clear evidence of its position as a relic of mankind's distant ancestry.[53] Jones claimed that Elliot Smith was now retreating on the question of the tarsier's zoological affinities, if not on the more general question of human evolution.[54] That the Old and New World monkeys had independently evolved the characters that identified them as monkeys was clear evidence of the power of convergence. If we accept that they "have led the same lives, felt the same needs, made the same adaptations and evolved the same type of structure," we should not be surprised "that an animal conforming to the popular conception of a Monkey has been developed twice."[55] The apes were merely Old World monkeys that had specialized in accordance with a "pithecoid trend" that led away from the primitive form retained by mankind. Only in those characters not affected by this trend did the ape show a detailed similarity to the human form, a view that led Jones to quote Klaatsch to the effect that "the less an ape has changed from its original form, so much the more human it appears."[56]

Jones was certainly willing to invoke the parallel acquisition of similar new structures by two different groups. He even tried to dismiss the evidence for the similar protein structure of ape and human blood by arguing that biochemical convergences were no less likely then their anatomical equivalent.[57] Yet despite his frequent references to convergence, he did not always rely on this effect to explain the similarities between mankind and the apes. The similarities were definitely adaptive, involving "a limited homeomorphy of purely adaptive features such as are common to animals that, being once arboreal, have become more or less terrestrial . . . more or less upright in posture, and more or less bulky in their general build."[58] In many cases, however, these were not independently acquired specializations, but independently retained primitive characters, which were still useful in the new lifestyle. Thus Jones assumed that a wide chest was a primitive mammalian feature, retained by any group that found it useful, including the brachiating apes and fully upright mankind.[59] The apes represent a retrogressive line of development stemming from the tarsioid ancestor, degenerating because of the "phylogenetic vice" of brachiation—a mode of progression that just coincidentally required the retention of characters similar to those retained by the fully upright ancestors of mankind.[60]

Jones wrote little in support of his theory for some time after *Man's Place among the Mammals*. In the early 1940s he began to speak out more openly on the philosophy of nature which underpinned his interpretation of human evolution. A short book entitled *Design and Purpose* made clear his commitment to a holistic and teleological view of nature and his belief in a spiritual purpose for human life. *Habit and Heritage* was an equally brief defense of the by now almost completely discredited Lamarckian doctrine of the inheritance of acquired characters. Jones had never concealed his distaste for natural selection, or his feeling that Lamarckism was a more flexible mechanism of adaptation for explaining how similar habits could bring two different forms closely together. His was no theory of nonadaptive orthogenesis, in which internally programmed trends forced a group of related species to evolve consistently in the same direction. Instead, he had always attributed the independent acquisition of similar characters to adaptation, and his position like that of many Lamarckians thus emerges as a consequence of the deep-seated preference for a theory in which habit was the driving force of evolution. For Jones, similar habits were capable of generating identical structures, so powerful was the ability of the inheritance of acquired characters to modify structures in accordance with their use. Nevertheless, to express such views in the 1940s, just as the Modern Synthesis of genetics and Darwinism was emerging triumphantly from the chaos of earlier decades, was to mark Jones off as an eccentric, no longer in touch with the realities of the times.

A more complete statement of Jones's philosophy came in his *Trends of*

Life in 1953, in which parallel evolution was linked both to Lamarckism and to vitalism. In the same year he published *Hallmarks of Mankind,* a final defense of his views on human evolution. Here he argued that the great apes themselves are not a unified group, but each has evolved separately from the monkeys, their similarities being due to adaptive convergence.[61] He also implied that his own theory was the only viable alternative to polygenism. The differences among the human races were so great that they could not have been produced by a recent divergence. The races either had totally separate origins or they were the products of a longstanding divergence in a stock that had retained its basically human character over a vast period.[62]

In *Man's Place among the Mammals,* Jones singled out W. K. Gregory for particularly strong criticism. Gregory hit back in defense of the ape ancestry of mankind with his appropriately titled *Man's Place among the Anthropoids* of 1934. Like Elliot Smith in 1919, he complained of Jones's exaggerated use of convergence to explain the detailed similarities between mankind and the apes. Certainly, it was important to distinguish between characters that revealed ancestral heritage and those acquired recently in response to new habits, but it was Jones himself who could not apply this distinction in a rational manner. It was "asking a good deal of convergence to produce such astonishing identity of general plan as we find in comparing the humerus of man with those of the chimpanzee and gorilla."[63] In the case of certain marsupials and placentals, convergence of structure was obvious enough, but numerous features allowed the naturalist to recognize the fundamentally different ancestries of the convergent forms.[64] The similarities between mankind and the apes were too great to be explained by the similarity of their habits of locomotion; they must indicate a common heritage in a more generalized ape ancestor.

Perhaps the most pithy response to Jones's theory came from another American defender of the ape theory, E. A. Hooton, who was quite prepared to accept parallel evolution on a smaller scale, but regarded the tarsioid theory as a ludicrous extension of the idea.

> To Wood Jones the anthropoid apes show many pithecoid or catarrhine specializations which make him refuse to recognize them as relatives, although man occasionally in individual cases exhibits these same features. The narrow specializations of Tarsius, whom he regards as our nearest primate relative, divergent from human development as they are, disturb him

no whit. He strains at the anthropoid gnat and swallows the tarsioid camel.[65]

While developing his response to Jones's attack, Gregory was forced to defend his position against the most important convert to the idea of a long-distinct line of human evolution, his own chief at the American Museum of Natural History and one of the greatest vertebrate paleontologists of the early twentieth century, Henry Fairfield Osborn. In early works such as his *Men of the Old Stone Age,* Osborn had adopted a fairly conventional view of mankind's origin from the apes. In the 1920s, though, he began to turn against this approach. He burned his fingers badly in 1922 by announcing the discovery of a fossil anthropoid tooth in North America, and claiming that *Hesperopithecus* might even be ancestral to mankind.[66] To the amusement of many critics—Wood Jones included—it soon turned out to be the badly worn tooth of an extinct species of pig. Osborn had been misled because he was already committed to a new and increasingly popular theory of the location of human evolution, which placed the key events in central Asia.[67] The assumption that the plains of Asia had once been continuous with those of North America suggested the possibility that the hypothetical ancestors of mankind might have migrated to the latter continent. The central Asian theory was, in fact, the inspiration for Davidson Black's later discovery of Peking man, but at this stage the lack of any fossil evidence for human ancestors in the region forced Osborn to postulate a yet unknown line of evolution leading toward mankind. He soon began to suspect that this unknown line may have already acquired its distinctly human character as early as the Pliocene, in effect joining Keith on this issue. Finally, in 1927 he announced his belief that the human line started far enough back in time to ensure that mankind and the apes were entirely separate stocks.[68]

Osborn now believed that mankind and the apes had diverged from an arboreal ancestor of Eocene times, although unlike Wood Jones he made no effort to present the tarsier as an illustration of this ancestral form.[69] Indeed, his whole interpretation of human evolution stressed the importance of an early transition to ground-dwelling on the plains of central Asia. The separation of the ape and human lines had occurred in the Oligocene, somewhat later than in Jones's theory. Using some of the arguments already developed by Keith, Osborn argued for the appearance of an already large-brained and essentially human form by the

Pliocene. Unlike Keith, though, he believed that the ancestors of this form would represent a distinct human line going so far back that the apes were bypassed altogether. He recommended rejection of Gregory's "ape-human theory" and recognition that the similarities between the two lines of evolution were due either to a shared preservation of ancestral characters or to the independent convergence of the apes toward human characters. Despite their similarities, the two lines had evolved significantly different habits: our own ancestors were alert ground-dwellers who soon began to use primitive tools, whereas the apes had remained sluggish creatures adapted to brachiation in the forests.[70] Osborn wrote:

> I regard the ape-human theory as totally false and misleading. It should be banished from our speculations and from our literature not on sentimental grounds but on purely scientific grounds and we should resolutely set our faces toward the discovery of our actual pro-human ancestors. In my opinion, the most likely part of the world in which to discover these "Dawn Men" as we may now call them, is the high plateau region of Asia.[71]

As a consequence of his new position, Osborn changed his attitude toward the known fossil hominids, and now published a new survey of the field entitled *Man Rises to Parnassus.* His purpose was to eliminate from our ancestry the "bar sinister of ape descent" and to advocate our descent from "dawn men" rather than "ape-men."[72] Osborn had originally been suspicious of *Eoanthropus,* with its apelike jaw, but after visiting Smith Woodward in London in 1921 he was converted to the cause of Piltdown man and now began to see *Eoanthropus* as a relative of the large-brained Pliocene ancestor of the modern races.[73] The apelike jaw still meant that the Piltdown race was not directly ancestral, though, and Osborn predicted the discovery of more modern forms in the Pliocene, with a prominent chin.[74] *Pithecanthropus* was seen as a case of arrested development, preserving a much earlier type into Pleistocene times in the isolation of Southeast Asia.[75] Not surprisingly, the apelike Neanderthals were dismissed from the ancestry of the modern races. They were a genuine human type, but had not evolved so far as the others because they had been exposed to a less stimulating environment.[76] Eventually the "central Eurasian empire of the Neanderthals" had been wiped out by the migration eastward of more highly evolved forms from the central Asian heartland.

Osborn ended his survey with an account of the modern human races in which he declared that their diversity was such that an unbiased zoologist would divide them into several genera, each with a large number of species.[77] This diversity indicated their great antiquity. The three primary stocks, the Caucasian, Mongolian, and Negroid, had "diverged from each other during the Age of Mammals, even before the beginning of the Pleistocene or Ice age," the Negroid retaining the most primitive character.[78] As with Keith and even Wood Jones, the antiquity of the human races was linked to the supposed antiquity of the human type itself. For Osborn, the study of racial origins was a vital matter in the intellectual life of the world, since their different characters ensured that the races would respond to political institutions in different ways.[79] He argued that each race had its "soul," and that it was vital for the best characters of each to be properly developed. On this basis he threw his weight behind the eugenics movement, and was elected president of the second International Congress of Eugenics held in 1921.[80]

Osborn insisted that his extension of human antiquity had a purely scientific foundation, whatever its social and intellectual consequences. While noting the anatomical similarities between mankind and the apes upon which Gregory had built the theory of an ape origin, he now insisted that such evidence was not enough. The general laws of mammalian evolution, as discovered by paleontologists, must also be taken into account, and these laws made the idea of an ancient line of human evolution far more plausible. Osborn's new position was consistent with the general theory of evolution he had derived from his own studies of the fossil record. He believed that each family consisted of multiple evolutionary lines advancing in parallel over vast periods toward a common goal. As he pointed out, the paleontologist can find fully developed members of the horse family in the Pliocene, and it was this knowledge that had first led him to postulate fully human types in the same epoch.[81] The supposed parallelisms between ape and human evolution were also consistent with his theory of orthogenesis, which allowed for evolutionary lines to inherit tendencies from a common ancestor that would drive both in parallel through a long trend of similar evolution.[82]

Nevertheless, it is difficult to escape the thought that Osborn steadily moved toward an extension of human antiquity as he began to realize that such a position would be useful in the broader debates in which he was involved. His strongly developed sense of racial diversity could easily be supported by arguing that the races themselves were of great antiquity. Osborn was also deeply concerned with the defense of evolutionism against the backlash of fundamentalist religion that had led to the trial in 1925 of John Thomas Scopes for breaking Tennessee's law against the teaching of evolution. He published a book on this topic in 1926, making clear his own commitment to the belief that evolutionism has a spiritual message to offer and that mankind has unique spiritual powers.[83] Osborn realized that for many people the ape-human theory was one of the most objectionable aspects of evolutionism, and he hoped that by giving mankind a less bestial ancestry he would make the general theory of evolution easier to defend. Osborn almost certainly shared the feeling of distaste for the apes, as illustrated by his reference to "the bar sinister of ape descent."[84] As E. A. Hooton put it: "Professor Osborn's opinion seems to be compounded from a desire to compromise with Christianity and an aristocratic aversion from the anthropoid apes."[85]

Gregory responded to Osborn's new theory with a barrage of articles defending the orthodox view of the relatively recent evolution of mankind from the apes. He too noted the convenient parallel between the new interpretation and the more traditional misgivings about human ancestry, and complained that newspapers and preachers were being misled into thinking that Osborn had rejected evolution altogether.[86] Despite the public controversy, it is unlikely that Osborn's arguments had any more effect than Wood Jones's on the scientific community. It must be remembered, however, that the now popular presapiens theory—which required a fully human form to have been in existence by the beginning of the Pleistocene—carried the obvious implication that the human line must have become distinct from that of the apes much earlier than once supposed. Several supporters of the presapiens theory thus came close to the position supported by Osborn, although few went so far as Wood Jones.

Marcellin Boule, who had started the move to eliminate the Neanderthals from human ancestry, maintained in his *Fossil Men* that the lines leading toward mankind and the apes had arisen separately out of the Old World monkeys.[87] Boule relied heavily on the concept of evolutionary parallelism, although he said little about the mechanism responsible. Louis Leakey also supported the view that the human and ape lines had split apart as early as the Oligocene. Like Osborn, he suggested that "a true man-like common ancestor of the various human genera was already in existence" in Miocene times.[88] Leakey said even less than Boule about the forces responsible for shaping these parallel developments, and since both men concentrated more on the later phases of human evolution, their work had little impact on this debate. The tarsioid theory and the less extreme variant adopted by Osborn thus remained scientific heresies throughout the 1920s and 1930s. But they were heresies with just enough support to require the defenders of the orthodox view to take them seriously, especially as they arose not so much from a headlong challenge to orthodoxy, as from an extension of orthodox principles to what most scientists considered to be ridiculous lengths.

The tarsioid theory was soon forgotten in the 1950s, as the renewed interest in the Australopithecines increased anthropologists' confidence in the hypothesis of our descent from the apes. There were a few belated attempts to defend the exclusion of the apes from human ancestry, including Wood Jones's own *Hallmarks of Mankind.* Curiously, Robert Broom, a longstanding supporter of the ape theory because of his interest in *Australopithecus,* had severely qualified his support in his classic monograph of 1946. He then argued that the human line sprang from a pre-*Dryopithecus* stock much earlier than he had at first supposed, a view that he conceded was closer to the position of Osborn and Wood Jones.[89] Broom's coauthor, G.W.H. Schepers, went even further, launching a blistering attack on the idea that we could be descended from the "specialized and degenerate anthropoids."[90] He praised Jones's "clarity of thought" and argued for an independent line of human evolution from the Eocene, with the apes and *Pithecanthropus* being degenerate side branches. This argument was coupled with explicit support for Lamarckism and the theory of inbuilt biological trends leading to orthogenesis.[91] Another unlikely source of support came from William L. Straus, who coauthored the paper establishing that the La Chapelle-aux-Saints Neanderthal specimen had been crippled by arthritis. In 1949 Straus too opted for a modified version of the Osborn thesis, in which he invoked parallelism to explain the separate origin of mankind and the apes from a preanthropoid, monkeylike stock.[92]

These last flurries of support show that the principles upon which the tarsioid theory had been built still held an attraction for a few naturalists in the postwar years. Yet the trend was setting strongly against any revival, and from the early 1950s onward the almost universal acceptance of the early Australopithecines as human ancestors muted opposition to the theory of ape descent. The increasing strength of the Modern Synthesis between Darwinism and genetics also undermined what little credibility was left to the evolutionary mechanisms of Lamarckism and orthogenesis, which had been the twin pillars of support for the tarsioid theory. More recent work on the genetic similarities between mankind and the apes has closed the gap between them even further, leaving little or no room for any alternative interpretation of their evolutionary relationship.

Chapter Six Polytypic Theories

Several of the anthropologists and anatomists who extended the antiquity of mankind beyond the normally accepted limits were motivated by a desire to create a time scale for substantial racial diversification within the human stock. They believed that the degree of diversity among the living races was so great that it could only have been built up over a vast period. In the case of Osborn, this exaggerated sense of racial differences led to the claim that by normal biological standards the human races had diverged enough to become distinct species, perhaps even distinct genera. Such an interpretation of the races had been common among the polygenists of the nineteenth century, who had coupled a strong emphasis on the anatomical differences among the racial types with an effort to minimize the significance of interbreeding among them. Explicit polygenism was no longer popular in the early twentieth century, although Osborn was not alone in his belief that the racial differences were equivalent to the species distinctions recognized elsewhere in the animal kingdom. Most of the anthropologists who emphasized the significance of race believed that the differences were brought about by a longstanding divergence from a common ancestor. But if the races were really distinct species that just happened to be able to interbreed, was it possible that—instead of diverging from a common ancestor—they had *converged* toward one another from quite separate origins in the great apes? Polytypic theories of human evolution were based on this assumption, and although never very popular, they continued as an irritant to orthodox anthropologists throughout the early twentieth century. Like the tarsioid theory, the polytypic approach was all the more annoying because it took certain principles upon which almost everyone was agreed and extended them to absurd lengths. Since most anthropologists accepted the importance of racial differences, and since many biologists accepted the possibility of parallel evolution, polytypic theories were an embarrassing reminder of what could happen if such beliefs were taken too far.

Polygenism had been advocated by French anthropologists such as Paul Broca and Georges Pouchet, Broca in particular arguing that not all human races could interbreed successfully.[1] Their views were not linked to any clearly defined theory of how the various human species had evolved, however, although Broca favored a system of multiple branches entwined together. The German anthropologist Carl Vogt considered the question of multiple origins carefully in his *Vorlesungen über den Menschen,* translated as *Lectures on Man* in 1864. Vogt was a confirmed polygenist, arguing that there was a greater difference between a European and a Negro than between two ape species, such as the gorilla and chimpanzee, so either the two races were distinct species or the apes were not.[2] He accepted the evolution of mankind from the apes via now-extinct intermediates such as the Neanderthals,[3] but held that branching evolution under the influence of different climates could not account for the diversity of modern races.[4] Vogt rejected the whole Darwinian approach to divergent evolution. There was no primordial form from which all life began: instead, there had been a multitude of different original types, each with its own potentialities for future development.[5] He did not directly apply this concept to the human races, but it did serve as a model for a theory in which the races had their origins in different primate ancestors.

Vogt pointed out that the great apes themselves did not form a single evolutionary hierarchy. The orang, chimpanzee, and gorilla each stood as the head of its own sequence of lower forms, indicating the existence of three parallel lines of ape evolution.[6] He then linked this point to his interpretation of the human races as distinct species: "If in different regions of the globe anthropoid apes may issue from different stocks, we cannot see why these different stocks should be denied the further development into the human type, and that only one stock should possess this privilege."[7] The human forms produced by further development of the different ape stocks would originally have been quite distinct, but

> by constant working of his brain man gradually emerges from his primitive barbarism; he begins to recognize his relation to other stocks, races and species, with whom he finally intermixes and interbreeds. The innumerable mongrel races gradually fill up the spaces between originally so distinct types, and notwithstanding the constancy of characters, in spite of the tenacity with which the primitive races resist alteration, they were by fusion slowly led toward unity.[8]

Vogt had an almost teleological view in which several branches had independently passed through the same sequence of forms up to the human level. Whatever the source of the original branches, the common trend was so powerful that the human products were close enough to interbreed, once their intelligence allowed them to recognize their morphological relationship.

The translation of Vogt's book was issued by the Anthropological Society of London, which championed the polygenist cause in Britain. Most of the early evolutionists were suspicious of polygenism, however, and would certainly have regarded Vogt's polytypic theory as too clear an expression of predetermined or goal-directed progressionism. For the Darwinians, the essentially human character of the modern races must stem from a common ancestor: it was unthinkable that different apes exposed to different conditions could have independently converged toward the human form. However, it would be possible for a Darwinian to believe that the branching evolution that had generated the modern races could have gone so far that the end products were on the point of becoming distinct species. As long as interbreeding was still possible, though, the process of speciation was not complete, and it never would be now that improved transportation allowed constant intermingling of the once-diverse populations.

In his well-known paper delivered to the Anthropological Society in 1864, A. R. Wallace tried to resolve the monogenist-polygenist debate by appealing to the extreme antiquity of mankind and hence of its racial subdivisions.[9] In effect, Wallace accepted a degree of parallel evolution by suggesting that the racial characters had already become established before the final evolution of the human brain. In the distant past, the races had been adapted by natural selection to the conditions of their local environment, but when intelligence became important, selection acted only on brain size, leaving the racial characters unchanged. At this point, Wallace accepted natural selection as the driving force of human evolution at all levels, a position he soon repudiated in favor of supernatural guidance.

Wallace at first assumed that all racial characters were adaptive, but Darwin almost immediately pointed an additional factor out to him. Although of no adaptive value in itself, skin coloration might be correlated with resistance to certain diseases, and could thus be selected for in certain areas.[10] T. H. Huxley accepted this suggestion, and was generally favorable to the claim that the races were formed by selection. He admitted that adaptation to the local environment did not take place quickly, but appealed to the extreme antiquity of the human species to avoid this problem.[11] Darwin does not seem to have shared this sense of human antiquity, and was far more cautious about the ability of natural selection to create the whole range of racial characters. His solution was to invoke sexual selection, which is why so much of the *Descent of Man* is given over to a general consideration of this mechanism. Darwin considered the evidence for the races being distinct species, and admitted that the inability to interbreed did not always correspond to the naturalists' intuitive sense of the boundary between species.[12] The most weighty argument against polygenism was not the extent to which the races could interbreed, but the fact that they graduated into one another through intermediates even where no interbreeding seemed to have occured.[13] The races were at best subspecies, and Darwin argued forcefully against Vogt's theory that separately originating forms of mankind had converged together.[14]

Thanks to the influence of the chief evolutionists, polygenism gradually became less influential in Britain, although many anthropologists continued to adopt what we should today call an inflated view of the significance of racial differences. In Germany, though, the leading evolutionist, Ernst Haeckel, had openly espoused the polygenist cause. Proclaiming that the racial differences were fully equivalent to those among species in the animals, Haeckel also argued that the woolly-haired species of mankind (Ulotrichi) had retained more of the apelike form from which all had evolved. While thus accepting a common ancestry for the whole of mankind in the apes, he believed that several species of ape-men had separately evolved into true men, citing as evidence the existence of several quite distinct groups of languages.[15] Presumably each species, as it separately acquired enough intelligence to begin articulate speech, had invented its own language. Here, in a more extreme version of Wallace's position, the racial divergence of the human stock had occurred before the transition to the fully human stage, and for Haeckel it was doubtful if the black races had ever completed the process.

The most prominent advocate of a polytypic theory of human origins in the early years of the twentieth century was Hermann Klaatsch, who had studied with Haeckel's close friend, the comparative anatomist Carl Gegenbaur.[16] Originally an evolutionary morphologist, Klaatsch first made a name for himself by proposing an unorthodox theory of the origin of vertebrate limbs. By the turn of the century he had begun to concentrate on the problem of human evolution, and again provoked controversy by suggesting that mankind had evolved not from the apes but directly from a more primitive form.[17] He also proposed that the crucial development had taken place in the isolated environment of Australia, with the aborigines being a remnant of the earliest human form. He spent the years from 1903 to 1907 in Australia attempting to confirm this view and also visited the Trinil site in Java, where he contracted malaria, which may have hastened his death. On his return to Europe he was appointed professor of anthropology and anatomy at Breslau, where he gained a reputation for producing controversial descriptions of new human fossils. In his work on Hauser's skeleton discovery at Combe Capelle, he described it as a new species, *Homo aurignacensis hauseri*. This individual (which most authorities regarded as a member of the Cro-Magnon race) was clearly modern in form. Klaatsch now became convinced that the differences between the Neanderthal and Aurignacian types was so great that they must be the products of entirely separate lines of evolution. He announced his polytypic theory in 1910, and before his death in 1916 had prepared a book summarizing his views under the title *Der Werdegang der Menschheit,* published posthumously and translated into English as *The Evolution and Progress of Mankind.*

Klaatsch presented his theory in an extensive paper in the *Zeitschrift für Ethnologie,* and backed it up with summaries in more popular German and English periodicals.[18] The basis of his argument was similar to that used by Wood Jones: it was necessary to distinguish carefully between those characters that were recently acquired adaptations and the more fundamental, nonadaptive characters that gave a true indication of evolutionary ancestry. Two species might appear to be closely related, but if their resemblances were all at the superficial, adaptive level, and if there were also more fundamental differences between them, it might be necessary to sever the presumed relationship and accept that the similarities had been produced by convergence in two quite separate lines of development. On the basis of his detailed studies of Neanderthal skeletons, and now of Hauser's Aurignacian specimen, Klaatsch decided that there were indeed fundamental differences between the two types of mankind. These were centered principally in the structure of the upper limbs, especially the ratio of the longitudinal and transverse diameters of the head of the humerus.[19] In overall appearance, the limbs of the Aurignacian man were graceful and slender, whereas the Neanderthal type was rough and thickset.

By itself, this distinction was no more remarkable than Boule's characterization of the Neanderthals, but Klaatsch extended its significance in two ways. Whereas Boule stressed only that the heavier limbs of the Neanderthal type linked it more closely to the great apes, Klaatsch made a distinction between the orangutan and the African apes and declared

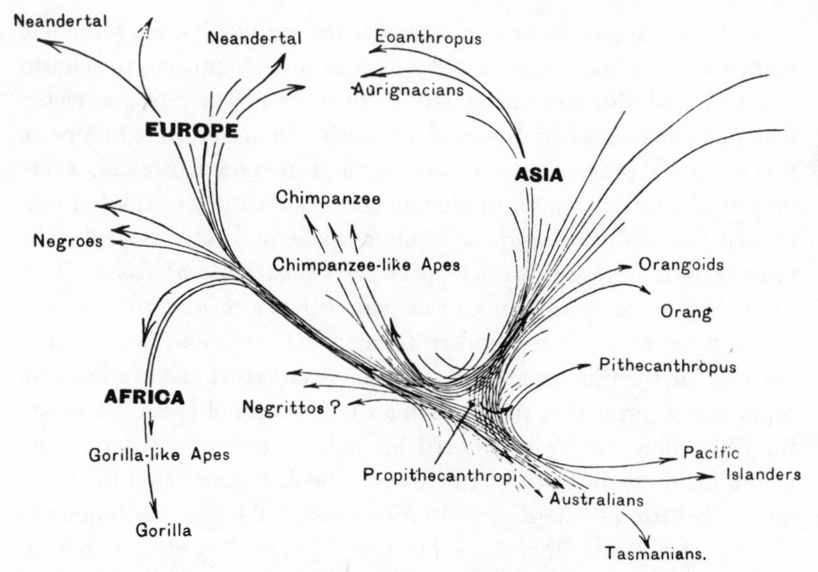

Figure 12. Hermann Klaatsch's scheme for the evolution of the apes and the human races. From H. von Buttel-Reepen, *Man and His Forerunners* (1913), p. 73. The newly discovered *Eoanthropus* (Piltdown man) has been added as a human branch close to the Aurignacian race.

that the Aurignacian man resembled the orang, whereas the Neanderthals could be linked to the gorilla and chimpanzee. There were, in effect, two fundamentally different types of mankind, the orangoid and the gorilloid, the distinction indicating a substantial evolutionary gulf between them. The superficial similarities that allowed the Neanderthals to be ranked as a form of humanity were the product of convergence, produced when two separate lines of development in different parts of the world had independently acquired a large brain and the ability to use tools. In the same way, the great apes were not a natural group, but the end products of two separate lines evolving toward massive jaws and a posture adapted to forest-dwelling. Klaatsch was thus able to sustain his original view that mankind had a more primitive structure than the apes, retained while our brain increased in size. It was the apes that had developed the more specialized characters as their ancestors had moved into the forests of Asia and Africa. Far from being the forerunners of mankind, the apes were degenerate and overspecialized side branches from the two main stems of human evolution.

Klaatsch now accepted that *Pithecanthropus* was close to the ancestral form of mankind and the apes (see fig. 12). The hypothetical "Propithecanthropi" had migrated throughout the Old World and had divided into two main groups, one evolving in Asia, the other in Africa.[20] Each of these groups had evolved toward a human form, but each had thrown off side branches degenerating toward the ape character. The apes were "unsuccessful attempts and dashes forward toward the goal of the definite creation of the human race—submerged branches of the primeval humanity which, in adapting themselves to special conditions of life, were compelled in the struggle for existence to sacrifice important parts of their organization, while a more favoured collateral branch, in quiet progressive evolution, developed into a human race."[21] The Propithecanthropi already had basically human limb proportions, and once having spontaneously acquired the human foot structure, became fully erect and capable of moving on to tool use and brain growth.[22] The apes represented those members of both groups that had not participated in the acquisition of an erect posture. As a consequence of this failure they had been driven into the trees, where their upper limbs had become overdeveloped and a changed diet had caused a massive increase in the jaw muscles, thereby limiting any increase in brain size.[23] A similar process of success and failure had produced the Neanderthal race of mankind and the gorilla in Africa, the Aurignacian race and the orang in the East. While the apes remained confined to the forests of Africa and Asia, the Neanderthals had migrated into Europe, but had been exterminated by the Aurignacians invading from the East.

As described so far, Klaatsch's position seems like a hybrid of the presapiens and tarsioid theories. Like Boule and Keith he wanted to push the Neanderthals off onto a branch of their own, separate from the evolution of the modern European race, and like Wood Jones he saw the apes as degenerate offshoots of a human stock that preserved a much older primate form. Had he restricted the gorilloid version of humanity to the extinct Neanderthals, the theory would have been controversial, but it would not have entailed a polygenist interpretation of the modern races. But Klaatsch was not willing to confine his assessment of gorilloid characters to the Neanderthals. In addition he went on to hint that the Negroid races of today share the same ancestry and have thus evolved separately from the Mongolian and Caucasian types.[24] Although developed only briefly in the 1910 paper, this polytypic theory of modern human evolution became a central thesis of Klaatsch's book *The Evolution and Progress of Mankind.*

Klaatsch accepted that the Neanderthals were truly human, with a large brain and a well-developed toolmaking ability. He suggested, however, that their brain had expanded more at the back than the front, so they had had little intellectual life.[25] The link between the Neanderthals and the black races was drawn in such a manner as to confirm the inferiority of the latter: "We should not be far astray in ascribing to Neanderthal man a good deal of the bestial aspect which the lower blacks seem to us to have."[26] Curiously, though, Klaatsch restricted his prejudice to the African races. His experiences in Australia had given him a profound respect for the normally despised aborigines, and he now suggested that their elaborate social life was typical of the earliest form of the Aurignacian race from which modern Europeans eventually developed. He was happy to promote a close link between Europeans and the black races of Asia and the Pacific, while enhancing the gulf between this group and the African blacks. Admitting that many were reluctant to adopt a complete polygenism, he went on:

> We cannot deny that the recent tendency of anthropology is not to support the idea of the unity of the race that had been suggested by religious and sentimental considerations. Modern science cannot confirm the exaggerated humanitarianism which sees brothers and sisters in all the lower races. The various types must be taken separately. The Australian aboriginals, the Samoans, and the Cinghalese are actually closely related to us, but a Zulu or a Herero is not.[27]

Klaatsch had only limited interest in whether the two human branches had given rise to distinct species—after all, he had reduced the species concept almost to absurdity by turning the Aurignacian or Cro-Magnon race into a distinct species. Nevertheless, his claim that the two human groups were as far apart as the Asian and African apes made clear the extent to which he was prepared to stand by a polytypic theory of human origins. He was working on the brain structure of the different races in the hope of confirming his theory when he died in 1916.

Klaatsch's theory seems to have gained few adherents apart from his own students, but it sparked off a good deal of controversy. The editor of his *Evolution and the Progress of Mankind,* Adolf Heilborn, claimed that everyone at the Cologne congress of anthropology in 1910 was excited by the new idea, and the rapid publication of an English summary in *Nature* supports this opinion.[28] Nevertheless Keith, in an immediate response to the *Nature* article, claimed that few Germans had actually accepted the theory. Keith dismissed Klaatsch's anatomical comparisons as "flimsy" and immediately put his finger on the weakest point by claiming that the demands made on "convergence phenomena" went beyond the limits of rational speculation.[29] A reply by Klaatsch's student, Gerhart von Bonin, admitted that the extent of convergence implied by the independent evolution of two human stocks was a problem, but insisted that the evidence for the ape-human parallels could not be dismissed.[30] Keith again responded, pointing out that glandular effects could produce major changes in a race's anatomy quite rapidly, and arguing that the skeletal characters upon which Klaatsch relied were highly variable because of the impression made by the muscles on the bones. Keith concluded: "Nothing is impossible in nature, but there are some things which are highly improbable. A multiple origin for a single species is one of the most improbable, and, so far as the human species is concerned, there is no need to suppose a multiple origin."[31] Keith continued to oppose the polygenetic approach for some time, although eventually his own strong feeling for racial differences led him in this direction. Within a short time, however, he had abandoned his belief in the rapidity of human evolution and had accepted the Neanderthals, at least, as a distinct species. It is not impossible that Klaatsch's views on this latter point may have played a role in stimulating this revolution in Keith's opinions.

Despite Keith's dismissive attitude, a few anthropologists took Klaatsch seriously. H. von Buttel-Reepen followed an account of Schwalbe's interpretation of the Neanderthal species with a detailed summary of Klaatsch's views on this issue. Noting that the theory implied that the apes were descended from mankind, rather than vice versa, he admitted that it would have to be taken with a grain of salt, but observed that the more human appearance of the young ape was in agreement with the theory.[32]

Two British anthropologists lent immediate support to Klaatsch's interpretation of the separate origin of the Neanderthals. In 1911, John Gray published an article in support of convergent human evolution, arguing that "if divergent varieties come to live in the same environment, convergence will result." Thus, if different ape species threw off ground-living varieties, each would converge toward the human form.[33] Gray was an early supporter of the presapiens theory, citing the Galley Hill fossil as evidence that the Aurignacian type had evolved long before the Neanderthals. He adopted Klaatsch's views on the extreme separation between the two types, differing only in his opinion that the Aurignacian form was derived from the chimpanzee rather than the orang. The brachycephalic races of Asia might be derived from the orangoids, but confirmation of this supposition was needed from the fossil record.[34] More support came from W.L.H. Duckworth, lecturer in anthropology at Cambridge, in a 1912 survey entitled *Prehistoric Man*. Although aware of the controversial nature of Klaatsch's position, Duckworth pointed out that authorities such as Osborn accepted that some domesticated animals, including the horse, might have been derived from more than one wild species. If this was so, it might be possible to modify Klaatsch's polytypic theory to make it more acceptable.[35]

A polytypic theory even more extreme than Klaatsch's was developed independently by the Italian anthropologist Giuseppe Sergi. Widely respected for his study of the Mediterranean race, Sergi went on to develop highly unorthodox views in his *Le origine umane* of 1913.[36] An early advocate of the exlusion of the apes from human ancestry, he argued for the parallel evolution of the ape and human stocks over a vast period. He also extended this extreme form of parallelism to the human stock itself, suggesting that even here a number of independent lines of evolution had given rise to the different human forms that we know as the races of modern mankind. The Neanderthals were dismissed as a separate genus, *Palaeoanthropus,* which had been overtaken by the development of other human types in many parts of the world. The modern races were divided not just into species, but also into genera, each genus being the product of an independent line of parallel evolution in its own area. The genus *Notanthropus* comprised the species that Sergi believed to have evolved in Africa, including *N. eurafricanus* (the basis of the Mediterranean race), *N. afer* (the African blacks), *N. pygmaeus* (the pygmies) and *N. australis* (the Australian aborigines). Another genus *Heoanthropus* included all the "species" evolved in Asia, some of which, by hybridization, had contributed to the creation of the European races. Sergi even went so far as to accept genera of mankind evolved in North and South America, *Hesperanthropus* and *Archaeoanthropus,* the latter a supposedly ancient form of mankind discovered in Argentina by Florentino Ameghino and almost universally rejected by other anthropologists.[37] Sergi held that the races presented absolutely fixed characters that could not be changed by the environment, and argued specifically against Franz Boas's demonstration of the mutability of racial character as shown by immigrants into America.[38] Although he accepted that some races had been created by hybridization between two "species," he was suspicious of the monogenists' claim that all forms of mankind can interbreed successfully.[39]

Sergi's theory implied that the races had evolved independently on every continent. His work was enthusiastically cited as late as 1929 by the French anthropologist Georges Montandon in support of the theory of ologenism.[40] This theory started from the assumption that life originally appeared over the whole surface of the globe, all future developments being somehow latent within the first forms. Evolution consisted of the periodic splitting of each line into advanced and retrogressive forms, with the event taking place simultaneously over the whole of the earth's surface. Montandon applied his idea to mankind by supposing that *Pithecanthropus* had a global distribution, and had undergone repeated diversifications to produce the various extinct and living races. Since the theory entirely eliminated the geographical element in speciation, it was not exactly a form of polygenism. Montandon claimed that his position was opposed equally to both monogenism and polygenism. Not surprisingly, few took his idea seriously, although it was published in the prestigious journal *L'Anthropologie.* Montandon's article was, in fact, preceded by an attack on the polygenism of Klaatsch and Sergi by Henri Vallois.[41]

If the anthropologists of the 1920s were unwilling to consider polygenism or multiple human origins, a handful of medical practitioners were prepared to explore the evidence. The leader here was F. G. Crookshank, whose *The Mongol in Our Midst* of 1924 offered a popular account of the race question. Crookshank noted that as early as 1844 Robert Chambers's *Vestiges of Creation* had suggested that the lower races corresponded to stages in the development of the highest, or white race.[42] Later, the work of J. Langdon Down had revealed the Mongoloid character of certain imbeciles, providing apparent confirmation of this hypothesis. Crookshank insisted that this trait was not just a medical problem: the Mongol character appeared in individuals who ranged from mentally retarded through to intelligent types who simply had a mentality different from that of the normal European.[43] The character was clearly a reversion, and thus provided evidence that the Mongols had contributed something to Europe's racial history in the distant past. Yet the effect was not known in Negroes, he claimed, indicating their quite different racial heritage.[44] Furthermore, there were distinct analogies between the racial types of mankind and the various ape species, including the typical posture adopted when squatting.[45] All this suggested that Klaatsch had been right in postulating the separate origin of the races from the Negro-gorilla and Mongol-orang stocks—to which Crookshank added a third group, also hinted at by Klaatsch in his later work, the white-chimpanzee stock.[46]

As befitted a medical man, Crookshank offered few hints as to *how* the evolutionary process had separated the ape and human types in each case; it was merely his intention to show that some such hypothesis offered the best means of accounting for the medical and anatomical evidence. In a greatly enlarged edition of his book in 1931, Crookshank noted that it had been translated into German by G. E. Kurz of Münster, a pupil of Klaatsch, and added accounts of Kurz's work on racial anatomy to support the theory.[47] Although critical of the anthropologists' refusal to consider the evidence he had brought forward, Crookshank seemed quite pleased with the reception it had received from the general public. A few anthropologists did begin to take polygenism more seriously in the 1930s. Klaatsch's belief that the apes were degenerate offshoots of early mankind was no longer accepted, but still left open was the possibility that different ape species had independently given rise to the various human races.

This modified version of Klaatsch's position can be seen in the early work of the American anthropologist Earnest Albert Hooton. The title of Hooton's 1931 survey, *Up from the Ape,* indicates his commitment to the theory of an ape ancestry for mankind. But he was also a firm believer in the possibility of parallel evolution, and this belief enabled him to explore the idea that offshoots from different ape species had each adopted an upright posture and thus evolved into a form of mankind. In the conclusion of an early article published in 1925, Hooton had already hinted at this interpretation, arguing for two or three fundamental humanoid stocks and pointing out that both Africa and Asia would have fitted the environmental needs for human evolution.[48] Like Klaatsch, he accepted the link between the black races and the African apes, and between the Mongols and the orangutan. In *Up from the Apes,* Hooton argued against Osborn's central Asian theory of human origins and, with a veiled reference to Montandon's ologenism, expressed his own preference for a much wider geographical range over which the transformation took place.[49]

> It seems probable that at this time [the upper Miocene] some of the more progressive apes took to the ground in Africa, others in southeastern Asia, possibly still others in Europe. In short, the probable cradle of humanity, the site of our prehuman ancestors' first ventures on the ground, is not some single hallowed spot or Garden of Eden, but the whole broad area through which the progressive great apes ranged. A terrestrial habitat and an upright biped gait were not God-given attributes of a single Adam and Eve among

> the great anthropoids. It is difficult to avoid the conclusion that of the diverse families and genera of giant anthropoids developed in the Miocene period, several may have taken to the ground in different areas and at various times. Some of these attained a semi-human status and some achieved complete humanity; some have survived to the present day and some have fallen by the wayside. Evolution operates not upon one single line and one single species, but upon multiple lines, some converging and some diverging, and upon large groups of animals. Nor is this process restricted to a single continent.[50]

This position certainly implies that the races had originated as distinct species or even genera, and Hooton accepted that despite the common agreement to include all the races within *Homo sapiens* the differences were enough to count as species anywhere else in the animal kingdom.[51]

Hooton's theory of how the separate forms of mankind had evolved was—like Klaatsch's—one of adaptive convergence rather than orthogenesis. Acquisition of an upright posture within a suitable environment was the key factor that had modified the various ape species in the same direction. Arthur Keith gave a very different interpretation of parallel human evolution in the later part of his career. In his *Antiquity of Man,* Keith had greatly extended the age of the human species, in part as a means of giving time for the diversification of the races. During the 1920s he had been forced to back away from his position on the antiquity of mankind, but since he felt unable to abandon his views on the significance of race, he now had to consider the possibility that the racial distinctions were even older than the species itself. Without explicitly abandoning the inclusion of all the races in the same species, he proposed that a number of racial types had evolved in parallel up to the fully human level. He expressed this theory most clearly in a 1936 address to the British Speleological Association, reported in *Nature.*[52] Like Hooton, Keith was now impressed with the wide geographical range of early mankind and was convinced that the races had evolved in their modern locations. Citing Weidenreich's views on the Mongoloid character of Peking man, Keith also noted the Australoid features of *Pithecanthropus* and the Negroid character of Leakey's Kanam fragment. The persistence of these racial types suggested that there had been a parallel evolution of the human form from the *Pithecanthropus* level in several parts of the world. Keith was puzzled about the causes that could have brought about such a parallel evolution, but suggested a form of orthogenesis brought about by inbuilt genetic trends. This approach "implies that the future of each race is latent in its genetic constitution. Throughout the

Pleistocene period the separated branches of the human family appear to have been unfolding a programme of latent qualities inherited from a common ancestor of an earlier period."[53]

Keith's polytypic scheme was more restricted than Hooton's, and carried no implication of multiple origins in the apes. The reference to Weidenreich identifies a closer parallel, since he too believed that racial variants in the *Pithecanthropus* / *Sinanthropus* group (now *Homo erectus*) had been preserved in different areas as each population had made the transition to the *Homo sapiens* level. Weidenreich was certainly no polygenist. He saw the differences between the Peking and Java forms as equivalent to the racial divisions within *Homo sapiens,* and used the names *Sinanthropus* and *Pithecanthropus* only for the sake of convenience.[54] He also explained the parallel evolution of the races as a response to the increase in brain size, and accepted that some interbreeding took place between the races at all stages in the process. Keith also appreciated that *Sinanthropus* and *Pithecanthropus* were similar, but it is not clear that he would have included them in the same species.[55] The quoted passage shows that he was thinking of parallel lines of evolution with no interaction between them, the parallelism being due solely to an inbuilt tendency. Keith's was thus far more a polytypic theory, since it gave each race a distinct origin at the lower level of development. Nevertheless, we can see why some later critics regarded Weidenreich's theory—and the modernized version of it promoted by Carlton Coon—as a kind of polygenism. There is an uncomfortable similarity to Keith's explicitly polytypic scheme, and once the racial differences are pushed back into an earlier species, it becomes all too easy to imagine that some races have lagged behind others in their achievement of the fully human status.

In the postwar years, the general trend of anthropology was away from the exaggerated emphasis on racial differences, and the Modern Synthesis in evolution theory made parallelism seem much less plausible. As late as 1948, however, the British plant geneticist R. Ruggles Gates applied his own interpretation of evolution theory to anthropology and came up with another version of polygenism. Significantly, Gates wrote his book *Human Ancestry from a Genetical Point of View* in America, under the influence of Hooton and Coon.[56] Although most geneticists had turned away from the concept of directed or parallel mutations, Gates was convinced that mutations could indeed occur consistently in the same direction within a number of related species. They would thus provide a mechanism to explain the orthogenetic trends reported by many paleontologists.[57] While accepting the sequence *Pithecanthropus*-Neanderthal-*Homo sapiens* as the basic outline of human evolution, Gates insisted that many independent lines had made their way through this sequence to form the various modern races.

Gates relied on Weidenreich's work for the racial characters of *Pithecanthropus* and *Sinanthropus,* but severely criticized the German anthropologist for his willingness to abandon the view that these early hominids were distinct species. He accused Weidenreich and many others of being led astray by the dogma that all mankind belongs to one species, a dogma created by Linnaeus in accordance with eighteenth-century views on the rights of man.[58] So anxious was Gates to develop a separate ancestry for the white and black races that he invoked the now almost totally ignored *Eoanthropus* as the basis for a bizarre elaboration of Klaatsch's scheme. The Piltdown find represented the start of an orangoid line of evolution in Europe leading toward the Caucasian race, whereas the Neanderthals formed a separate, gorilloid line aimed at the blacks.[59] Just as many anthropologists were beginning to move back toward the Neanderthal-phase-of-man theory, Gates revived the idea of a Neanderthal extinction, with the modern black races being merely parallel branches of the same gorilloid family.[60]

Gates was convinced that the sterility criterion for distinguishing between species had now been shown to be worthless. Throughout the plant and animal kingdoms, forms universally recognized as distinct species could interbreed. The fact that the various human types could interbreed was thus no bar to them being considered as distinct species. Commenting on the view that the Neanderthals constituted only a subspecies of *Homo sapiens,* Gates sneered: "To such straits are biologists reduced when they try to adhere to the absurd conception of interspecific sterility as the sole criterion of specific distinction. Innumerable attempts have been made to found the conception of species on a single criterion. They have all broken down, and none more hopelessly than that of interspecific sterility."[61] In effect, Gates had synthesized the positions of Keith and Widenreich: the races could interbreed, yet they were species whose parallel evolution was entirely due to the occurrence of similar mutations in accordance with an inbuilt orthogenetic trend.

Despite Gates's position as a geneticist, it would be difficult to imagine a theory more at variance with postwar trends in evolutionism and anthropology. His excursion into the field of human evolution is of interest only because it shows that a genetical explanation of orthogenesis and parallel evolution was, at least, conceivable. Coon and Hooton continued to promote their ideas of parallel racial evolution in the post-war years, but their approach was increasingly out of touch with that advocated by the evolutionary biologists.[62] Turning their backs on Gates's interpretation, population geneticists and systematists were now becoming convinced that the breeding population was a vital unit in the evolutionary process. In effect, the species was defined as the breeding population, making any theory based on interbreeding "species" a contradiction in terms. The trend in both genetics and evolution theory was away from the notion of predetermined or orthogenetic trends, and back toward acceptance of the natural selection of random variation. The taxonomic deflation that accompanied these developments eliminated the vast range of specific and generic names that had been applied to the fossil hominids, and emphasized the comparatively trivial nature of the geographical differentiation within modern humanity. In these circumstances polygenism becomes unthinkable, along with any theory of polytypic evolution. The only source of support for the parallel appearance of human forms in various parts of the world comes from the few naturalists who have stood out against modern Darwinism, especially Leon Croizat.[63]

Part Three The Causes of Human Evolution

Chapter Seven Brain, Posture, and Environment

The preceding chapters show the great diversity of opinion on the actual course of human evolution. Starting from the most obvious link with the great apes, opinion became fragmented into a host of differing positions. Scientists disagreed over the nature of our prehuman ancestors, over the time at which a separate human line of evolution emerged, and even over the possibility of a multiple origin for the human species. They further disagreed over the causes of evolution in general and human evolution in particular. The scientists' views on the causes at work in evolution necessarily influenced their attitudes toward the rival phylogenies. Indeed, they constructed the phylogenies at least partly in order to vindicate rival interpretations of evolutionary causation. In this final part of our study, we investigate the wide range of debates over the causes of human evolution.

The Introduction raised the question of the relative importance to theories of human evolution of prevailing ideas on the mechanism of evolution or of narratives specifying a unique combination of circumstances that affected our ancestors. In the early twentieth century, scientists were clearly concerned with issues directly related to evolutionary mechanisms. Those anthropologists who argued for long-separate or multiple lines of human evolution were attracted to mechanisms such as Lamarckism and orthogenesis that allowed a greater degree of convergent or parallel development. These issues are the subject of chapter 8. In addition, however, a debate ranged widely over the means by which our ancestors had been set on the course that so extensively separated them from their primate relatives. As we understand it today, the story of human origins requires the specification of evolutionary novelties that distinguished our ancestors from those of the apes, novelties that are in turn explained by appealing to the advantages gained by our ancestors in coping with the particular environment to which they were exposed. We do not take it for granted that a greater intelligence is automatically an advantage in all circumstances, since we appreciate the need to explain why the apes did not participate in this trend. But it has not always been clear to paleoanthropologists that their theories must depend quite so exclusively upon a series of unique events linked into a narrative structure.

Much late-nineteenth-century thought on human origins was still influenced by a faith in the inevitability of natural progress, which seemed to justify the assumption that growing intelligence was the central theme of evolution. For such a view, it was obvious that some member of the animal kingdom would eventually reach the level at which rational thought became possible, and there seemed little need to specify why one line of primate evolution had reached this goal first. Only with the breakdown of linear progressionism in the early twentieth century did it become necessary to think more carefully about why the branches of primate evolution have gone in different directions. Even then, though, one could not completely escape the influence of progressionism: many still assumed a general trend toward brain development in the primates, so that the real problem was to explain why the apes had somehow branched away from the main line of evolution. Although one could appeal to the different environments to which human and ape ancestors were exposed as a means of accounting for the divergence, there was still a reluctance to ascribe too much significance to this factor. Early twentieth-century theories of human origins thus frequently combined the narrative style of explanation with invocations of general evolutionary principles that were supposed to explain important aspects of the process.

Darwin's theory of evolution was not based on a progressive trend, and sought to explain the origin of each form in terms of the circumstances to which it had been forced to adapt in the course of its history. Darwin himself made a notable effort to break away from the progressionist assumption and to specify evolutionary novelties acquired by our ancestors and not by those of the apes. But general interest in human evolution almost inevitably focused on those characters by which we have advanced beyond the animals and on which we pride ourselves. The opponents of Darwinism stressed the unique nature of our mental and moral faculties to defend the traditional belief that they could only have arisen by divine creation. The evolutionists responded by trying to show how the growth of the brain—from which all the higher characters were alleged to flow—could be explained in natural terms. Consequently they stressed the progressive nature of evolution, so that the move toward higher levels of organization seemed inevitable. The linear evolutionism of cultural anthropologists such as Tylor must also have encouraged this tendency. Instead of treating human evolution as the product of unique circumstances, this approach made the final step up to a human level of intelligence seem merely the culmination of a general trend. Moral significance was sought in the supposedly goal-directed character of the progressive trend, in effect transferring human values into nature. Paradoxically, the old idea of a special status for mankind might have been better preserved by supposing that the breakthrough to human status was a unique event, a view that is precisely what Darwinism implied.

In an effort to break with the most obvious form of progressionism, Darwin tried to define a unique step in human evolution by appealing to another character that was popularly supposed to distinguish mankind from the rest of the animals: our upright posture. He argued that it was not just an abstract sign of nobility, but a practical link in the chain that led to the increasing use of tools. No longer used for climbing in the trees, the hands were free to make tools, and the use of tools stimulated the growth of intelligence by a sort of feedback loop: more intelligence gave better tools, more tool use led to higher intelligence. The use of tools for hunting and food-gathering fitted the Darwinian requirement for a change that conferred adaptive benefits, as well as specifiying a unique anatomical change in the course of human evolution. Our ancestors would have lost their apelike teeth only after they had developed more sophisticated tools—a point that significantly helped the plausibility of the Piltdown finds.

Stephen Gould has pointed out that Friedrich Engels used a similar argument in an essay written in 1876 to justify the claim that labor played a vital role in the emergence of mankind.[1] Engels endorsed the view that the achievement of an upright posture was the first breakthrough, and led to the use of tools and the consequent development of the brain. For him, this view was an antidote to the belief that brain growth was in some mysterious way the driving force of evolution—a belief that he took to be a reflection of idealist philosophy. Gould admits that Engels's work had no influence on scientists, but Darwin's support was enough to gain the idea an important following. Unfortunately, neither Darwin nor Engels realized that a simple appeal to tool use was not enough to break the shackles imposed by the assumption that the brain must lead the way. They still imagined that the advantages of tool-using were so obvious that our ancestors would automatically have been led to adopt this habit. This view left them unable to explain why the ancestors of the apes had failed to participate in the same humanizing trend.

The situation became even more complicated in the early twentieth century, when Arthur Keith and others began to argue that an upright posture had already been achieved by our gibbonlike ancestors in the distant past. The transition to bipedal locomotion on the open ground now became the key breakthrough that had freed the hands for the making of tools. But why had some of the apes left the forests, while others remained? Were some naturally more intelligent or progressive than others, and thus able voluntarily to recognize the advantages of walking and tool-using? Such an appeal to an extra increment of intelligence must presuppose an earlier trend toward brain growth, and once again throws the emphasis on explaining why some of the apes did not participate to the same extent in the trend.

Alternatively, one could suppose that our ancestors were forced out of the trees by a climatic change that brought about a dramatic reduction in the area covered by forests. Sheer necessity would thus cause some of the apes to attempt a life on the plains, thereby precipitating the switch to upright walking and the use of tools. Such a view abandoned the idea of a progressive trend and threw the emphasis onto a purely contingent environmental change that forced a new mode of life onto our hapless ancestors. We shall see, however, that many of the paleoanthropologists who included a change of habitat as an important component in their theories managed to smuggle in the progressionist assumption through the back door. They insisted that, even when faced with a changing environment, it was still the more intelligent and enterprising apes that had responded in a positive manner, leaving their duller relatives languishing in the ever-retreating forests.

Early Speculations

The paleontologists of the mid-nineteenth century had developed a progressionist interpretation of the history of life on the earth, although many felt that the progress represented the unfolding of a divine plan rather than a natural transmutation of forms. Louis Agassiz was a leading exponent of the transcendentalist school of thought and a staunch opponent of evolutionism. He believed the Creator had chosen the human form as the goal toward which the development of the vertebrate type was aimed, via a series of upward steps. In some respects, the evolution theory of Robert Chambers's *Vestiges of the Natural History of Creation* of 1844 was an extension of Agassiz's view, since it accepted that transmutation occurred in accordance with a progressive plan of creation. Chambers shocked everyone, though, by making the process continue through to mankind, and by exploring the implication that our higher mental powers are simply the product of an enlarged brain. This was an early version of the belief that evolution is a progressive trend that would inevitably sweep up to the human level and beyond. James Dwight Dana's theory of cephalization also stressed the growth in brain size through the history of life, although only with some misgivings did he accept an evolutionary interpretation of the trend. Dana believed that the process of brain increase had reached its final goal in mankind, a view echoed as late as 1884 by John Cleland. The majority of post-Darwinian naturalists, however, abandoned the concept of a final goal, and treated the progress of intelligence as a general trend affecting many lines of evolution, as in Othniel C. Marsh's "law of brain growth."[2]

Some evolutionists openly welcomed the teleological implications of a theory based on a trend toward inevitable progress. For Lamarckians such as Edward Drinker Cope, such a trend indicated that the emergence of the human mind was still a central feature in the divine plan of creation.[3] To a surprising extent, the more radical thinkers who adopted an evolutionary perspective remained oblivious to the teleological implications of linear progressionism. They felt that their calls for social reform would appear more plausible if presented as the natural extension of a universal process. This attitude is clear in the work of Gabriel de Mortillet and Karl Vogt.[4] De Mortillet's sequence of archaeological periods certainly generated a linear scheme of human evolution, the whole being linked to French socialism. Vogt was also a noted radical, expelled from his native Germany in 1848, yet his *Lectures on Man* accepted the parallel evolution of the human form from separate ape origins. Vogt seems not to have appreciated the need to specify *why* the trend should have been so powerful that it could affect apes living in different parts of the world. He assumed that the advantages of being human were so obvious that the progressive trend could be taken for granted even at this late stage in the history of life.

E. B. Tylor's cultural evolutionism also carried a political message based on the inevitability of progress. Many anthropologists shared the belief that progress was slow, but inevitable, and were thus forced to confront the question of why some branches of mankind have advanced less than others. In the end, even Tylor succumbed to the prevailing belief that some races were constitutionally less able to develop civilization than others. The white races, formed when mankind at last migrated to the more vigorous climate of the temperate regions, had thereby been stimulated to develop "the powers of knowing and ruling which give them sway over the world."[5] This was the standard way of accounting for white supremacy, but it was an explanation that could not be applied to the earliest phases of human evolution, which were assumed to have taken place in the tropics. The cultural evolutionists seem to have believed that mind and culture developed together through mutual interaction, and this belief accounts for the linking of their approach to the Lamarckian mechanism of the inheritance of acquired characters.[6] The earliest cultures stimulated the growth of intelligence, which was transmitted to future generations, and this process allowed higher levels of culture to be developed. The process began an indefinite cycle of interactions, which could be stimulated by more challenging conditions, but which did not appear to have been initiated by a change in the environment.

Few of the evolutionary anthropologists considered the question of why the feedback loop between mind and culture had become established in mankind's ancestors, but not in those of the apes. Tylor still accepted an enormous gulf between the apes and even the lowest race of mankind, and one of the American cultural evolutionists, Daniel Brinton, openly appealed to an evolutionary saltation.[7] Even Herbert Spencer, with his strong interest in biological evolution, showed no awareness of the need to specify conditions that would explain why mankind had advanced so much further than its closest relatives. In both the *Principles of Psychology* and the *Principles of Sociology,* Spencer argued that the human mind emerged through the replacement of primitive reflex actions by more rational modes of behavior. He believed, as mentioned earlier, that from primitive mankind onward, this "development of the higher intellectual faculties has gone on *pari passu* with social advance, alike as cause and consequence."[8] This is a clear statement of the principle behind the Lamarckian feedback loop, yet how was the loop first established? Spencer's fascination with the general theme of evolutionary progress seems to have prevented him from seeing the significance of this question. Social evolution was certainly a higher level of development than its biological counterpart, but the one graded insensibly into the other, and he saw no need to specify why one branch of biological evolution had moved so much more rapidly into the higher phase.

The closest that Spencer ever came to this question was his citation of his American disciple, John Fiske.[9] In his *Outlines of Cosmic Philosophy* of

1874, Fiske had sketched in a scheme of universal evolution which differed from Spencer's mainly in its claim that the growth of altruism was a central feature of biological progress. Although he tried to minimize the distinction between the apes and the lowest forms of humanity, Fiske appreciated the need to explain how the new phase of human mental and moral evolution had begun. He based his account on the emergence of the social habits, which in turn he attributed to the growing need for a family life to sustain the infant through its long period of helplessness.[10] Unfortunately, Fiske also suggested that the longer growth period was required because of an increase in intelligence, thereby making the argument circular. Elsewhere, he appealed to an unexplained variation in the degree of sociability among the Miocene ancestors of the apes and mankind.[11]

Of Darwin's own followers, it was George John Romanes who became most deeply concerned with the development of human intelligence. Again, though, fascination with the general principles of mental development blocked any serious thought on the question of how the trend was initiated. Romanes took it upon himself to defend Darwin against the attack led by St. George Jackson Mivart, who had argued that certain human faculties were unique to mankind, and could not have developed from an animal origin. One of the most important distinctions was thought to be the use of language, and Romanes went out of his way to emphasize the extent to which—in his opinion—animals did make use of some form of language. Accepting that the much greater development of sign-making was the foundation of human superiority, he observed that the apes may have been blocked off from this route to progress simply by the lack of suitable vocal organs.[12] Yet in his main discussion of this issue, Romanes was deliberately evasive. He suggested, in opposition to Darwin himself, that our ancestors were already largely human in form *before* the development of speech began. *Homo alalus* was already an erect, toolmaking, social creature, and it was social interaction that stimulated the development of language and the refinement of the human mind.[13] Romanes was evidently not interested in how this intermediate stage had been reached.

All of the writers mentioned above believed that social or cultural life was crucial for the development of human mental powers. Implicit in their view of human evolution was the assumption of the existence of a threshold, at which the process of biological evolution gradually began to give way to a new level of development stimulated by the feedback loop between culture and intelligence. The normal process of adaptive evolution, in the long run, produced a slow increase in intelligence, but at a certain point in the scale, social activity began to give a greater impetus to mental development. They showed little interest in the question of why one line of biological evolution entered this new phase before the others, because their faith in the progressive nature of the process convinced them that it was only a matter of time before one line or another reached the threshold. Since many of the more advanced lines of animal evolution were in principle capable of reaching this goal, little was to be gained by asking why one line had, in fact, got there ahead of the others.

This point was explicitly made in an article written in 1895 by the American sociologist and Lamarckian Lester Frank Ward. He devoted the bulk of this article to a critique of the various suggestions that had been offered as a means of drawing a distinction between mankind and the lower animals. First, however, Ward raised the question of how mankind had acquired its exalted status, only to dismiss it as largely uninteresting because the human form had no important distinguishing characters. Two propositions upheld this view: "The first is that if the developed brain had been awarded to any of the other animals of nearly the same size as man, that animal would have dominated the earth in much the same way as man does. The other is that a large part of what constitutes the physical superiority of man is directly due to his brain development."[14] Like some biologists, Ward emphasized that by anatomical standards, the human form was not a particularly advanced one. He went on:

> It is difficult to conceive of a being entirely different in form from man taking the place that he has acquired; but if any one of the structurally higher races possessed the same brain development it would have had the same intelligence, and although its achievements would doubtless have been very different from his, they would have had the same rank and secured that race the same mastery over animate and inanimate nature.[15]

According to Ward, the erect posture was merely a direct consequence of our intelligence, achieved only at a very late stage in our evolution.

If Ward was articulating a view taken for granted by other sociologists and anthropologists, this assumption would explain the general lack of interest in the question of human origins. They were concerned in an abstract way with the transition from the biological to the social phases in the process, but they did not believe that any special characteristic enabled the ancestors of mankind to reach the threshold first. Perhaps, as Fiske speculated, some chance increase in the social instincts of our ancestors had played a role, but this was not the real cause of human evolution. If our ancestors had not been given this advantage, some other form would eventually have reached the threshold and the new level of evolution would have taken off from a different starting point. The emphasis was on the inevitability of the transition, not the accidental characters that determined which line of biological evolution made the actual breakthrough.

By contrast, the majority of Darwinians were suspicious of the idea that social activity could initiate the last phase of human evolution. They knew that many of the higher animals live in social communities, and found it difficult to believe that our earliest ancestors could have differed significantly in this respect. In the *Descent of Man,* Darwin spent much time in accounting for the origin of the moral sense, but he was convinced that our apelike ancestors were already social beings.[16] By the end of the century, many sociologists had come to accept that instincts inherited from our animal ancestors played an important role in human social life. The increasingly complex nature of social and moral activity was a product of the interaction between these instincts and the growing intelligence: it was an effect, not a cause of human evolution. Darwin argued that any social animal would acquire a moral sense once its intellect had developed to a certain level.[17] The crucial problem was to explain why human intelligence had developed further than that of any other social animal, so as to permit the emergence of the moral sense. A Darwinist would thus be forced to look for other factors that might have switched our ancestors onto the path of intellectual development. The most obvious character that could have played this crucial role was the use of the hand for toolmaking, which in turn could be linked with the freeing of the forelimbs from locomotion as a consequence of walking upright.

Darwin himself laid the foundation for this theory in the *Descent of Man.* Here he noted that if a primate ventured from the trees onto the ground, it would have to become either quadrupedal (like the baboon) or bipedal. The clue to why our ancestors had taken the latter course lay in the structure of the hand, which was less efficient than that of the apes for climbing, but far better at more delicate tasks such as toolmaking.[18] Darwin argued that the hand could not have been perfected for toolmaking if it were still used for locomotion, so it would have been an advantage for our ancestors to adopt the upright posture to free the hands.

Once started, the process would continue to completion, since a more erect posture would always give a higher degree of freedom to the hands. The free use of the hands was thus "partly the cause and partly the result of men's erect position."[19] The modern apes stand partly erect, showing that the intermediate stages in the transition were viable. Darwin seems to assume that the increased use of tools would have encouraged the growth of intelligence, although he does not explore this point in detail.

One consequence of the ability to make artificial weapons would be the loss of the great canine teeth used by the apes for defensive purposes.[20] Darwin attacked a suggestion made by the Duke of Argyll that the intermediate stage between ape and human would have been too helpless to survive, since the fighting abilities of the ape would have been lost before the making of artificial weapons could compensate.[21] Darwin insisted that the animal characters of our ancestors would be lost only in proportion to the compensating gain in mental powers. The belief that the great teeth of the apes would be retained long after the growth of the brain had begun would later emerge as an argument for the validity of the Piltdown discoveries. Darwin also pointed out alternatives to this interpretation, however. The female apes do not possess such large canines, yet are well able to survive. In some circumstances, then, the loss of natural defensive ability might not have been a hindrance, and Darwin mentioned the possibility that the early stages of human evolution might have taken place on an isolated island such as Australia, New Guinea, or Borneo, where there would have been few predators.[22] Elsewhere, he expressed a general preference for Africa as the cradle of mankind, while noting the impossibility of making any firm prediction now that the discovery of *Dryopithecus* had shown that the apes once ranged over Europe.[23] Whatever the exact location, Darwin favored a tropical habitat with plentiful supplies of fruit for our earliest ancestors.

Ernst Haeckel, whose views on human evolution Darwin regarded highly, also favored the acquisition of the upright posture as the first step in the process. In his *History of Creation,* Haeckel identified two evolutionary innovations that had made the human race what it is today: the upright posture and articulate speech.[24] The latter was the more important, since from it stemmed the higher mental faculties—in this respect Haeckel agreed with Mivart and Romanes. Yet the upright posture was the first step, and the original ape-man had been upright but speechless: *Pithecanthropus alalus.*[25] Haeckel was not altogether consistent in this view, however, since elsewhere he suggested that primeval mankind had been only half erect.[26] This observation did not prevent him from hail-ing Dubois's discovery of *Pithecanthropus erectus* as confirmation of his original prediction.[27] Unlike Darwin, though, Haeckel did not really explore the reasons for the upright posture being the first step. Since speech rather than toolmaking was crucial for the growth of the intellect, the value of the erect posture was not altogether clear, hence Haeckel's qualification of his prediction.

Darwin's more realistic evaluation of the benefits conferred by an upright posture was extended by two of his followers, Alfred Russel Wallace and E. Ray Lankester. In his 1864 paper on the origin of the human races, Wallace argued that the development of the brain was the last step in human evolution, implying that our ancestors achieved a human physical form, presumably including the upright posture, first.[28] At this stage, Wallace still believed that natural selection would produce the final improvement in intelligence, but he abandoned this position in his 1870 paper "The Limits of Natural Selection as Applied to Man." Here he maintained that a number of our mental and physical characters could have been formed only by direct supernatural intervention in human evolution. He specifically included the hand, which he now claimed had latent powers not exploited by savages and hence "all the appearances of an organ prepared for the use of civilized man, and one which was required to render civilization possible."[29] By the time he wrote his *Darwinism* of 1889, Wallace had slightly softened his position, and now argued only that certain mental powers required a supernatural explanation. He accepted that the erect posture, with the consequent freeing of the hands, was the crucial difference between mankind and the apes. This character had been acquired very early in human evolution, and was the key to a line of development that had ultimately enabled mankind to conquer the world.[30]

E. Ray Lankester was a prominent second-generation Darwinist who made substantial efforts to convey the implications of evolutionism to the general public. In an article "From Ape to Man," originally published in the *Daily Telegraph,* he emphasized the importance of the opposable thumb as the starting point of mankind's emergence. He argued, though, that it was the ability of the hand to explore the environment, rather than toolmaking, that promoted the growth of intelligence.[31] The erect posture was an essential factor allowing the perfection of the hand. In another article, Lankester supported the view that the emergence of mankind was a comparatively recent event. The frequency with which modern humans suffer from hernia showed that the walls of the abdomen had not yet fully adjusted to the upright posture.[32] He also argued that the human foot could not have been developed from a structure like that of the chimpanzee or gorilla. The ancestral ape must already have had a plantigrade foot more like that of the gibbon.[33]

Emphasis on the role played by the adoption of an erect posture came independently from Charles Morris in America. In a paper published in 1896, Morris clearly developed the theme that the transition to walking upright had freed the hands and thus promoted the growth of human intelligence. He was less clear on how the change had been brought about, however, suggesting that our ancestors were forced to walk erect by changes in their limb proportions which prevented them from brachiating or adopting a quadrupedal gait.[34]

One of the most forceful statements supporting the priority of the erect posture was delivered by the Scottish archaeologist Robert Munro in a Presidential Address to the Anthropology Section of the British Association for the Advancement of Science in 1893. Munro spoke as though both the apes and mankind had begun from a quadrupedal posture, and had undergone major anatomical changes rendering the apes semierect and mankind fully erect.[35] The human foot had become, in effect, a tripod, supporting weight on the great toe, side and heel, an idea endorsed by T. H. Huxley.[36] Munro strongly criticized Wallace's view that a supernatural power must have guided the growth of the human mind. There was indeed a higher power involved, but it was the reasoning power of our ancestors themselves, now rising above the laws of physical nature.[37] This new level of intelligence was developed because of the agility of the hand, once it had been freed from the task of locomotion, although it was also promoted by the constant warfare between tribal groups.[38] Significantly, though, Munro said nothing about *why* our ancestors alone adopted the upright position and thus initiated the new phase of mental development.

This last point leads us back to Wallace's *Darwinism* and an important qualification of Darwin's own view of the circumstances in which the erect posture was adopted. Darwin had suggested a tropical environment with plentiful supplies of fruit as the most likely habitat for early mankind. Curiously, it was Wallace—who still accepted a role for supernatural intervention—who put his finger on the crucial weakness of this hypothesis. The upright posture, he pointed out, is linked to an even more important distinction between mankind and the apes: we are terrestrial, whereas they are arboreal. It was thus most unlikely that the transition to upright walking had taken place in a forest environment, to which the apes were still better adapted. Open plains or high plateaus in the temperate zone were the more probable location, with our ancestors adopting the life of hunters and trappers.[39] Arguing that the apes were comparative newcomers to Africa, Wallace went on to suggest that the most likely cradle of mankind was Eurasia, an observation that would fit in with the idea that the Mongolian race was intermediate between the two extremes of humanity, white and black.[40] Since much of Asia was still unexplored, it would also explain the lack of early human fossils. Wallace thus anticipated in many important respects the central Asia theory of human origins later popularized by H. F. Osborn and others. He had also shown that a complete account of human origins would have to include an explanation of our ancestors' evacuation of the forests—it could not simply assume that the transition to upright walking was prompted by the advantages to be gained from the subsequent use of tools.

The Brain Leads the Way

Wallace's critique suggests that Darwin himself had fallen into the trap of believing that the advantages of a human type of behavior—including toolmaking—were so obvious that the transition would take place automatically. While recognizing that the upright posture preceded and caused the growth of the brain, he had accepted that the technological and intellectual advantages of the move had somehow drawn our ancestors into the earliest phase of the process. His inability to explain why the apes—in the same environment—had not followed this lead suggests that he had not really escaped the logic of the unilinear progressionism so popular at the time. Wallace had realized that to account for the divergence between the ape and human stocks it was necessary to postulate a change of habitat as well as behavior. The apes had stayed in the trees and stagnated, whereas mankind had ventured out onto the open plains, where an upright posture was a clearer advantage and the arms were no longer needed for grasping branches.

One question remained unanswered, a question so vital that it still left room for fundamental disagreements about the cause of human evolution. If one assumes that toolmaking had boosted human intelligence, and that a move onto the open plains was the most likely origin of the bipedal locomotion that had freed the hands, *why* had our ancestors—and not those of the apes—adopted the new habits that led them out of the forests? Continuing the line of argument opened up by Wallace, one could suppose that they were forced out of the forests, perhaps by a climatic change that greatly reduced the area of woodland available. In this case, the emergence of mankind would be a purely contingent affair, depending on an almost unrepeatable coincidence of geological and biological processes. The alternative was to assume that our ancestors recognized the potentialities of the new environment and deliberately moved out of the forests to exploit them. In this case the most intelligent apes would have been the ones that made the move, whereas their less progressive relatives would have stayed in the trees to stagnate. The approach of transferring the cause of the change back to a difference in mental abilities reintroduces the possibility that an earlier trend toward brain development was, after all, the key to the origin of mankind.

The resulting debate was far more than a mere priority dispute between the supporters of brain growth and those of the upright posture. Those who stressed the importance of a general evolutionary trend toward increased intelligence admitted that the transition to bipedal locomotion was essential for the final growth of the human mind. The most advanced forms produced by the general trend were the ones that saw the value of walking on their feet and leaving the hands free for other purposes. Their initiative, rather than the upright posture that resulted from the transition, was the real key that opened the door to a new phase of brain growth. Many advocates of this view believed that the upright posture was in fact extremely ancient, an inheritance from the brachiating habits of earlier primates. The crucial step was our ancestors' decision to exploit the potentialities of this primitive character by using the foot for walking and the hand for toolmaking.

We saw in chapter 4 how the simple, linear scheme that derived mankind directly from the great apes was replaced in the early twentieth century by a more complex theory of branching evolution. We can now appreciate how this new approach could still be made consistent with a more sophisticated kind of progressionism. If one dismisses the great apes as a degenerate side branch, the implications of Darwin's claim that the upright posture promoted toolmaking and brain growth were subverted. The upright posture was an ancient character that merely had to be properly exploited to yield its benefits. Instead of being a crucial breakthrough, the development of bipedal locomotion was merely another product of the triumphant growth of the brain. Branching evolution was necessary, but only to make room for all the failures that had been unable to seize upon the sequence of opportunities offered to them by the steady increase in primate intelligence. To explain why the apes' ancestors had been less innovative than mankind's, one only had to postulate that they had already become too specialized, too committed to the trees. The supporters of the presapiens theory might hail their approach as a more sophisticated image of human evolution, in tune with the general principles seen elsewhere in the animal kingdom. But a vestige of the old progressionism remained, because they could not shake off the belief that the line of brain growth leading to mankind was central to nature's purpose.

The important feature of this position was thus not the temporal priority of brain growth over posture change, but the claim that at every stage it was the increasing capacity of the brain that initiated the anatomical modifications leading toward the human form. By contrast, an environmentalist interpretation of human evolution stressed that the brain growth produced by toolmaking was a consequence of the new habits forced upon our ancestors when they were driven out of the forests. Even if the arboreal life of the early primates had preadapted them for walking upright, so that only a relatively minor change in foot structure was required, an external stimulus was needed to force new habits onto the earliest ancestors of mankind.

Misia Landau has called these two scenarios the "escape" and the "expulsion" from the trees, and has stressed the different images conjured up by these metaphors.[41] "Escape" implies that our ancestors used their superior intelligence and initiative to leave the evolutionary trap of the forests and venture into a whole new world. "Expulsion" portrays our ancestors as hapless vagrants compelled to abandon their home by forces outside their control. Both images may well appeal to deep-seated cultural values and archetypes, including the capitalist ideal of individual enterprise and the biblical notion of the expulsion from the Garden of Eden. Whatever the traditional support for the latter, Landau notes that in fact the escape image was far more popular. Even those anthropologists who accepted a role for environmental change frequently appealed to the more progressive character of the individuals who responded positively to the challenge. The progressionist ideal of the brain leading the way thus dominated the thinking even of those who saw the move to a new environment as a vital step in human evolution.

If the theory that the brain led the way was indeed linked with a more general progressionism, it would throw doubt on Landau's further suggestion that the transition to ground living was seen as a crucial breakthrough leading to the separation of mankind from the rest of nature.

True, many early twentieth-century paleontologists saw the achievement of bipedal locomotion as a vital step in the growth of human intelligence. Yet Elliot Smith and other supporters of the brain-first theory saw this transition as the culmination of a long trend toward brain growth that had originally been fostered by life in the trees. Although lifted above the animals by the final stage in his evolution, mankind was still a product of nature, our most cherished values the goal of a universal trend.

The distinction between escape and expulsion can itself be seen as an expression of the much wider debate in evolution theory over the relative significance of internal and external forces. Many naturalists at this time sought to restrict the scope of environmental factors, and the appeal to a general trend toward brain growth served this purpose. Even the claim that initiative was the force that led animals to explore new environments can be found in the more general evolutionary debates. It was precisely this factor that Lamarckians such as Samuel Butler used as a means of constructing a more humane alternative to the trial-and-error materialism of natural selection. Paleoanthropologists may not always have referred explicitly to these more general debates, but we should be wary of assuming that there was no interaction between the issues as they were raised at the two levels.

A temporary product of the belief that the growth of the brain led the way in human evolution was renewed support for Darwin's brief suggestion that our ancestors would have retained fighting canine teeth until they were rendered superfluous by the invention of artificial weapons. This conclusion seemed reasonable as long as it was accepted that we had evolved directly from the great apes, and it certainly helped to create a climate of opinion in which the Piltdown fraud would be taken seriously. It is sometimes assumed that the same factor was responsible for the early suspicion with which Raymond Dart's discovery of *Australopithecus* was greeted. This assumption fails to take into account the presapiens theory, however, and the growing belief that the erect posture was an ancient primate characteristic. In fact many authorities found the Piltdown jaw a bit too apelike (not surprisingly), and so tended to push *Eoanthropus* off onto a side branch marked by an anomalous development of the jaw. Two of the naturalists who supported the retention of an apelike jaw in the prewar years—Elliot Smith and Sollas—backed away from this position in the 1920s and were prepared to give *Australopithecus* considerable significance as a clue to human evolution. If our early ancestors were more generalized apes, there was no reason to suppose that they had the specialized teeth of the modern great apes. Nor was the upright posture claimed by Dart for *Australopithecus* a problem, since many authorities now felt that it was the modern apes' slouching gait that was the true evolutionary novelty.

The real culprit (if we can use the term) in the case of *Australopithecus* was the rapid development of the view that Asia, not Africa, was the cradle of mankind. Had Dart's fossil turned up in Asia, it is probable that paleontologists such as H. F. Osborn would have hailed it as a vital piece of evidence, especially as Osborn now suspected that many human characters were more ancient than those of the apes. An upright creature with human teeth and a more progressive brain was just what Osborn was looking for, but he was looking in Asia and had convinced the majority of his contemporaries that there the important evidence would emerge. The central Asia theory was, as we shall see, a product as much of the overzealous application of geographical principles as of the fascination with brain growth.

Before exploring these debates in detail, I should note that some of the scientists who figured prominently in the discussions of human phylogeny were not deeply concerned with the question of causation. Gustav Schwalbe and Eugene Dubois, although they supported Haeckel's views on the transition from the apes to mankind, said little about the underlying causes.[42] For them, evolutionism was merely the construction of genealogies based on morphological relationships and the few available fossils. They were interested in the course of human evolution, not its cause. Even Marcellin Boule and Arthur Keith contributed little on the initial stages of the process. Both became increasingly concerned with later episodes such as the emergence of the Neanderthals and the modern races. Boule was committed to a general theory of parallel evolution that left no room for the concept of a key transition affecting a single line of development. Even though Keith pioneered the view that the erect posture was an ancient character, he commented rarely on the transition to a terrestrial habitat. One reason why the evolutionary biologists who pioneered the central Asia theory of human origins were able to carry the day so easily may have been that the anatomists and anthropologists who did so much of the work on human fossils had simply failed to take up a position on this issue.

In 1909 and 1910, W. J. Sollas was president of the Geological Society of London and used his annual addresses to expound his own views on human evolution. The 1909 address argued that the main developments had probably taken place outside Europe and anticipated the theme of *Ancient Hunters:* the importance of racial migrations. In 1910 Sollas went on to survey the theoretical debates that were already beginning to rage as the linear progressionism of Schwalbe came under fire. Noting that Klaatsch and others were now excluding the apes from human ancestry, Sollas objected that such theories created entirely hypothetical lines of evolution, and he defended the general outline of the sequence from the apes to mankind.[43] But what had actually caused the further developments that led to the human form? Sollas noted the importance of the erect attitude and even ventured part way toward an environmentalist explanation of the transition. The apes were almost erect, but had not completed the transition because they had remained in the forests. If resources had somehow become limited in the forests, however, "it would be very strange . . . if some ancestral members of this intelligent group had not endeavoured to escape the pressure of the environment by invading the plains or open country. Such a change of habitat would almost inevitably necessitate either the assumption of an erect attitude or a return to a more perfectly quadrupedal state."[44] The baboons illustrated the latter alternative, whereas man had taken the other direction, "most likely because he had already begun to make important use of his hands." The erect attitude freed the hands to participate in the development of the higher mental functions.

The suggestion that the hands were already coming into use before the invasion of the plains allowed Sollas to back away from the environmentalist implications of his argument. He had introduced his discussion in the following manner: "I may perhaps be permitted to state, merely as a confession of faith, my belief that the really fundamental change, underlying all the rest, was the increasing growth of the intellectual powers, and this I regard as an ultimate fact as difficult of explanation as any other ultimate fact, such as the origin of variations or even of life itself."[45]

He reconciled this principle with his remarks on the importance of the erect posture by emphasizing that it was intelligence that stimulated the move onto the plains. "It was not, in all probability, that the erect attitude was first acquired as a necessary preliminary, but that an increasing appreciation of the powers of the hand led to a more frequent adoption of that attitude, and that the transformation of both organs [brain and hand] probably proceeded *pari passu.*"[46] The ultimate cause was thus the growth of the brain that allowed some of the ancestral apes to recognize the advantages of the new lifestyle, a cause that Sollas freely admitted was beyond his comprehension.

Since he still accepted the evolution of mankind from the apes, Sollas believed that the earliest participants in this process were still formidable creatures, well able to take care of themselves. He criticized Klaatsch's suggestion that we developed from less powerful ancestors in the protected environment of Australia. "If we abandon the view that Man commenced his existence as a puny creature retaining the primitive characters of the lemurs, and ascribe his origin to a point on the ancestral tree not far below that from which the chimpanzee and gorilla branched off, we may then attribute to him at the beginning a strong bodily frame and a dentition well fitted for the purpose of offence and defence."[47] The freeing of the hands then led to the use of sticks as primitive spears, so that "the massive jaws and fighting teeth can now be dispensed with, and may safely undergo a regressive development with adaptation to purely alimentary functions."[48] Sollas also noted Keith's view that the brow ridges of the apes were derived from the strength of the jaw muscles, and deduced that these ridges would be reduced at the same time. This last point, however, illustrates the problematic character of the Piltdown finds. When Sollas later claimed that the apelike jaw of *Eoanthropus* exactly fitted this prediction, he carefully had to forget that the Piltdown skull showed no trace of brow ridges.[49]

Sollas's prediction was echoed in more general terms in Grafton Elliot Smith's 1912 Presidental Address to the Anthropology Section of the British Association for the Advancement of Science, which is perhaps the most elaborate statement of the view that the brain led the way in human evolution. After the Piltdown discovery, Smith claimed that he too had predicted the combination of a large brain and an apelike jaw, although his position had actually been stated in rather more general terms.[50] His name was so closely associated with the theory of brain development, though, that even Keith accepted him as the true source of the prediction.[51] Arthur Smith Woodward's original description of *Eoanthropus* does not refer to the use of teeth for defense, although in the British Museum's *Guide to the Remains of Fossil Man* he agreed that the teeth would only degenerate after intelligence had increased.[52] As a paleontologist, though, Woodward attributed all evolutionary trends, including the one toward brain growth, to nonadaptive orthogenetic factors, thereby undermining the whole point of progressionism.[53] It was thus Elliot Smith's account, incorporated with modifications into his *The Evolution of Man* of 1924, which remained the primary source of the theory that brain growth was the key to human evolution.

Although Smith's 1912 address included the general notion that the refinement of the facial features might lag behind the expansion of the brain, this aspect was not his real concern. As the leading authority on cerebral anatomy, he was convinced that the brain had indeed led the way throughout the evolution of the mammals. His main purpose was to attack the claim that the achievement of an upright posture was a key breakthrough permitting the subsequent development of human intelligence. In the preceding year he had even suggested that the first appearance of the cerebral cortex (which he called the neopallium) was responsible for the more skilled limb movements that had prompted the evolution of the mammals from the reptiles.[54] It was by an extension of this process of building up the cortex that the human mind was produced. Like many subsequent evolutionists, Smith tried to pretend that his position had been endorsed by Darwin himself. He specifically criticized Munro (and also Sollas) for trying to switch attention from the brain to the upright posture as the key to human evolution.[55]

The starting point for Smith's attack on the claims made for the upright posture was the view, usually attributed to Keith, that such a posture had already been developed by the primates in their early, gibbonlike phase. Keith had argued that the gibbon's mode of progression by brachiation required that the body move into a vertical position, suspended from the arms.[56] The consequent production of the necessary anatomical adjustments in the trunk would thereby preadapt the body to the posture adopted when the legs are used for support on the ground. Curiously, Keith himself seems to have had little interest in the switch from brachiation to walking, although he did endorse the view that the brain led the way.[57] He also commented on the advantages gained by adopting a terrestrial lifestyle, which would open up the whole surface of the earth to human occupation.[58] Yet Keith's major works on human evolution concentrated on the later phases of the process, and did not explore these topics in detail.

By adopting the view that the erect posture was extremely ancient, Smith was able to taunt Munro: "If erect attitude is to explain all, why did not the Gibbon become a man in Miocene times or even earlier?"[59] The answer to this question derives from the view that the gradual specialization of the brain was responsible for initiating all major developments in primate evolution.

> . . . such advances as the assumption of the erect attitude are brought about simply because the brain has made skilled movements of the hands possible and of definite use in the struggle for existence. Yet once such a stage has been attained, the very act of liberating the hands for the performance of

> more delicate movements opened the way for a further advance in brain development to make the most of the more favourable conditions and the greater potentialities of the hands.[60]

The earlier gibbons did not immediately become human because their brain was fitted only for exploiting the hand in tree-climbing. Only after further brain development did they become intelligent enough to exploit the hands more efficiently and thus realize the advantages of becoming terrestrial. In Smith's view, "it was not the adoption of the erect attitude that made Man from an Ape, but the gradual perfecting of the brain and the slow upbringing of the mental structure, of which erectness of carriage is one of the incidental manifestations."[61]

Smith stressed the importance of the arboreal mode of life for the early stages in primate brain development, a theme taken up by Wood Jones in his *Arboreal Man.* Smith explained how tree-climbing had forced the primates to abandon the reliance on the sense of smell that had characterized the early mammals, and to depend instead on vision. This development in turn had stimulated their inquisitiveness and their ability to use their hands, both of which encouraged further growth of the brain. A popular book by Dorothy Davison later expanded on Smith's vision of the brain developing in the trees, comparing the arboreal environment to a Montessori school, where everything was designed "to provide an adaptable brain with exactly the stimulus needed to develop it."[62] The trend of brain development led from the tree shrews through the tarsiods to the monkeys and the apes. In their original forms, both the monkeys and the apes had been far more generalized in structure than their modern counterparts, which had been sidetracked from the main trend of primate evolution by overspecialization. Apart from these dropouts, the main trend had continued inexorably, as the stimulating environment of the trees exerted its effect on those primates who remained flexible enough to respond.

Smith accepted that the final step in the development of mankind involved the transition to a new and even more stimulating environment on the open plains. Yet he believed that this transition had been initiated as a result of the extra brain power gained by certain primates from their long development in the trees. Already fully erect, and with hands quite well adapted to grasping, our ancestors finally acquired sufficient intelligence to accept the challenge of a terrestrial mode of life.

> In one group the distinctively Primate process of growth and specialization of the brain, which had been going on in their ancestors for many thousands,

> even millions, of years, reached a stage where the more venturesome members of the group—stimulated perhaps by some local failure of the customary food, or maybe led forth by a curiosity bred of their growing realization of the possibilities of the unknown world beyond the trees, which had hitherto been their home—were impelled to issue forth from their forests, and seek new sources of food and new surroundings on hill and plain. The other group, perhaps because they happened to be more favourably situated or attuned to their surroundings, living in a land of plenty, which encouraged indolence in habit and stagnation of efforts and growth, were free from this glorious unrest, and remained apes.[63]

Although an environmental change may have encouraged our ancestors to move out of the trees, Smith makes it clear that their extra intelligence and initiative were crucial to the transition.

In another passage of his 1912 address, omitted from the reprinted version in *The Evolution of Man,* he makes this point even more explicitly.

> Once the Simian ancestor of Man began to anticipate the consequences of his acts and put this knowledge and the growing appreciation of the powers of his hands to useful purpose, for using weapons, or even making them, the erect attitude would become a regular habit, so as to emancipate his hands entirely for their new duties. The realization of his ability to defend himself upon the ground, once he had learned the use of sticks and stones as implements, would naturally have led the intelligent Ape to forsake the narrow life of the forest and roam at large in search of more abundant and attractive food and varieties of scene. . . . Thus we have come to realize the steps by which a growing brain makes it possible and desirable for the most intelligent of the Apes to forsake the purely arboreal life and seek a wider sphere of activity upon the earth: they emerged from their original forest home, and in troops invaded the open country, led no doubt by the search for a more plentiful supply or a more appetizing variety of food.[64]

Since greater intelligence was the key that opened the way into the new habitat, the further development of the brain after the transition to the ground was, in a very real sense, a continuation of the trend already established in the trees.

Smith emphasized that one of the characters needed to ensure the development of the brain throughout this whole process was the retention of a primitive bodily structure. Although primitive characters were often portrayed as a sign of degradation, a high degree of specialization

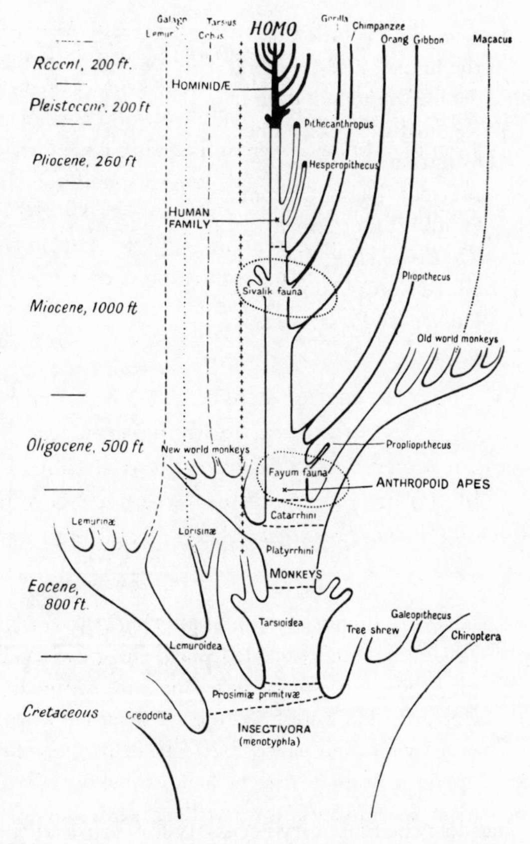

Figure 13. Elliot Smith's scheme of relationships for the primates, showing the "main stem" of evolution leading toward mankind, with the apes and lower human forms as side branches. From Grafton Elliot Smith, *The Evolution of Man,* 2d ed. (Oxford: Oxford University Press, 1927), p. 3.

was in fact the true indication of evolutionary weakness, since it circumscribed all future development.[65] The ancestors of mankind took advantage of their humble status, remaining in the trees where flexibility of structure and behavior was a vital factor in stimulating the growth of the brain. Evolution was thus a branching process, yet one branch—that in which brain growth predominated—was clearly the most important. Smith added an almost teleological progressionism to the foreword of his *The Evolution of Man* entitled "Man's Pedigree." Here he uses diagrams to show how the apes and lower forms of mankind such as the Neanderthals had diverged from the "main stem" which ran through to the modern Nordic race (see figs. 13 and 14). This view was certainly not a simple-minded progressionism, since the central line of brain development was made possible by the retention of primitive anatomical characters. The lemurs, monkeys, and apes had become specialized in various ways, and

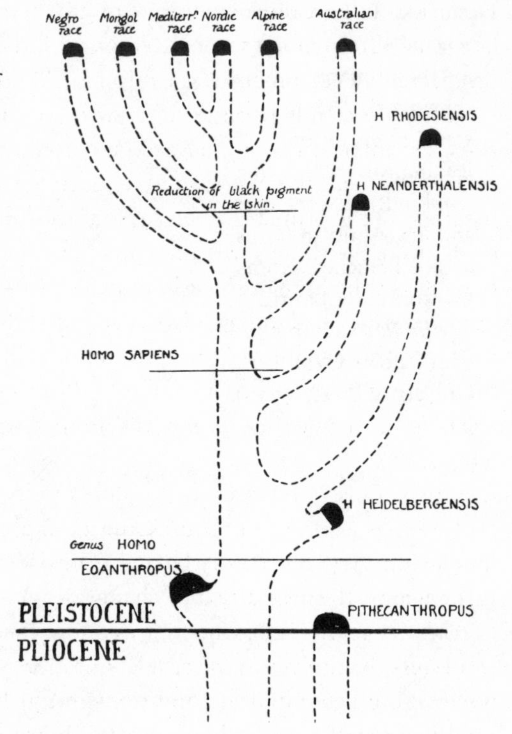

Figure 14. Evolution of the human races: a more detailed representation of the final stages of human evolution shown in the previous diagram. From Grafton Elliot Smith, *The Evolution of Man,* 2d ed. (Oxford: Oxford University Press, 1927), p. 2.

had lost the plasticity necessary for future advance in the most important character of all.[66] The Nordic race was placed at the apex of the main stem because in some respects it was more primitive than the other races, for instance in its retention of straight hair.[67] Nevertheless, the use of a term such as "main stem" to denote the line of brain development shows that Smith regarded the divergent forms as evolutionary failures, which had lost their chance to become human because they had taken the easier alternative offered by bodily specialization.

The teleological character of Smith's theory is discussed in chapter 9. Many paleoanthropologists accepted the claim that an additional increment of brain power had allowed our ancestors to escape from the forests into the even more stimulating environment of the plains. Smith's explanation of how this extra intelligence had been gained, however, was open to many interpretations. By presenting this last step in human evolution as the culmination of a longstanding trend affecting the whole history of the primates, Smith had opened up the possibility that the trend itself might be the determining factor in the creation of human nature. The transition to bipedal locomotion might become merely an incidental factor, a shifting of the scene against which the preordained unfolding of intellectual development was exhibited. When pushed to this extreme, evolutionary trends became so fascinating that some paleoanthropologists lost interest in the change of environment as a key step in human origins. Some, like Wood Jones, extended the trends to include all members of the primates, even those already deflected onto side branches unconnected with our own ancestors. The apes—like the less successful branches of the human stock in Smith's theory—became parallel developments which had independently acquired the characters by which they mimic the human form. Others suspected that the growth of the primate brain might be the product of an inbuilt orthogenetic trend, destined to reach its goal whether or not the results were of any adaptive value (chapter 8).

Smith himself refused to go along with those who used directed evolution as a means of eliminating even a generalized ape form from human ancestry. His theory balanced the inevitability of brain development against the necessity for a change of environment to allow the trend to reach its goal. Those who followed him in this moderate position had to address themselves to the crucial question of how critical the stimulus of changing geographical conditions might be in prompting the most intelligent apes to quit their ancestral home in the trees.

Asia or Africa?

Smith himself seems to have believed that a changing environment played no significant role in the transition to bipedalism. The open plains merely passively offered themselves as a source of new opportunity for those apes with enough intelligence to exploit them. Since no change in the environment was required to force our ancestors out onto the plains, any location where plains and forests lay side by side was a possible cradle for the human race. Smith himself defended Darwin's claim that Africa was the most likely area in which the transition had actually taken place. Others, however, were beginning to suspect that the environment played a far more active role. Darwinists had always accepted that gradual changes in the climate produced by geological activity were an important stimulus to animal evolution. Was it possible that the emergence of mankind could have been triggered either by the opening up of a new habitat or by pressure created when the ancient forests were reduced in size? The most likely source of such a pressure seemed to be the opening up of the plains of central Asia as a result of the uplift that created the Himalayan mountain range. Along with other considerations derived from the principles of biogeography, this source gave rise to the widely popular theory of a central Asian origin for mankind. Many of this theory's strongest advocates managed to retain a role for initiative, however, by emphasizing the mental qualities of those apes which responded to the environmental challenge in a creative way.

The possibility of an environmental stimulus had been hinted at by the French anthropologist L. Manouvrier in his discussion of the newly discovered *Pithecanthropus*. Accepting Dubois's claim that here was a small-brained hominid that already stood upright, Manouvrier suggested that the transition to ground dwelling may have been prompted by the destruction of the forests in Southeast Asia, perhaps by volcanic activity.[68] No one else took such a localized effect seriously, and the chief line of support for the environmentalist approach came from those naturalists and paleontologists who looked to central Asia as the cradle of mankind. Two strands of thought interacted to create this theory. One stressed the generally stimulating effect of northern regions on the evolution of all mammals, including the primates. The other looked to a distinct worsening of the climate in central Asia during recent geological epochs because of the uplift of the Himalayas. Although often mixed together, these two approaches are actually rather different, one placing the center of evolution permanently in the more stimulating environment of the North, the other implying a critical episode in which a group of ancestral apes was trapped by the rising mountains and forced to abandon the forests in which they had evolved.

The theory of a northern or "Holarctic" center of mammalian evolution was popularized by the Canadian-born paleontologist William Diller Matthew, in his "Climate and Evolution" of 1914.[69] Matthew argued that many of the inferences drawn from the geographical distribution of organisms were false, because they had been based on the assumption that the location in which a primitive form was found represented the center of evolution from which more advanced types had radiated outward. In fact, he argued, the advanced forms are to be found at the true center of dispersal, their less developed ancestors (if they survived at all) having been pushed out to marginal areas.[70] This point led many anthropologists to decide that *Pithecanthropus* was just such a relic, driven into Southeast Asia from a more northerly center of human evolution. Matthew himself applied a similar argument to the modern human

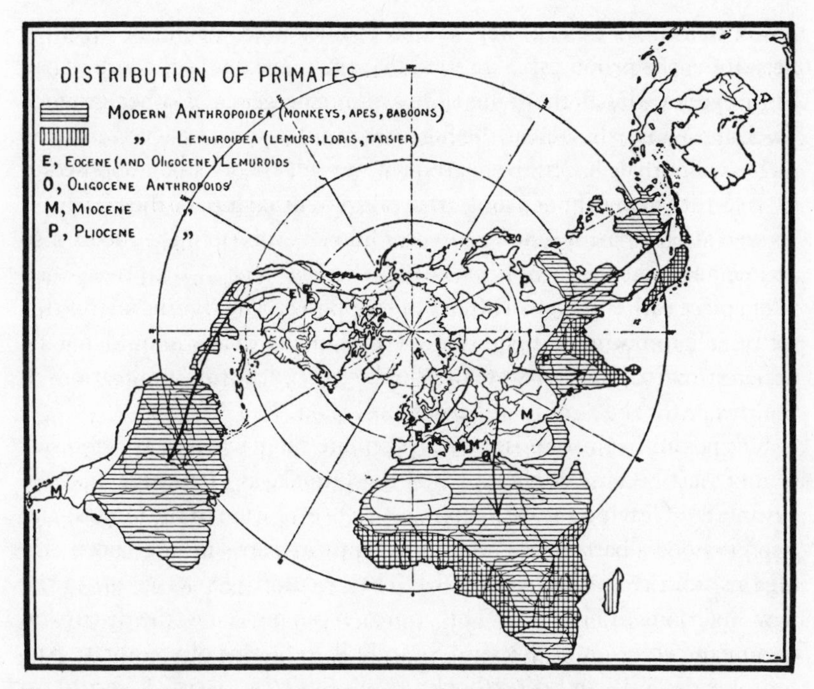

Figure 15. Map showing the dispersal of the modern primates from a hypothetical Holarctic center. From W. D. Matthew, *Climate and Evolution* (New York: New York Academy of Sciences, 1939), p. 46.

races, with the Caucasian race occupying the center of evolution and dispersal in northern Asia.[71] But his main point concerned the general process of mammalian evolution throughout the Tertiary. He argued that in the course of the earth's history the planet's overall climate had alternated between periods when warm, moist conditions were widespread and intervals in which a harsher climate had overtaken many northern regions. The last warm spell ended in the Eocene, and since then the ever drier and colder climate of the Holarctic region had been the center for all mammalian evolution (see fig. 15).[72] The remaining areas of tropical forest in the South had become the refuge of the earlier and more primitive types.

Matthew's theory certainly pointed a finger at central Asia, and his assumption of progressively worsening conditions was just what was needed as the stimulus that might have driven our ancestors from the trees to the ground. But it was a theory of general mammalian evolution, and could hardly apply to the relatively recent transition that was supposed to be the crucial step in the evolution of mankind. Applied literally to the primates, it implied that the move out of the forests had begun very early, at the same time as the emergence of other ground-dwelling mammals such as the horse family. Matthew believed that the earliest mammals had all been arboreal, but his theory stood opposed to Elliot Smith's concept of a long arboreal apprenticeship for the primates. As well as postulating a new center for human evolution, the theory also carried an immediate implication that the key steps in the process had taken place earlier than had often been supposed. The significance of this factor in promoting support for the presapiens theory should not be underestimated, but few were prepared to accept the full implications of the theory for the timing of human evolution.

One possible compromise was to postulate a somewhat later deforestation of Asia because of the uplift of the Himalayas. This view had the advantage of leaving the earlier phases of ape evolution in the forests, and also provided a barrier that would prevent those apes in more northerly regions from retreating southward. They would have to face up to the new environment or perish. This approach can be seen in the discussion of human evolution by Richard Swan Lull, professor of vertebrate paleontology at Yale, in his textbook *Organic Evolution,* 1917. Referring to Matthew's theory, Lull designated continental elevation as the "impelling cause" of human evolution, and pointed to the drying up of central Asia in Miocene and late Pliocene times as a force that would "compel the descent of the prehuman ancestor from the trees, a step which was absolutely essential to further human development."[73] Lull drew a comparison with the evolution of the first terrestrial vertebrates. The primitive amphibians had been forced to move across dry land as the lakes and ponds of the Devonian had gradually dried up. In the same way, the ancestors of mankind had been forced to migrate from one dwindling patch of forest to another, until eventually they were able to live permanently on the plains. An almost identical account of the process was given by the geologist Joseph Barrell.[74] Although Lull was not above calling in orthogenesis when the fossil evidence seemed to require it, he ended his book with a firm statement of the principle that all great developments in the evolution of life have been stimulated by climatic changes.[75]

One of the leading advocates of the central Asia theory in the 1920s was Henry Fairfield Osborn. As early as 1900, Osborn had used the similarity of fossil faunas in Europe and western North America to predict that Asia would turn out to be the center of mammalian evolution. This idea lay dormant until 1921, when Roy Chapman Andrews led the first of a series of Central Asiatic Expeditions to Mongolia. Over the next few years, these expeditions made fossil discoveries that seemed to confirm Osborn's predictions, although no fossil hominids were found.[76] During the mid-1920s, Osborn's own views on human evolution changed dramatically, as he began to argue for an ancient line of hominids entirely separate from the apes. In a popular article written to support the central Asia theory in 1926, that theory was already beginning to urge him toward a new position on the antiquity of mankind. He was now using the term "dawn man" rather than "ape man," and suggesting that the plains of central Asia had provided an ideal climate to foster the emergence of the human form much earlier in the Tertiary than was normally supposed.

Osborn made extensive use of Matthew's theory to back up his claims, linking the origin of mankind to the view that Asia had long served as the center of mammalian evolution.[77] Throughout the Tertiary, he argued, the uplands of central Asia had provided the stimulating environment needed for the evolution of the dawn man, whereas lowland Asia supported the forests in which the apes had retained their arboreal lifestyle.

> This high plateau country of central Asia was partly open, partly forested, partly well-watered, partly arid and semi-desert. Game was plentiful and plant food scarce. The struggle for existence was severe and evoked all the inventive and resourceful faculties of man and encouraged him to the fashioning and use first of wooden and then of stone weapons for the chase. It compelled the Dawn Men—as we now prefer to call our ancestors of the Dawn Stone Age—to develop strength of limb to make long journeys on foot, strength of lungs for running, and quick vision and stealth for the chase. . . . [The result was that] while the anthropoid apes were luxuriating in the forested lowlands of Asia and Europe, the Dawn Men were evolving in the invigorating atmosphere of the relatively dry uplands.[78]

Osborn accepted Matthew's view that human evolution was coordinated with the general trend of mammalian evolution. The drying-out of central Asia had occurred as early as the Oligocene, forcing all the mammals of that time to face the choice of adapting to the new conditions or moving south to remain with the forests.[79] Most orders of mammals had contained some more progressive types that faced up to the challenge of the new environment. The dawn men were simply the primate component of this general episode in mammalian evolution. Although he was disgusted by the supposed link between mankind and the apes, Osborn was quite willing to see parallels between human and animal evolution, as long as those parallels were applied to lines that had developed separately since the early Tertiary.

There was an element of environmental compulsion here, since the mammals, including the primates, had been forced either to adapt to the new, more arid conditions or to give up a major part of the earth's surface. Yet they had faced a "choice" of whether to adapt or migrate south, and this option allowed Osborn to modify the implications of his theory by emphasizing the mental superiority of those primates which had responded in a positive way. The dawn men were derived from the alert, progressive types which could sense the opportunities of the new environment, whereas the more sluggish ancestors of the apes had chosen to stay in the forests. The earliest humans were more intelligent, not just because they were evolving in a more stimulating environment, but because they were derived from the best elements of the primate stock, creamed off at a very early stage in the process. In the 1927 paper announcing his final rejection of an ape ancestry for mankind, Osborn stressed the psychological differences that had always distinguished the ape and human stocks.[80]

In *Man Rises to Parnassus,* Osborn adopted a slightly different position. He now insisted that the arboreal phase in human evolution had never been profound or extensive, implying that some of the primates had always been at least partly terrestrial.[81] This view undermined the whole logic of the "expulsion from the trees" scenario, since the earliest ground-dwellers would have been natural candidates for exploring the newly opened plains. The upright posture was clearly very ancient, but it could hardly have been derived from brachiation, and Osborn was thus left with the problem of explaining how the dawn men had begun to walk upright. He tried to minimize the significance of this question by insisting that postural adaptations could be acquired quite easily in response to new habits. Osborn is best known for his theory of orthogenetic evolution, and he was certainly prepared to invoke parallelism to explain the similarities between the apes and mankind. But the switch to an upright posture required a dramatic change in the direction of evolution, not rigid linearity, forcing him to fall back on a hypothesis originally developed at a much earlier stage in his career: organic selection.

In the late nineteenth century, Lamarckians such as Samuel Butler had stressed that the inheritance of acquired characters would allow the rapid response of bodily structure to newly acquired habits. Since this view would allow the animals' initiative to direct their behavior, and hence their evolution, the Lamarckians urged their theory as an antidote to the trial-and-error materialism of natural selection. Osborn himself had at first been tempted by Lamarckism, and although he soon abandoned it, he seems to have felt the need to retain a role for habit as the driving force of evolution. Organic selection was a mechanism introduced independently by Osborn and by James Mark Baldwin in the 1890s to allow a role for initiative in evolution *without* Lamarckism.[82] As Osborn explained in *Man Rises to Parnassus,* the mechanism worked if one assumed that the individual's body could rapidly adapt itself to a new posture required by new habits.[83] Such acquired characters were not inherited directly, but would allow individuals to adapt temporarily until equivalent genetic variations came along to fix the posture in the stock. Osborn cited postural differences among the modern races to illustrate this effect. Although he did not explain why our earliest ancestors had acquired the new habit of walking upright, the implication seems to be that organic selection would have enabled the transition to be effected quite easily.

Osborn's willingness to resurrect an idea first proposed long ago shows how far he was prepared to go in promoting his belief that our ancestors' superior intitiative allowed them to respond to any challenge. The fact that organic selection was originally introduced to complement the Lamarckians' emphasis on initiative as the driving force of evolution is evidence that there were parallels between the debates on human evolution and on general evolution theory. Although the paleoanthropologists who stressed the role of initiative in human evolution seldom directly referred to use-inheritance, their attitude reflected a view of nature that had once been an integral component of Lamarckism.

The Central Asiatic Expeditions gained wide publicity and achieved notable results in archaeology, but they did not locate the hominid fossils that Osborn had predicted. It was Davidson Black's discovery of Peking man that for a time seemed to highlight the central Asia theory, although the dating of *Sinanthropus* was far too late for Osborn's new time scale. Black, a young Canadian who had trained with Elliot Smith, went to China because he was convinced that Asia was the cradle of mankind. In 1925, just before the first hominid fossils were located at Chou Kou Tien, he published a paper entitled "Asia and the Dispersal of Primates," giving his own views on human evolution. Black was strongly influenced by Matthew, and accepted the idea that primitive forms would be pushed into remote regions as more advanced types appeared in the main center of evolution.[84] *Pithecanthropus* and the European Neanderthals were primitive forms of early mankind that had been expelled in opposite directions from the Asian center where higher types were evolving.[85] Black accepted the view that the rise of the Himalayas in Oligocene and Miocene times had dried out the central Asian region, and that the resulting disappearance of the forests had coincided with the differentiation of the ape and human stocks. Because of his commitment to Matthew's theory, Black saw northern Asia as the original home of the primates. It was the ancestors of the apes that had migrated southward in an effort to retain their original habitat. The ancestors of mankind had simply remained in the main center of evolution, where they produced a series of ever-higher types in response to the stimulus of the new conditions.

Black was convinced that from the very beginning the ancestors of mankind were distinct from those of the apes. One group of early primates was of a conservative, early-maturing type. They were

> unable or unwilling to modify their mode of living to suit the necessities of an altering environment. They therefore became faced with the alternative of migration or extinction. To this group belong the ancestors of the modern great anthropoids, whose migrations in their attempt to retain their environment and escape the pressure of competition with more progressive types, has determined the present distribution of the recent members of the Simiidae.

The progressive group, by contrast, had a long childhood which allowed the development of a larger brain.

> By reason of the resulting increase in mental capacity, the individuals of this group were both able and willing to alter their mode of life to suit progressive environmental conditions. To this group belonged our direct protohuman and human ancestors, who in turn were overwhelmed and forced to migrate elsewhere by the constant pressure of competition with more progressive generations within the broad extent of their original dispersal centre.[86]

Once again, the idea of an environmental compulsion had been converted into a form of brain-led progressionism by an appeal to the superior initiative of those individuals who responded to the challenge. As a student of Elliot Smith, Black probably believed that the earliest primates had developed a large brain in the trees, but it was the variation of intelligence among them that had decided which would continue to progress and which would stagnate.

A number of influential paleontologists endorsed the central Asia theory with varying degrees of enthusiasm. W. K. Gregory, although strongly opposed to Osborn's extension of human antiquity, accepted the general idea of an Asian ancestry in his *Origin and Evolution of Human Dentition* of 1922. He believed that the precursors of mankind were Miocene apes that had already taken to the ground, but cautioned against the assumption that their homeland had been geographically extensive. The lack of Pliocene hominid fossils showed that they had not ranged widely over the plains, implying that "in some restricted and more or less isolated Palaearctic region they were in the course of differentiation from ground-living apes inhabiting the border regions between forests and plains."[87] Gregory seems to have evaded the question of whether our ancestors were driven out of the forests or left of their own accord.[88] The sentence quoted above hints at a less dramatic interpretation, in which a ground ape began to exploit an ecological opportunity offered by its local environment, gradually moving farther out onto the plains.

Although a loyal supporter of Piltdown man, Arthur Smith Woodward eventually conceded that the transition to the ground may have taken place in Asia.[89] In general, though, he stressed the trend toward primate brain development, not the achievement of bipedalism, as the chief cause of human evolution. Marcellin Boule also expressed little interest in the development of an upright posture, but favored Asia as the cradle of mankind on the basis of the distribution of fossil primates.[90] Boule was careful to qualify his support by noting that one could not altogether rule out the as-yet unexplored continent of Africa.

It was, of course, in South Africa that Raymond Dart, another student of Elliot Smith, discovered the Taungs skull in 1924 (see fig. 16). Dart did not consider himself to be a paleontologist, and there is little to suggest that he had deliberately chosen to go to South Africa because he hoped to confirm the African origin of mankind. He soon appreciated, however, that the *Australopithecus* skull might represent the elusive missing link. Dart's original description emphasized that *Australopithecus* was an upright biped, whose hands "were already assuming a higher evolutionary role not only as delicate tactual, examining organs which were adding copiously to the animal's knowledge of its physical environment, but also as instruments of the growing intelligence in carrying out more

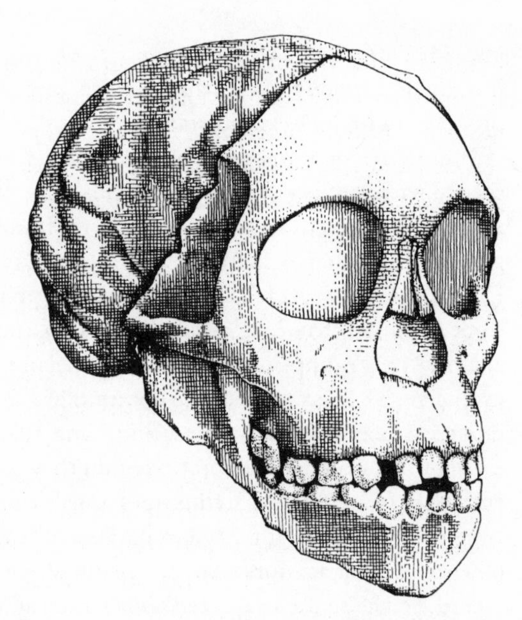

Figure 16. Sketch of the endocranial cast and face of the Taungs specimen of *Australopithecus* discovered by Raymond Dart. From Grafton Elliot Smith, *The Evolution of Man,* 2d ed. (Oxford: Oxford University Press, 1927), p. 11.

elaborate, purposeful movements, and as organs of offence and defence."[91] He implied that *Australopithecus* was already using primitive tools and thus had no need to develop the fighting canine teeth of the modern apes.

Dart had evidently anticipated Osborn's views on the importance of the open plains as the only environment that could stimulate the growth of human intelligence. He believed:

> There has been a tendency to overlook the fact that, in the luxuriant forests of the tropical belts, Nature was supplying with profligate and lavish hand an easy and sluggish solution, by adaptive specialization, of the problem of existence in creatures so well equipped mentally as living anthropoids are. For the production of man a different apprenticeship was needed to sharpen the wits and quicken the higher manifestations of intellect—a more open veldt country where competition was keener between swiftness and stealth, and where adroitness of thinking and movement played a preponderating role in the preservation of the species.[92]

There was no evidence that southern Africa had ever been a jungle, and the Australopithecines were two thousand miles from their nearest ape cousins. The inference to be drawn was that our ancestors had actively ventured out into an entirely new environment rather than merely holding their ground as the environment changed around them. There was no hint of external compulsion here, only the deliberate exploitation of a new way of life by a generalized ape that had more intelligence than its stay-at-home relatives. While admitting that the Australopithecines did not have a much larger brain than an equivalent-sized ape, Dart insisted that their mentality was sufficiently more advanced to make all the difference in the way they responded to their surroundings: "Their eyes saw, their ears heard, and their hands handled objects with greater meaning and to fuller purpose then the corresponding organs in recent apes."[93] Like Elliot Smith, Dart thus saw brain development as the key to the change of habitat that stimulated the last phase of human evolution.

Although some anthropologists, led by Keith, saw little of interest in Dart's discovery, it would be easy to overestimate the hostility of the overall reaction. Elliot Smith included a substantial discussion of the new fossil in the second edition of his *Evolution of Man* in 1927. Smith had always supported Darwin's original prediction of an African location for human evolution, and he now proclaimed *Australopithecus* as confirmation of it.[94] Here was an ape that already showed signs of the mental development that would lead toward mankind. As Dart's former teacher, he fully endorsed the skill with which the endocranial cast had been revealed to show the slightly enlarged brain.[95] Although he doubted Dart's claim that the creature had walked upright, he admitted that it could correspond to the earliest phase of the move out of the trees so crucial for further brain development.[96] Clearly, Smith had now faced up to the implications of his view that the ancestral form was only a generalized ape, and had realized that there was no need for the earliest members of the human stock to have had the enlarged teeth of the modern apes.

Dart's interpretation was also supported by W. J. Sollas in a brief note to *Nature* in 1925.[97] In the following year, Sollas published a more detailed account of the Taungs skull, again confirming that it had characters approaching those of the Hominidae. In the discussion following this paper, D.M.S. Watson supported its conclusions, and Sollas added a claim to the effect that in his *Ancient Hunters* he had predicted that the first step in human evolution was the emergence of an early primate onto the plains as a hunter.[98] In his book, Sollas had certainly emphasized that the earliest humans were already hunters, but he was perhaps going too far in saying that he had anticipated a form like *Australopithecus*. Still, the fact that he was prepared to adopt Dart's discovery in this way shows how easily it could be fitted into current views on human evolution, now that Piltdown was falling into the background. One cannot help feeling that if the fossil had been found in Asia, Osborn too would have hailed it as confirmation of his own views on the importance of a terrestrial habitat.

At a more popular level, the importance of *Australopithecus* was also emphasized in the 1925 edition of H. G. Wells's *Outline of History.*[99] Some years later, Wells's far more authoritative *The Science of Life* (written in collaboration with Julian Huxley) included a substantial discussion of the transition to a terrestrial life. There was no suggestion that increased intelligence led the way to bipedalism. Instead, they emphasized that a change of habitat is normally made only under pressure of necessity. If this rule applied in the case of human evolution, the most likely explanation was that our ancestors were trapped in central Asia by the rising Himalayas and were thus forced to adapt to a deforested environment. In these circumstances there would be intense selection favoring those individuals who were capable of changing their way of life.[100]

The authors of *The Science of Life* conceded that another explanation was possible. Perhaps our ancestors were already more terrestrial than those of the apes, inhabiting open glades and forest margins, and therefore they might have pursued an increasingly terrestrial lifestyle as a typical evolutionary specialization. A similar idea had already been suggested by Gerrit S. Miller, who had objected to the "destruction of the forests" scenario by pointing out that drastic changes in the environment all too often lead to extinction. Far more important was the "local adaptive radiation" that led individuals of some species to exploit slightly different habitats in the neighborhood of their original home.[101] By such a process, some of the earlier, less specialized apes might have been led to adopt the habit of walking upright to explore the open spaces near the trees. This view was similar to the approach we have already seen hinted at by W. K. Gregory, although Miller thought that the achievement of an upright posture was a much slower process. Gregory at first favored an Asian origin for mankind, but Wells and Huxley pointed out that this theory, by eliminating the element of climatic compulsion, would make Africa an equally plausible candidate. This option in turn would bring *Australopithecus* back into the limelight, a point that Gregory himself would eventually come to appreciate.

By the time he collaborated with Wells on *The Science of Life,* Huxley was already moving toward the revised Darwinism that would eventually form part of the Modern Synthesis in evolution theory. The concept of local adaptive radiation fitted in very well with this approach to speciation, and allowed the first step in the divergence of humans from apes to be explained within a Darwinian framework. Instead of being the product of an overall trend toward brain development, or a blinding flash of adventurous inspiration, the transition to bipedalism became merely a new adaptation, exploiting an ecological opportunity offered by the local environment. In the 1940s, the majority of paleoanthropologists came to accept the bipedalism of *Australopithecus* as an adaptive step, achieved without the benefit of a bigger brain. *The Science of Life* thus strikes a refreshingly modern note.

The acceptance of *Australopithecus* as a possible human ancestor coincided with the discovery of new fossil evidence in the late 1930s, confirming that this type had been small-brained yet fully upright. These fossils had been discovered by Robert Broom, who had been an early convert to Dart's interpretation of the Taungs specimen. Yet Broom himself was very much a product of the older approach to evolution theory: he accepted the significance of a trend toward brain development, and adopted an explicitly teleological explanation of this and other evolutionary trends. Broom's opinions confirm that the original *Australopithecus* discovery was not incompatible with the progressionist interpretation of human origins. Although the later Australopithecine discoveries would often be interpreted as evidence for the priority of bipedalism, the Taungs fossil could equally well be seen an an important step in the final phase of brain expansion. Broom himself found no difficulty in linking even his later discoveries with his teleological view of evolution. This fact suggests that these later fossils did not play a crucial role in the destruction of the old progressionism. It was the theoretical revolution in which Huxley participated—not the fossils—that finally convinced most biologists that an explanation of human evolution based on brain expansion was unsatisfactory. A Darwinian view of the evolutionary mechanism required the specification of an adaptive novelty at the start of the hominid line, and the early adoption of an upright posture by the Australopithecines was soon seen as an ideal candidate for this role in the story of human origins. Broom's new Australopithecines provided useful evidence for this argument, but without the theoretical revolution they would probably have been integrated into the traditional picture along with the Taungs fossil.

Chapter Eight Trends in Human Evolution

Although most authorities agreed that the transition to bipedalism initiated a new phase in human evolution, many saw this transition as the culmination of an earlier trend toward the enlargement of the primate brain. Our ancestors inherited their upright posture and their agile hands from a much earlier stage in primate evolution, and their decision to walk upright was taken deliberately, as soon as their intelligence was able to recognize the advantages of moving out into the open. Although their higher intelligence distinguished them from their ape cousins, it was the end product of a longstanding trend. Elliot Smith in particular stressed the importance of this trend, portraying it as the central theme in primate evolution, and dismissed the side branches as failures which had allowed themselves to be sidetracked by excessive anatomical specialization.

Smith seems to have been unable to free himself completely from the influence of an earlier view of evolution, in which the human form marked out a goal toward which all progress was aimed. Yet the trend that he postulated could, to some extent, be explained in naturalistic terms. Increased intelligence was a product of the interaction between the primitive primate anatomy and the stimulating environment of the trees, in which eye, hand, and brain cooperated in a harmonious way to increase both dexterity and intelligence. In theory, all primates could participate, and all had indeed progressed at least some way along the scale of mental development until sidetracked by physical specialization. The assumption of a common trend toward brain growth laid the foundations of several of the controversies that racked the world of early twentieth-century paleoanthropology. If many different forms could participate in the same trend, because they were exposed to the same environment, was it not possible that anatomical similarities usually attributed to common descent might have been independently acquired in several different lines of primate evolution? This was precisely the point made by Wood Jones in his tarsioid theory of human origins. Convergence, because of evolution producing the same results in similar environments, explained the resemblances between mankind and the apes. A similar point was made by Klaatsch and others to support a polytypic theory of human origins: several different apes had independently recognized the advantages of becoming human. Once the concept of a trend driving evolution in a particular direction had been introduced into the discussion, there were always a few anatomists and anthropologists who were prepared to allow the trend so much power that it would seriously compromise what the orthodox regarded as the most rational interpretation of the evidence for human evolution.

Both the tarsioid and polytypic theories violated the Darwinian principle that evolution has been an essentially haphazard process of branching and divergence. But so did Smith's notion of a main stem of evolution aimed at mankind. The heresies merely brought out into the open an element of progressionism inherent in the orthodox picture. Jones made it clear that the similarities produced by convergence indicated a powerful role for habit as the guiding force of evolution, and openly proclaimed his preference for a Lamarckian over a Darwinian mechanism. There was no shortage of paleontologists who would confirm that the fossil record appeared to reveal linear trends in the evolution of many groups of animals. The existence of such trends lent credibility to both the heretical views of human evolution and their anti-Darwinian implications. It was a long-established tradition to regard evolutionary trends as evidence for non-Darwinian mechanisms such as Lamarckism and orthogenesis.[1] But in one sense, the anthropologists who appealed to the paleontologists for support got more than they bargained for. Many of the trends thought to exist elsewhere in the animal kingdom were nonadaptive—supposedly the product of internal, orthogenetic forces that drove evolution in a particular direction without regard for the well-being of the organisms concerned. In extreme cases, such trends generated overdeveloped structures such as the antlers of the Irish elk, leading to racial senility and extinction. Was it possible that some of the trends in human evolution might display this alarming character?

Many preferred not to confront this disturbing prospect, and even Henry Fairfield Osborn, a leading exponent of orthogenesis in mammalian evolution, refused to apply it to the origin of mankind. A few bold spirits did explore the possibility, however, and were even prepared to treat the growth of the brain itself as the result of such a nonadaptive trend. Not surprisingly, though, the harsher aspects of orthogenesis tended to be ignored in this context, and the trends were given an almost teleological interpretation. Those paleontologists who attributed the trends to an inbuilt tendency for living things to vary in particular directions adopted an almost preformationist outlook. Others explored the possibility that the purposeful activity of the living body itself would somehow be able to make use of nonadaptive variations forced upon it by the germ plasm.

These issues raise once again the vexing question of the relationship between convergence and parallelism. Although these terms were sometimes used interchangably, we continue to use "convergence" solely in the context of adaptive changes.[2] In such cases, similar structures are produced independently in different lines of evolution solely because both groups of organisms have been exposed to a similar environment and have adopted a similar lifestyle. By solving the same problem in the same way, they have been compelled to acquire the same structural modifications. Darwin admitted the possibility of a limited amount of convergence, but always denied the ability of this effect to produce a fundamental identity of structure in organisms that were originally quite different. In order to allow similar habits to generate indentical structures from different starting points, the supporters of convergence frequently appealed to Lamarckian use-inheritance, which was thought to bring about more consistent adaptive modifications than natural selection.

"Parallelism," on the other hand, was the term more frequently used to denote the phenomenon by which independent lines of evolution followed identical or similar courses when no adaptive benefit was conferred by the changes. In effect, parallelism was the product of orthogenesis, of a tendency—perhaps inherited from a remote common ancestor—that pushed variation consistently in a particular direction whether or not the results were beneficial. Osborn believed that there was often an adaptive phase of the orthogenetic trend, but he could never explain how this correlation with the environment came about, and he agreed that in the end such trends normally went beyond the point of utility and contributed to extinction. Other paleontologists saw the trends as totally nonadaptive, with the mechanism controlling variation being totally out of touch with the environment, so that the trends almost inevitably led to the overdevelopment of structures, racial senility, and extinction. If an exception to this rule was to be made in the case of the trend toward primate brain growth, it would create an almost explicitly teleological form of orthogenesis.

The preceding chapter showed that many paleoanthropologists invoked the transition from the trees to the plains as an important episode in the story of human origins. This aspect of their theories is readily susceptible to the analysis in terms of narrative structure favored by historians of anthropology such as Misia Landau. Yet several of the theories contained elements that did not fit easily into this pattern. For Elliot Smith, the transition to the plains was the culmination of a longstanding evolutionary trend that had produced a steady, one might almost say predictable, increase in brain size. To the extent that he was prepared to construct a historical account of human origins, in which our ancestors' changed habits or environment helped to separate them from the lower primates, his theory can be analyzed as a narrative. But to the extent that human characteristics were seen as the inevitable product of a trend that was central to all primate evolution, the narrative pattern is not in evidence. A story, like a historical explanation, depends upon the identification of unique circumstances that shaped the course of events in an unpredictable manner. If the present state of affairs can be deduced or predicted from a general law of development, then we have not a historical narrative, but a genetic or deterministic explanation of its origins. Elliot Smith himself certainly presented the trend toward brain growth as the main driving force of primate evolution, and this aspect of his theory was deterministic rather than historical. A significant proportion of modern human character was produced by a purposeful evolutionary process, not by a unique event separating us from the rest of the primates. To analyze Smith's theory in terms of narrative structure alone, even with the evolutionary trend playing a role in the narrative, is to present a one-sided image of a theory in which the narrative and deterministic modes of explanation were interwoven in equal parts.

The deterministic style of explanation becomes even more strongly apparent in some of the theories discussed below, since the evolutionary trends take on an even more extensive directing role. Although Wood Jones accepted that a transition to ground-living had occurred in human evolution, he was not very interested in this late stage of the process. His real concern was to establish that the human and ape stocks had been separated since the earliest phase of primate evolution. Because they shared an arboreal environment, both lines of evolution had subsequently been affected by the same trend, leading to the duplication of many features including a comparatively large brain. For those who adopted an explicitly orthogenetic view of the trends in primate evolution, the development of the brain had gone on inexorably, whatever the environment, driven by an inbuilt biological force which could not be deflected by external factors. This was particularly so for cases in which a number of lines were thought to have reached the fully human status independently, as in the polytypic theories. In these cases, the deterministic style of explanation has taken over almost completely from the narrative. The most important question to be answered in a study of such theories concerns the nature of the evolutionary mechanisms thought to be capable of producing the rigid trends.

Convergence and Lamarckism

We have seen two sources of inspiration for the belief that convergence has played an important part in primate evolution. One was the attempt to separate the line of human ancestry from that of the apes; it required a strong element of convergence to explain how the two lines of evolution had developed striking similarities in many aspects of their anatomy. Wood Jones's tarsioid theory was the best-known product of this movement, and Jones openly appealed to Lamarckism in support of his claims. The other source of inspiration was the ever-present heresy of polygenism, which required an explanation of how a number of separate lines of ape evolution had been able to move independently toward a human form. Early polygenists such as Vogt and Klaatsch were never very clear on the question of the evolutionary mechanism involved, but seem to have favored convergence rather than orthogenesis.

Wood Jones's theory of parallel evolution in apes and humans owed its origins in part to Elliot Smith's views on evolutionary trends. As we saw in the previous chapter, Smith appealed to the stimulating character of the arboreal environment to explain the constant tendency of the primate brain to increase in size. The claim that progressive evolution occurs because of the individual organism's creative response to the challenge of its environment had been a standard theme of late nineteenth-century Lamarckism. Many paleontologists had appealed to use-inheritance to explain what appeared to be linear trends of specialization in the fossil record. In effect, Smith followed this model in his theory of brain development. Without ever stating that the extra intelligence acquired by the individual was inherited by its offspring, he managed to imply that the stimulus of the environment acting consistently on individuals over many generations was responsible for the trend. He had thus set up a functional process that would increase primate brain size throughout the period of arboreal life—and would continue in the yet more stimulating environment of the plains once the human stock had become intelligent enough to make the transition.

Smith's real interest was the effect of the trend on the main stem of primate evolution leading toward mankind. He accepted that there had been some parallel evolution in the various branches of the primates, but he did not seek to explain the parallelism as the result of the same functional trend affecting all the branches. For him, once the monkeys and apes had been sidetracked from the main trend by physical specialization, they ceased to be of interest as far as brain development was concerned. Wood Jones took up Smith's interest in the environment as the stimulus creating an evolutionary trend and generalized it to give a mechanism of parallel evolution based explicitly on Lamarckism. He believed that the trend affected brain growth *and* physical adaptation, and would affect all lines of primate evolution exposed to the same environment. Thus even the side branches would continue to advance in a manner paralleling the development of the human stock, and the similarities between humans and apes might be independently acquired in stocks that had been separate for a vast period.

In his *Arboreal Man,* Jones supported Smith's views on the importance of retaining a primitive, and hence more flexible structure, and on the stimulating character of arboreal life for the development of the primate brain (see chapter 5). When he began to argue for the complete separation of the ape and human stocks, he said little at first about the evolutionary mechanisms involved. His "The Origin of Man," first published in 1918, extended Smith's views on the perils of overspecialization by dismissing the apes altogether as a degenerate side branch, whereas the human line was supposed to have developed directly from the primitive tarsioids. Nevertheless, Jones had to accept that in the course of their long evolution in the trees mankind and the apes had acquired similar characters which had misled many anatomists into believing that they were closely related. He dismissed the similarities as "superficial" adaptive convergences.[3] They had only been taken so seriously because the Darwinists had refused to admit the power of convergence to produce similar results in different stocks that had adopted the same habits.

Jones's chief purpose was to emphasize the longstanding gulf between humans and apes, not to explain why the two branches had originally separated. The unequal potential of the two lines was not a result of the factors that had caused that separation. Change of environment was not involved, since both had developed in the trees and had benefited—if unequally—from this stimulating environment. Even in their brains, the apes had evolved at least partly in the same direction as ourselves. Jones believed that the apes had been held back because at an early point in their history they had overspecialized for life in the trees. Nevertheless, he was forced to spend most of his time discussing the characters that apes and humans have in common, since his chief concern was to show that these similarities could have been produced independently. The final acceleration of brain growth no doubt occurred when the early humans descended from the trees, but Jones had no incentive to present this step as the key to the separation between humans and apes. In his most detailed defense of his theory, he briefly raised the question of *why* our ancestors had made this move to a new environment, but professed himself unable to answer it.[4] He may have preferred Elliot Smith's idea of deliberate adventure out onto the plains, but Jones's emphasis was firmly on the high antiquity of the separation between the human and the less adventurous ape stocks.

Jones's *Man's Place among the Mammals* of 1929 not only contained a more extensive defense of his thesis, but also included a discussion of the evolutionary mechanisms that were thought to be involved. In chapter 6, Jones argued explicitly for the transmission of acquired characters. He conceded that there was no point in reviving the tragic controversy surrounding Paul Kammerer, the Austrian naturalist at the center of the "case of the midwife toad."[5] Kammerer had tried to provide experimental evidence for the inheritance of acquired characters, and the exposure of a possible fraud in his work had greatly encouraged the geneticists in their campaign against Lamarckism. But Jones insisted that the indirect evidence for the effect from paleontology and other areas was so great that the geneticists' dogmatic rejection could be regarded as unscientific. This was a common attitude among paleontologists during the period before the Modern Synthesis drew the various fields of biology together again in the 1940s. Jones also made it clear why the Lamarckian mechanism was so important for his thesis. The inheritance of acquired characters would allow a more plastic conception of adaptive change than would Darwinism, thereby explaining how the effects of habit could influence the whole anatomy of the organism.[6]

In the following two chapters Jones discussed the phenomenon of convergence in more detail. He again argued that the correlation of all the parts involved in an adaptive modification required something more subtle than the selection of chance variations. There must be a harmonious interaction of the whole organism with the environment. In these circumstances, similar habits could affect the fundamental structures of originally different organisms, producing a striking similarity in their basic anatomies. The resemblances between many of the Australian marsupials and the equivalent placentals was an illustration of this effect.[7] It was necessary for the comparative anatomist to eliminate all the adaptive convergences in order to recognize the slight but crucial differences that indicated the separate heritage of the similar forms. Jones also cited numerous examples of parallel evolution drawn from paleontology to support his claim that the phenomenon was widespread thoughout the history of life. Some of these trends were nonadaptive, however, suggesting that in his anxiety to combat Darwinism Jones was prepared to seize upon evidence that did not altogether fit in with his own commitment to adaptive convergence.[8]

Jones returned to the defense of Lamarckism in his *Habit and Heritage* of 1943. In the previous year, his *Design and Purpose* had expounded a holistic view of the organism and a teleological view of evolution. Finally, in 1953, Jones summed up his philosophy in his *Trends of Life.* Here he openly supported vitalism as a foundation for his teleological philosophy. He insisted that purpose was to be found not only in the intentions of an external Designer of nature, but also in the activities of living organisms themselves, which directed the evolution of species through the influence of habit on structure.[9] This was exactly the line of argument used by Samuel Butler and the earlier generation of Lamarckians to uphold the moral value of their theory as a bulwark against Darwinian materialism. Jones's later writings suggest that his view of evolution rested on a foundation built from the anti-Darwinian attitudes of the late nineteenth and very early twentieth centuries. Needless to say, such opinions had long gone out of fashion by the times Jones came to develop them explicitly in the 1940s and 1950s.

The only other paleoanthropologist to offer such an active critique of Darwinism was Robert Broom. In his *The Coming of Man: Was It Accident or Design?* of 1933, Broom argued forcefully against the belief that the natural selection of fortuitous variations could produce the many purposeful trends observable in the history of life. He dismissed genetic mutations as the products of unsanitary laboratory conditions, of no value in evolution.[10] Citing the American neo-Lamarckian E. D. Cope as "the most philosophical of the palaeontologists of the nineteenth century," he declared: "Though I have for many years been much more inclined to favour Lamarckism than Darwinism, I have never had much hope that the matter could be proved by experiment. Evolution, as the palaeontologist sees it, goes on so slowly that he hardly expects to see any appreciable change in his lifetime."[11] This was a standard argument used by most of the Lamarckian paleontologists, although Broom also mentioned examples of nonadaptive, orthogenetic trends.[12] In the case of human evolution, he accepted that use-inheritance would allow increased activity to stimulate the evolution of the brain,[13] and stressed the importance of the Lamarckians' insistence on the role of mental powers in the direction of evolution.[14] There must be something in Lamarckism, he claimed, although it was not the whole answer, and in the end it would be necessary to call in a supernatural guiding agency. Although Broom himself did not include these ideas in the 1946 survey of his new Australopithecine discoveries, his coauthor, G.W.H. Schepers, openly hinted at his preference for "the explosive hypothesis of the inheritance of acquired characters."[15] Broom repeated his own views in a more popular format as late as 1950.[16]

Because of his work on the Australopithecines, Broom is often assumed to have been a leading supporter of the theory of an ape origin for mankind. It is curious, then, to find that his views on the mechanisms of evolution were similar to those of Jones. It must be remembered, however, that in his 1946 account of his discoveries he abandoned his earlier commitment to the ape-origin theory and openly supported an interpretation more in line with Jones's position, a move fully endorsed by Schepers.[17] Evidently there is a link between Lamarckism and the attempt to separate the apes from mankind, but not a direct one. In effect, Broom's and Jones's thinking converged from two different, but related, starting points. Both were deeply interested in the need to preserve a vision of nature that would uphold the spiritual values once regarded as mankind's chief distinguishing mark. Jones began from the assumption that this goal could only be achieved by dividing mankind from its most bestial cousins, the apes. He seems to have believed that the Darwinians had welcomed the theory of an ape ancestry as a means of demolishing the traditional status of humanity. As he explored the implications of his rejection of their position, he realized that he needed a theory of the evolutionary mechanism that would itself undermine the whole edifice of materialism built around natural selection. Broom adopted a more conventional approach, arguing from the start that the whole process of evolution was incompatible with Darwinism. Since evolution itself was purposeful, he did not believe that it would be demeaning to link mankind directly with its closest animal relations. Only gradually did he come to see that, given his belief in the purposeful nature of many evolutionary trends, he was in a position to join Wood Jones in rejecting the close link to the apes.

Henry Fairfield Osborn was an earlier convert to a position resembling that of Jones. Although Osborn had long been known as a leading advocate of parallelism in mammalian evolution, he was reluctant to apply the orthogenetic element of his theory to the origin of mankind. In his 1927 paper announcing his belief in an Oligocene dawn man, he referred only to convergence to explain the similarities between ape and human anatomy.[18] Although no longer a Lamarckian, he was still prepared to give habit an important role as a guiding force in human evolution, as we saw in chapter 7. He cited the work of Gregory and Morton on the similarities between the gorilla and human foot, presumably to indicate that the partly terrestrial habits of the apes would produce a parallel transformation of this organ. It must be remembered, though, that Osborn was anxious to stress the differences between the habits and the habitats of the two stocks. In another paper he mentioned the possession of similar "family and special characteristics" as the source of parallel adaptations in mankind and the apes, although it is difficult to see how this aspect could explain anatomical convergences.[19] Osborn also tried to minimize the significance of the problem by insisting that many of the similar characters were merely inherited from a common ancestor in the distant past.

Although Osborn agreed with Jones that the human and ape stocks had been distinct throughout the late Tertiary, their views on the causes of human evolution were very different. Osborn argued that the earliest humans had a distinct, terrestrial lifestyle, and was thus forced to be rather casual in his references to adaptive convergence. Obviously there had to be convergence in the human and ape stocks (unless both were affected by an orthogenetic trend), but Osborn could hardly follow Jones's efforts to show how similar habits could produce similar results in forms that were only distantly related. If he could not explore the topic of convergence in detail, however, Osborn was in a far better position than Jones to account for the original separation of the stocks, and his theory thus falls far more naturally into the narrative style of explanation. Nevertheless, Osborn's reliance on the long-term effects of a stimulating environment to shape the higher characters of mankind reveals the extent to which his thinking in this area followed his customary reliance on rigid evolutionary trends. By postulating an early adaptive radiation of the primates, followed by a linear trend of specialization in each branch, Osborn related his ideas on human origins to his general theory of mammalian evolution, a theory whose origins lay in the neo-Lamarckism of an earlier generation of American naturalists.

Of those paleoanthropologists who invoked a less extreme version of parallel evolution, several seem to have shared Jones's lack of interest in the cause of the original separation. Marcellin Boule, for instance, accounted for both the apes and the extinct hominids by postulating an almost explicitly teleological trend forcing all branches of primate evolution toward the human form. In *Les hommes fossiles,* Boule argued that the ape and human stocks were separate offshoots from a stem represented by the Old World monkeys.[20] The apes, and the apelike forms of humanity such as *Pithecanthropus,* were all parallel lines of evolution that had striven to reach the same goal.

> Not only at its origin, but even toward the end of its evolution, the human branch would thus have had as neighbours, and in a sense as rivals, other branches of the higher Primates, the offsprings of a common stem. Various ape-like forms, starting from the first anthropoid stages, would seek, under the influence of the same environment or of the same needs, to evolve toward types of greater perfection. Several of these forms may have been able to surmount the stage where the living Anthropoids seem to stand, and to have acquired some of the higher characters that Man possesses to-day; but the direct descendants of our primitive ancestors alone would seem to have reached the end of this race towards the goal of progress.[21]

Boule's suggestion that needs (*besoins*) direct evolution may be a hint at Lamarckism, and certainly suggests that he was thinking of convergences produced by adaptive evolution. However, his willingness to treat the path toward humanity as the road to progress illustrates the teleological character of his thinking on this issue. He seems to have believed that the advantages of becoming human were so obvious that many different primates had tried this avenue of development, only one succeeding completely. Despite Boule's reputation as an early advocate of an image of branching evolution, this passage shows that he saw the branches being driven in a basically similar direction, rather than diverging toward their own specializations.

A rather similar position emerged in the writings of another prominent supporter of the presapiens theory, Louis Leakey. In his *Adam's Ancestors* of 1934, Leakey argued not only for the great antiquity of modern humanity but also for the existence of a parallel line of evolution that had generated a distinct and more primitive family of humans, the Palaeoanthropidae.[22] Since this family included the Neanderthals, it had obviously developed at least up to a level in which toolmaking had become important, despite its retention of several characteristic apelike features. Leakey was thus advocating a form of parallel evolution capable of pushing two long-separate lines along the path toward a human level of intelligence. He said little, however, about the evolutionary mechanisms that might be involved. At this time Leakey was predominantly an archaeologist, with an expert knowledge of the complex pattern of paleolithic culture developments. It seems probable that his belief in parallel lines of cultural evolution was simply translated into a theory of parallel biological evolution, and the advantages to be gained from technology counted as the force that guided the lines in the same direction.

In a similar manner, the advocates of a polytypic theory of human origins also had to invoke an element of parallel evolution that had raised each of the modern races to human status from separate ape ancestors. Karl Vogt had pioneered this interpretation in his *Lectures on Man* of 1864, although he had said little about the forces that had produced the parallel developments. He assumed that apes in different parts of the world would begin to enjoy the "privilege" of becoming human, which seems to rule out environmental adaptations.[23] Presumably, the advantages of a growing intelligence were thought to be so obvious that several different apes took this route of development, somehow acquiring the physical character of humanity as a result.

The leading exponent of polytypic human evolution in the early twentieth century was Hermann Klaatsch, who had gained his early reputation as an advocate of the view that the apes did not represent a stage in human evolution. Although he mentioned the theory of parallel evolution in his early work, his main concern had been to emphasize the primitive character of the human form.[24] Significantly, he insisted on the need to bring in Lamarckian use-inheritance to explain the adjustment of the skeleton to an upright posture.[25] When he announced his theory of polytypic evolution in 1910, he was forced to spend most of his time defending his controversial identification of the orangoid and gorilloid types of humanity. The apes were presented as human failures that had been sidetracked by adaptation to special conditions—an indication that the path toward humanity was supposed to be a natural one.[26] In his *Evolution and Progress of Mankind,* Klaatsch developed his theory further, arguing that the long arms of the apes were an overspecialization for life in the forests.[27] Curiously, he also suggested that the apes' brain would have reached human proportions had it not been restricted by the enormous facial muscles.[28] He thus implied an inherent trend toward brain growth, although Klaatsch was certainly aware of the significance of mankind's transition to upright walking. The human foot was the crucial organ here, since our climbing ancestors had already adopted an upright posture in the trees. Perhaps the unique character of the human foot was a "spontaneous" or accidental variation that just happened to fit some individuals for upright walking.[29] The mutation would have had to occur in several different ape species for the polytypic theory to be plausible. Klaatsch did not emphasize this point, but it seems to imply a capacity for parallel genetic modification inherited from the common ancestor of the ape-human lines. Once given this new character, however, the ancestral human species had all followed a similar path of anatomical adjustment to the new means of locomotion and cultural development.

Klaatsch did not discuss the evolutionary mechanism responsible for bringing about these parallel developments, although he may well have preserved his early interest in Lamarckism. His theory drew upon the idea of convergent adaptations to bipedalism, but his appeal to an accidental variation to define the new direction of development implied a nonfunctional "trigger" for the process. A similar combination was exploited by John Gray, an English supporter of Klaatsch's polygenism. Gray argued that several ape species had thrown off varieties capable of taking to the ground, with the result that convergent evolution had transformed them into the races of mankind. As to how the varieties were originally "thrown off," Gray only speculated about the "separation and isolation of an accidental variation of the stock or germplasm."[30] Again, the assumption was that any ape adopting a terrestrial lifestyle would automatically be transformed into a human, but the teleological implications were undercut by the appeal to chance mutations to initiate the trends. If the supporters of convergent evolution were forced to invoke chance mutations to start the trends, it is hardly surprising that other anthropologists would explore the possibility that the trends themselves might be the product of a nonfunctional mechanism.

Nonadaptive Trends and Orthogenesis

The possibility that significant aspects of human physical and mental character could be the product of trends that were essentially purposeless was obviously a threat to the optimistic image of evolutionary progressionism. Yet the prevailing belief that such trends could be observed throughout the development of the animal kingdom was strong enough to ensure that at least some paleoanthropologists took the idea seriously. Whatever Osborn's qualms about the application of orthogenesis to human origins, others were quite willing to use his work on parallelism in mammalian evolution as the basis for extending the phenomenon into this new area. Thus the Italian anthropologist Giuseppe Sergi supported his polytypic theory of the origin of the human races with references to Osborn's "polyphyletic law" and the similar conclusions reached by other paleontologists.[31] Sergi said little about the underlying mechanisms that produced the trends, and only a few bold spirits tried to explore this question in any detail. In the process they almost invariably managed to lose sight of the well-known fact that most orthogenetic trends were supposed to end with extinction.

Sir Arthur Smith Woodward is best known for his descriptions of the Piltdown remains, but in fact he was a paleontologist specializing in vertebrate evolution and a leading advocate of orthogenesis. In 1913 he delivered an evening lecture at the meeting of the British Association for the Advancement of Science in which he pointed out that whenever new fossils were discovered, they confirmed the existence of general trends in evolution. He supported the theory of racial senility, which assumed that groups such as the Ammonites had reached a natural limit to their evolution and had then declined toward extinction. He also noted that extinction appeared often to be the result of an over-development of certain structures, as in the case of the Pleistocene deer known as the Irish elk, whose antlers were thought to have grown so large that they seriously interfered with the animals' movements.[32] Such trends suggested that the variation of a character somehow took on a momentum of its own that enlarged the character regardless of the consequences. Woodward was not afraid to speculate that the growth of the human brain itself might have been the product of a similar trend, but in this case just happened to give results that were increasingly useful at all levels of development. This trend was a general feature of primate evolution: "Compared with the rest of the body, the brain tended to be overgrown, and it seems reasonable to suppose that this overgrowth eventually led to the complete domination of the brain which is the special characteristic of man. As soon as an animal could feed and defend itself by craft, its teeth and other primitive weapons could degenerate."[33] Woodward was eager to apply this last point to explain the apelike jaw of *Eoanthropus*. He also argued that the original form of the human skull lacked heavy brow ridges, and that the production of such ridges represented an evolutionary trend shared by the ape and the Neanderthal lines.[34]

In his British Museum *Guide to the Fossil Remains of Man,* Woodward repeated his view that the human brain was merely the lucky product of the kind of trend that all too often led to extinction.

> Now, the study of many kinds of fossils has shown that when, in successive generations, one part of the body begins to increase in size or complication much more rapidly than other parts, this increase rarely stops until it becomes excessive. As a rule it passes the limit of utility, becomes a hindrance, and even contributes to the extermination of the races of animals in which it occurs. In the case of the brain, however, a tendency to overgrowth might become an advantage, and it seems reasonable to imagine that such an overgrowth in the early apelike animals eventually led to the complete domination of the brain, which is the special characteristic of man.[35]

Woodward made the point again in 1948 in his last account of the Piltdown remains, *The Earliest Englishman.*[36]

There is no evidence that Woodward wished to use his theory of orthogenetic brain growth to denigrate the power of the human mind, which he freely admitted had ushered in a new era of evolution. He was able to face up to the prospect of a nonfunctional explanation of brain growth because it fitted in with the overall theory of orthogenetic evolution to which he had dedicated much of his career. He made no effort to postulate a functional element in the creation of the orthogenetic trends, and in a more general discussion of this topic had openly appealed to physical forces built into the constitution of the germ plasm.[37] In the end, Woodward seems to have believed that the trends were preformed in the origin of living matter, programmed to unfold in accordance with some ultimate purpose. The final chapter of *The Earliest Englishman* portrays the sequence of trends as nature's experiments, each of which was aimed at the production of a higher type. The overdevelopment and extinction that all too often resulted in the end were presumably unfortunate byproducts of the Creator's loose method of controlling his universe. For Woodward, "man is not only the natural end of the procession of life which is revealed by the fossils: he seems to have come into being in accordance with the rules by which the various animals have succeeded each other during geological ages."[38]

Woodward's account of brain growth differed considerably from that offered by Elliot Smith, for all that the two agreed on the interpretation of the Piltdown skull. Smith certainly believed that the increasing size of the brain was caused by the functional value of intelligence, yet he too was prepared to allow a certain amount of parallelism in primate evolution. Although he repudiated Wood Jones's tarsioid theory, his rejection conceded a more limited amount of parallel evolution.

> There is abundant evidence of convergence in the three suborders of the Primates, but it is clearly the expression of the tendency of similar traits to develop in the various descendants of the same common ancestor, and therefore can hardly help those who refuse to admit the close connexion of the Lemuroidea with the Primates. Henry Fairfield Osborn has emphasized the fact that "the *same* results appear independently in descendants of the *same* ancestors."[39]

Despite his use of the term "convergence," Smith's reference to Osborn's law of parallel evolution suggests that he was thinking of a genetic predisposition for related forms to vary in a particular direction. He did not mention this possibility in his discussions of the functional theory of brain development, presumably because it would imply an entirely different form of directed evolution.

The most detailed study of parallelism in the primates was Wilfrid Le Gros Clark's *Early Forerunners of Man* of 1934, which was dedicated to Smith. Clark certainly accepted a general trend toward brain development, and followed Smith in postulating a main line of primate evolution culminating in the human species. "The line of evolution of the Anthropoidea has been marked by the successive branching off of specialized groups from a central stem in which a progressive expansion of the brain has been accompanied by the retention of a bodily structure of a remarkably generalized type. It is this main stem which culminated in the appearance of Man himself."[40] Clark differed from Smith, however, in arguing that the trend toward brain growth was not confined to the central line of development. The apes too had participated in the trend after their divergence, so that many of the similarities between the human and gorilla brain had evolved independently. With such an emphasis on the power of parallel evolution, it is not surprising that Clark showed little interest in the transition to ground-dwelling as a key step in human evolution.

The most striking aspect of Clark's theory, however, was his willingness to invoke nonfunctional or orthogenetic trends to explain parallel evolution. He accepted the widespread occurrence of nonadaptive characters, and stressed their importance in taxonomy. Such characters could be produced independently in different lines of evolution, and Clark insisted that parallel evolution had occurred on a scale that most evolutionists had been unwilling to admit. "The fact is that the minute and detailed researches which have been carried out by comparative anatomists in recent years have made it certain that parallelism in evolutionary development has been proceeding on a large scale and is no longer to be regarded as an incidental curiosity which has occurred sporadically in the course of evolution."[41] Parallel evolution resulted from an inherited tendency for related forms to vary in the same direction, a phenomenon frequently ignored because of the Darwinians' reliance on random variation. Clark believed that the striking resemblances between the brains of the Old and New World monkeys had been developed independently, and claimed that if parallelism could do this, it would be "difficult to place any limit on its possibilities."[42] The early primates "were at first characterized only by the fact that they incorporated potentialities for evolutionary development along certain definite lines, and they eventually gave rise to a group of mammals all of which are to be distinguished by a complex of structural features which they possess in common."[43] Indeed, the primate order would have to be defined not by the possession of certain characters, but by the evolutionary tendencies that eventually turned a group of quite generalized mammals into the primates we know today.[44]

There were a number of these evolutionary tendencies, and they could certainly not be seen as the result of adaptive pressures. The teeth, in particular, showed evolutionary trends that were nonfunctional, although the end results might sometimes be put to a useful purpose. Thus the specialized front teeth of various lemurs were used as a comb for the fur, although the structure had clearly arisen independently.

> This conclusion inevitably leads to the conception of an orthogenetic trend of evolution dependent upon an inherent tendency in the common progenitor to the production of similar features in divergent groups of descendants. For, although clearly the front teeth of the modern lemurs are used similarly in different groups for toilet purposes, there is no demonstrable reason for supposing that the characters of the furry coat in these animals

> are in any way so peculiar that they *inevitably* demand the elaboration of such a dental comb.[45]

Trends with a nonfunctional origin could thus produce a useful structure, a description of exactly what had happened in the case of the most important trend of all, the growth of the primate brain. Clark insisted that the large brains of the modern apes were acquired through a trend paralleling the growth of the brain in the ancestors of mankind.[46] The overall similarities between humans and gorillas were due to "common potentialities and tendencies for evolutionary development" inherited from a much earlier common ancestor.[47] Clark was, however, prepared to see the human brain as the "culminating point of an evolutionary tendency which is displayed to varying degrees among all the primates."[48] A general trend toward brain growth in all the branches of primate evolution paralleled the "main stem" of evolution aimed at mankind. The human brain was a culmination of the trend only in the sense that here the trend had gone further than in any other branch. Such an interpretation effectively undermined Smith's functional theory of brain development.

In his conclusion, Clark cited Osborn's work on parallel evolution and insisted that it implied orthogenesis: tendencies to vary in a particular direction built into the germ plasms of related species. He went on the express the view that with such a mechanism evolution became essentially predetermined.

> It seems certain that the instances of parallelism in the evolution of the Primates which have been brought to light in the preceding chapters are to be interpreted satisfactorily only by the conception of definite predetermined trends of development—that is, by the conception of Orthogenesis. This conception puts the onus of evolutionary progress more on the germplasm and regards the influence of the environment as of somewhat secondary importance. Hence it seems to intensify the mysteries of the germplasm, which (it implies) is endowed from the beginning with countless potentialities for evolution in definite directions. It becomes, therefore, increasingly difficult to conceive of evolution as being fundamentally merely a matter of action and reaction between the physico-chemical factors of the environment and those of a passive or at least a neutral and completely plastic organism. For this reason, Orthogenesis is apt to be dismissed rather abruptly as a "vitalistic" principle complicating in an unwelcome manner the mental pictures which biologists have striven to elaborate under the

> influence of mechanistic ideas. But if the mysteries of the living and evolving germ-plasm are even deeper and more enigmatical than we have been inclined to believe, it were better to recognize the fact.[49]

The reference to vitalism suggests that Clark was aware of the teleological implications of what he was saying. If the human brain was to be seen as the product of a nonfunctional orthogenetic trend, then only by seeing all such trends as an expression of an original purpose built into nature could any transcendental significance for human life be preserved. Despite the difficulties created by cases of racial senility, both Woodward and Clark seem to have been prepared to adopt such a view of orthogenesis. In the postwar years, of course, Clark accepted the Darwinism of the Modern Synthesis and abandoned his reliance on orthogenesis, although he was still prepared to follow George Gaylord Simpson in postulating a limited amount of parallelism because of homologous mutations in related lines of adaptive evolution.[50]

The American anthropologist Earnest A. Hooton also appealed to parallelism as a major factor in primate evolution. In his *Up from the Ape* of 1931, Hooton at first seemed to accept Elliot Smith's approach, since he stressed the role of initiative in stimulating our ancestors' descent from the trees.

> Let us suppose, then, that man's ancestors neither fell out of the trees, nor fell with the trees, but descended to the ground on their own initiative, leaving their more cautious and conservative anthropoid relatives glaring at them disapprovingly from the branches. These radical ancestors of ours saw and accepted the chance of a larger, more varied, and fuller diet; they wanted to live their lives more abundantly. A careful and dispassionate examination of the facts and probabilities of human evolution indicates that this crucial event was not the result of environmental accident, but rather a manifestation of that superior intelligence and initiative which, inherent in the proto-human stock, determined its evolutionary destiny.[51]

Yet when it came to explaining why our ancestors had this extra level of intelligence and initiative, Hooton repudiated Smith's approach and opted for a nonfunctional interpretation. He had already published an article attacking "functional theories of primate evolution," although it is significant that he exluded the arboreal origin of the erect posture from his list of targets.[52]

Hooton certainly did not believe that all primate evolution is nonadaptive, and in cases involving adaptive trends, he was quite willing to invoke Lamarckism. He reproduced Bernard Shaw's withering comments on August Weismann's anti-Lamarckian experiments, insisting that, whatever the lack of hard evidence, no biologist could seek to explain evolution without the inheritance of acquired characters.[53] But if many primate characters had a nonfunctional origin, an orthogenetic rather than a Lamarckian explanation would be required. In *Up from the Ape,* Hooton openly maintained that the various lines of human evolution shared "certain progressive and non-adaptive hereditary forces making for jaw reduction."[54] He even criticized Smith for arguing that the eye, hand, and brain had interacted together in promoting the growth of intelligence. This "beneficient evolutionary cycle . . . is nevertheless a vicious circle of reasoning," and it was necessary to postulate instead "certain inherent differences in the ancestors of man, on the one hand, and of the chimpanzee, on the other, which determined the development of a large brain and a large intelligence in one and a smaller brain and a lesser intelligence in the other."[55] The result was that the various apes sharing this more progressive tendency had moved out of the trees and evolved into the extinct and modern human races.[56]

The great problem with a theory of orthogenesis was to explain the origin of the tendency to vary in a particular direction. Hooton had nothing to offer on this score, but both Woodward and Clark assumed that the trends were built into the germ plasm, almost as though from the creation of life itself. Other paleoanthropologists, however, were looking for evolutionary mechanisms that would explain the setting up of trends without absolute predetermination. Some paleontologists interested in orthogenesis had suggested mechanisms based on the growth process of the individual organism. It was known that hormonal changes could affect growth to produce significant changes in adult structure, and if this effect could accumulate through inheritance over many generations, an evolutionary trend would result. This mechanism required the inheritance of acquired characters, although it was not based on use-inheritance and hence was not controlled by function. Its advantage was that, since the growth process was supposed to be purposeful, the changes could be accommodated in an orderly manner even though not originating through functional activity.

One of the most widely discussed efforts to link human evolution to variations in growth came from Louis Bolk, professor of human anatomy at Amsterdam. Bolk seized upon a point that had been noted by J. Kollmann in 1905: that the features of the young ape are considerably more human in appearance than those of the adult.[57] Kollmann had used this point to argue, rather implausibly, that the earliest humans had been pygmies. Bolk ignored this point and instead elaborated a general theory of human evolution based on a process of "fetalization" from an ape ancestor.[58] If growth was progressively retarded, our ancestors would mature at ever earlier stages in the developmental process, gradually losing the more bestial aspects of the adult apes. Man is, in effect, "a Primate-fetus that has become sexually mature."[59] The changing of the whole anatomy in a coordinated manner indicated to Bolk "that this fetalization cannot have resulted from external influences acting on the organism." The cause was internal, an "organic principle of development."[60] Although not the results of specific adaptive demands, the changes were coordinated by the growth process so that the end result was always functional. There was thus no need for Bolk to fear that the trend might lead to the production of nonadaptive structures and extinction.

Bolk divorced his theory of an internally controlled trend from the more disturbing aspects of orthogenesis by transferring control from the germ plasm to the growth process itself. The ultimate cause of human evolution was a systematic modification of the physiological process that regulates the growth of the body, originating in the endocrine system.[61] Growth can be influenced by hormones, and a cumulative alteration of the system producing the hormones might thus result in a retardation of the whole process, leading to neotony: the preservation of immature characters into adult life. Inhibiting hormones could be produced that would check growth in this way, although if their production in a modern individual were prevented by illness, ancestral characters might reappear. Bolk does not seem to have been interested in tracing the variation in hormone production back to a predetermined trend in the germ plasm. Rather, he assumed that the effect, once produced, was permanent in the sense that it could be transmitted to the next generation. Each generation thus built on a foundation already modified by its ancestors and extended the process one step further, resulting in a cumulative evolutionary trend.

Sir Arthur Keith strongly approved of Bolk's efforts to show that cumulative glandular changes could produce an effect such as fetalization.[62] From an early stage in his career, Keith had believed that the apelike features of the Neanderthals might have resulted from overactivity of the pituitary gland, as in modern cases of acromegaly.[63] He also became convinced that modern racial characters are the result of hormonal differences affecting growth in various ways.[64] In the course of the 1920s Keith returned frequently to the view that the growth process of the individual was important for the evolution of the species. The ability of hormones to affect growth in a coordinated way provided a means of explaining adaptive evolutionary trends without either Lamarckism or natural selection. Change was always functional because the growing embryo possessed an automatic mechanism for recognizing and responding to the needs of the organism.[65] Starting from the idea that hormonal disturbances could accumulate to give nonadaptive racial characters, Keith thus ended up with a theory in which the coordinated process of growth could generate functional trends, or at least trends that could be accommodated by the functioning body.

The crucial question in all this was, as in Bolk's theory: *how* do the changes in growth accumulate over many generations to create an evolutionary trend? Keith had always opposed Lamarckian use-inheritance,[66] but he was forced to consider the possibility that the effects produced by hormones might actually be transmitted from the parent to its offspring by the hormones themselves. The hypothesis that hormones acted as an additional vehicle of heredity was actively supported by Lamarckians such as J. T. Cunningham. Keith considered this possibility and was forced to reject it. He believed that the evidence was in favor of only detrimental characters being transmitted by external agents "damaging" the germ plasm.[67] How then did the accumulation of more positive growth modifications occur? Keith was forced in the end to fall back on the idea that the germ plasm was predisposed to vary in a particular way. He found references to the possibility that variation might be predetermined along certain lines in the writings of T. H. Huxley, and offered these as hints toward a future solution of the problem.[68] If such built-in trends worked through the production of hormones, the purposeful nature of the growth process would ensure that the resulting anatomical changes were functional, even though the trend was not initiated by any functional need.

In the second edition of his *Antiquity of Man,* Keith referred to Osborn's theory of "collateral" evolution and insisted that it was valid for mankind.[69] Several distinct branches of human evolution had independently developed a large brain, including the Piltdown, Neanderthal, and modern types. It was not necessary, he suggested, to suppose that the Pliocene human stock already had a large brain, only that "the bias or tendency to a cerebral increase was latent in the common Pliocene ancestor, and that this tendency obtained its first structural manifestation during the evolution of its descendants."[70] Keith thus joined

Woodward and Clark in attributing the growth of the brain to orthogenesis. He also argued that an independent evolution of certain features in several racial types implied that "an evolutionary bias may be latent in a stock."[71] In 1936 he supported a theory of the parallel evolution of the human races because of the "unfolding of a programme of latent qualities inherited from a common ancestor of an earlier period."[72] It thus seems that Keith was drawn ever closer to the idea that the actual direction of the trends had been built into living matter from its creation.

Parallel racial evolution was also included in Franz Weidenreich's theory of human origins. Here several racial types were supposed to evolve, at least partly independently, through *Pithecanthropus* and Neanderthal phases toward fully human status. Weidenreich seems to have believed that the racial characters themselves were nonadaptive,[73] but the overall trend represented a continuing adaptation to the erect posture and the enlargement of the brain.[74] He did not discuss the cause of this enlargement, made no mention of nonadaptive trends, and presumably accepted that all the races were responding to the benefits conferred by increased intelligence. It may be significant that in the 1920s Weidenreich had been an open supporter of Lamarckism.[75] He makes no reference to Lamarckism in his later works on human evolution, but his commitment to a form of linear progressionism may owe something to his earlier interest in the inheritance of acquired characters. He also speculated about the possibility of an additional trend toward relative brain growth that might result if modern mankind had evolved from an earlier giant form.[76] Just as in the various breeds of dogs, a reduction in absolute size entailed a relative increase in the ratio of brain to body size. At least part of mankind's enlarged brain might be explained if such a trend had occurred in human evolution, and Weidenreich believed that the latest fossil discoveries confirmed the existence of "giant early man." He did not speculate on how the trend might be produced, but for all his opposition to Bolk's theory, he may have shared the view that the forces affecting individual growth were important for evolution.[77]

In the postwar years, Weidenreich's views on parallel racial evolution were exploited in a very different context by the geneticist R. Ruggles Gates. Citing extensively from the works of Osborn and other paleontologists who had supported orthogenetic evolution, Gates insisted that the trends in human evolution were the result of parallel sequences of genetic mutations occurring repeatedly in the various human "species," whatever their adaptive value.[78] The fact that none of the prewar paleoanthropologists who had been interested in evolutionary trends had considered this mechanism illustrates the extent of the gulf between genetics and the older biological disciplines of anatomy and paleontology. A number of early geneticists had speculated about parallel mutations as an explanation of orthogenesis, but paleoanthropologists had continued to think in vaguely Lamarckian terms, or to postulate mysterious trends built into the germ plasm. Until Gates belatedly moved into the field of human origins, it had never occurred to the anthropologists that the newfangled laboratory work of the geneticists had anything to offer them.

By the time Gates applied his idea to a polygenist theory of human origins, he himself had lost touch with developments in genetics and evolution theory. Most geneticists were now convinced that mutations occurred essentially at random, and had come to accept that natural selection was the only mechanism that could guide evolution in particular directions. Gates repudiated the neo-Darwinism of the Modern Synthesis, which left him free to make an alliance with anthropologists such as Hooton, who were still steeped in the atmosphere of prewar evolutionary thinking. The new generation of paleoanthropologists was incorporating Weidenreich's work into a modernized version of the Neanderthal-phase theory, but without the element of racial parallelism. Evolutionary trends in the primates were not rejected—indeed, as Matt Cartmill emphasizes, the postwar consensus promoted by Le Gros Clark and George Gaylord Simpson defined the order in terms of a collection of such trends.[79] But the trends were now supposed to be largely adaptive, and thus compatible with Simpson's application of the Modern Synthesis to paleontology. The sudden loss of interest in orthogenetic trends represents one of the most obvious watersheds separating the paleoanthropology of the 1920s and 1930s from that of the modern era.

Chapter Nine Accident or Design?

The Coming of Man: Was It Accident or Design? Robert Broom asked in the title of his 1933 book. He himself had no doubt that some form of supernatural guidance was needed to explain progressive evolution. The Darwinian theory of natural selection was helpless to account for the purposeful trends that appeared at all stages in the ascent of life, and especially in human evolution. Few scientists now went so far as to appeal to the supernatural, but the majority of early twentieth-century paleoanthropologists were attracted to theories of human origins that were to some extent teleological. Many saw growth of primate intelligence as a definite trend that seemed to imply the almost inevitable appearance of the human mind. Yet if they were not to follow Broom in abandoning all hope of finding a natural explanation of this trend, something would have to be done to camouflage its more obviously teleological character.

Intelligence, of course, was not the only aspect of the mind that entered into the debates. Ever since Chambers's *Vestiges of Creation* had raised the question of human evolution, the origin of the moral sense had disturbed conservative thinkers. Chambers had certainly shaken things up by declaring that morality was the product of natural law, but he had said little about *how* progressive evolution generated moral feelings. Darwin had tackled this problem squarely in the *Descent of Man,* postulating an effect now known as group selection to explain how those tribes with a strong cooperative instinct survived at the expense of those less well organized.[1] According to Darwin, a moral sense resulted when the increased level of human intelligence was applied to the analysis of the social instincts generated by evolution. Even some of Darwin's supporters found this difficult to swallow, and many of his opponents insisted that the human spirit must have a supernatural origin. By the end of the century, however, a direct appeal to divine creation had become less popular. The triumph of evolutionism meant that everyone now had to see the moral sense as a product of nature, and the only hope of retaining a transcendental significance for morality was to regard the human mind as the inevitable product of a purposeful evolutionary process.

The cultural evolutionism of Tylor and Lubbock provided a framework within which a steady increase in both intelligence and sociability could be taken for granted. Biologists now agreed that many of the higher animals enjoyed some kind of social life, and it was easy to assume that progressive evolution promoted this kind of behavior. Perhaps our ancestors' adoption of a lifestyle based on hunting accelerated the process in human evolution, since the hunting group depended on cooperation for success. The temporary popularity of instinct theory in sociology suggests that many early twentieth-century thinkers accepted biological evolution as the source of human social activity. Professional sociologists soon abandoned this approach, but many paleoanthropologists continued to believe that evolution would reveal the true nature of social behavior and the moral sense. This seldom became a crucial issue in the theoretical debates, however, since most of the participants assumed that social instincts were the inevitable product of progressive evolution.

There were, in fact, two ways of thinking about the evolution of social behavior, although they were seldom clearly distinguished at the time. One approach stressed the practical advantages to be gained from cooperation, and then simply assumed that the cooperative instincts were the basis of what we call morality. This was a very Darwinian way of looking at the issue, whether or not natural selection was called in to explain how useful habits were converted to instincts. Efforts to specify the circumstances that would have favored social activity among arboreal primates or among early forms of mankind would fit into this category. A more popular alternative was to transfer the moral sense into nature itself by supposing that evolution actually worked by promoting cooperation or altruism, a view expounded by popular writers such as Henry Drummond and Peter Kropotkin.[2] Cooperation, not competition, was the driving force of nature, although to make it work without falling back on group selection it was necessary to invoke Lamarckism to explain how habits became instincts. This approach thus allied itself with the efforts of neo-Lamarckians such as Samuel Butler and E. D. Cope to emphasize how other "human" qualities such as initiative were important in animal evolution. Many Lamarckians believed that nature itself was a purposeful system designed by its Creator to encourage the development of those values cherished as essential to humanity.

Although a few anthropologists suspected that nature may have endowed us with more warlike instincts, the majority favored a more pacific account of our origins. It is thus legitimate to ask if their theories were part of an effort to humanize nature. Drummond and Kropotkin made no direct contribution to paleoanthropology: having adopted a benign image of nature, they took it for granted that the human moral sense was the goal of evolutionary progress. Lamarckism was popular among students of human evolution, however, and with it a desire to see nature as a more creative—and more ethical—system than Darwinism would allow. Elliot Smith certainly linked his theory of brain development with an attempt to undermine the harsher implications of evolutionism. There may be a parallel with the views of Drummond and Kropotkin here, but paleoanthropologists were forced to think more carefully about the relationship between mankind and the rest of the primates. They had to explain why the line of human evolution had come further than any other along the road of mental and moral development. Adopting a more harmonious view of nature was not enough, since it was necessary to point out specific events in human evolution that had raised our ancestors above the rest of the animal kingdom. If not the actual causes of human evolution, these events were at least conditions necessary to ensure that evolutionary progress would reach its goal.

This consideration leads us back to the more general question of how an almost teleological account of human evolution could be presented in a form that appeared to be compatible with the principles of natural science. The earliest, almost linear model of evolution from ape to human probably drew much of its inspiration from the morphological progressionism of the mid-nineteenth century, although it was certainly compatible with the linear view of cultural progress. By 1900 it was becoming recognized that the linear model was too obviously goal-directed, and efforts were made to depict primate evolution as a branching process more in line with what was known to have occurred throughout the animal kingdom. The branching model was designed, however, in such a way as to allow the preservation of a special role for mankind. As long as the "main stem" of evolution ran in the right direction, multiple efforts to achieve the goal and even a large number of failures could be tolerated. It was precisely the metaphor of nature "experimenting" to achieve its goals that allowed these evolutionists to believe that they had escaped the most obvious weaknesses of teleology. At the same time, the idea that progress could be maintained through key transitions in habitat and behavior—the switch to the open plains and hunting—also helped to conceal the element of linearity inherent in the notion of a main stem aimed at mankind. Few were prepared to admit that key events might be triggered by a purely environmental stimulus, but it was generally accepted that a satisfactory theory would have to compromise between progressionism and environmentalism. The narrative element in the theory appealed to a series of distinct events to separate the line of human evolution from the less successful branches that had tried to participate in the overall progression.

But what of "social Darwinism," that classic expression of the Victorians' belief in the harsher side of nature? Since there were few workers, at least in the field of anthropology, who accepted natural selection as the principal mechanism of evolution, we should not expect to see individual competition stressed as a factor in the origin of mankind. Rather more influential was the belief that tribal or racial competition might be important, although this belief was not confined to Darwinians and some of the most influential race theorists were neo-Lamarckians. Paradoxically, conservative thinkers who were convinced that nature is purposeful were quite willing to see racial conflict as the testing mechanism of evolution, designed to eliminate those products that did not make the grade. Another suggestion was that our ancestors' adoption of a hunting way of life had led to the creation of bloodthirsty instincts. Largely ignored at first, this idea was later taken up by Raymond Dart and was popularized by Robert Ardrey and others in the 1960s. Although often compared with modern sociobiology, this "anthropology of aggression" does not seem to have had the same clear-cut origins in the Darwinian theory of natural selection.

Progress and Harmony

As late as 1884, John Cleland, professor of anatomy at Glasgow, expressed the once-popular view that "man is a terminus, and not only *a* terminus, but *the* terminus of the advance of vertebrate life."[3] Few naturalists in the post-Darwinian era would have been prepared so explicitly to endorse the claim that mankind is the goal of the whole organic progression. Yet progressionism in the more subtle forms continued, and even a staunch Darwinian like Ernst Haeckel all too often portrayed evolution as a series of stages in the advance toward the human form. The theory of a simple progression from ape to mankind expounded by Haeckel's followers, including Gustav Schwalbe and Eugene Dubois, was certainly not couched in teleological language, but it was taken for granted that the human form would be the end product of the process. All that the anatomist needed to do was to work out the sequence of steps by which this form was created out of its nearest animal relative. There was no direct connection between this theory and the linear progressionism of the cultural anthropologists, but one cannot help noticing how neatly the sequence of cultural developments fitted in as a continuation of the morphological advance. Neanderthal man was culturally and biologically lower in the evolutionary scale than his modern counterpart. Progress toward humanity and progress toward Western civilization were complementary goals in a society that took its own superiority for granted. The artificiality of this whole model of evolution was frequently pointed out by the early twentieth-century advocates of the presapiens theory, who stressed that constant branching was a major feature of the general evolutionary process.

Despite the apparent sophistication of its appeal to divergent evolutionism, the presapiens theory was often used to imply that a number of branches moved with varying degrees of success toward the same goal. We have already seen this tendency at work in the writings of Marcellin Boule,[4] and Hermann Klaatsch regarded the apes as "unsuccessful attempts and dashes forward toward the goal of the definite creation of the human race."[5] The belief that the human form and by implication the human mind were goals toward which all the primates had been striving, was difficult to shake off. The apes had *tried* to become human, even though they had failed and been forced to settle for second best. The presapiens theory treated *Pithecanthropus* and the Neanderthals as closer, but equally unsuccessful side branches—in a sense, these types were even less successful, since by challenging the ancestors of modern mankind more openly they had caused their own extermination. The advantage of this position was that the more obviously teleological aspects of linear progressionism could be concealed by emphasizing how nature was forced to make many attempts to reach its goal in order to ensure that at least one would succeed.

This form of "trial and error" progressionism was still a far cry from the Darwinian view of evolution. For Darwin, there could be no single goal toward which all the species within a group must be striving. Each population had to make the best job it could of adapting to its own local environment, so the branches of evolution diverged away from one another instead of moving in parallel. Natural selection was equally difficult to reconcile with parallel evolution, since it was based on the assumption of random individual variation. By the early twentieth century, though, few naturalists took natural selection seriously, and even those most closely identified with a Darwinian view of evolution accepted only the general principle of divergence. W. K. Gregory came closest to this position, and has been called a "closet Darwinian" by one modern historian of paleontology.[6] Gregory certainly led the defense of the ape-origin theory, emancipating it from the older linear progressionism by emphasizing the transitions to new habits that had periodically turned human evolution into new paths of development. He was more willing than most of his contemporaries to accept that in a system of branching evolution there could be no single goal toward which all the branches moved. But even Gregory was strongly tempted to call in Lamarckism to explain how changed habits were reflected in morphological evolution.[7] In more popular works such as his *Our Face from Fish to Man* of 1929 he fell back into Haeckel's old technique of presenting vertebrate evolution as a series of steps toward the human form.

Other paleoanthropologists were more explicit in their support for the idea of a main stem of vertebrate, or at least primate, evolution aimed at the production of mankind. All side branches were treated as failed experiments in nature's striving for perfection. The points at which the branches diverged marked the periods at which the stock had been tested to pick out those individuals who remained true to the central theme of brain expansion. The metaphor of nature experimenting to ensure the eventual production of mankind was a common one. Sir Arthur Smith Woodward, whose work on the Piltdown remains ensured that he would be aware of multiple branches in the tree of evolution, wrote of the "several distinct approaches to modern man" produced by the orthogenetic trend toward brain growth.[8] In America, E. A. Hooton suggested that nature did not want to put all her eggs in one basket,[9] and used his polytypic theory of human origins to argue that "nature has conducted many and varied experiments upon the higher Primates, resulting in several lines of human descent."[10]

The concept of nature experimenting to reach its goals was also exploited by the leading exponent of a functional theory of brain growth, Sir Grafton Elliot Smith. We have already noted Smith's commitment to a main stem of evolution marked by "the steady and uniform development of the brain along a well-defined course throughout the primates right up to man."[11] The only hope of softening the teleological implications of a theory with such an emphasis on the linear character of the central theme was to imply that nature had to experiment in order to ensure success. Smith's use of the metaphor was not confined to the later phases of human evolution. Even the origin of the first mammals was characterized by nature trying out numerous experiments as soon as a new brain structure appeared in the ancestral Therapsid reptiles.[12] Later on, Asia and Africa were "the laboratory in which, for untold ages, Nature was making her great experiments to achieve the transmutation of the base substance of some brutal Ape into the divine form of Man."[13] Few passages could more clearly bring out the point that the experiments were designed to reach a preordained goal.

Of course, Smith could not admit the teleological aspect of his theory openly. He offered the excuse that "in attempting to attain conciseness of expression I have used teleological phraseology in many places merely as a matter of convenience, and not from any idea of accepting Teleology."[14] The reader must decide for himself or herself whether this explanation is adequate to account for the previous quotation. In one sense, perhaps, we can accept Smith's assurances. His theory of brain growth was based on a purely functional mechanism, not on orthogenesis, and he resisted the temptation to appeal to Lamarckian use-inheritance. In fact, Smith mentioned natural selection rather more frequently than most of his contemporaries, although it is clear that he saw selection being directed by intelligence and initiative, much along the lines of H. F. Osborn's "organic selection." He wrote of mankind constantly exposing itself to new conditions that favored the operations of natural selection and other forms of selection governed by the increased powers of intelligent choice.[15] Our arboreal ancestors' unspecialized anatomy left them free to develop large brains, which would enhance the potency of natural selection by generating an inquisitiveness that constantly led them into new situations.[16] Smith also held that sexual selection, depending upon individual choice, was responsible for the later refinement of human features.[17] Without appealing to Lamarckism, he thus shared the neo-Lamarckians' desire to see the mental activity of individual organisms as the directing factor in progressive evolution.

Smith was not impressed by the power of natural selection when it was driven by sheer necessity, since he believed that this process could lead only to anatomical specialization, sidetracking the species from the direct line of brain development. He saw the trees as a haven within which our early ancestors were safe from the pressures of competition on the ground.[18] When they finally ventured out onto the plains, their brains had grown enough to ensure that they could protect themselves with primitive weapons. The plains thus became a new stimulus for the growth of intelligence, whereas at an earlier stage they would have led to the trap of specialization. "The realization of his abilities to defend himself upon the ground, once he had learned the use of sticks and stones as implements, would naturally have led the intelligent Ape to forsake the narrow life of the forest and roam at large in search of more abundant and attractive food and varieties of scene."[19] Smith did not stress hunting as the key to the new lifestyle, although he realized that the development of better weapons would eventually allow big-game hunting to begin. The earliest humans were essentially gatherers. The increased level of their social activity was due to the cooperation needed for protection against predators. "Like most creatures who live in the open, the adoption of social habits is one of the surest means of protection; for the eyes and ears of each individual thus become the servants of the whole community, giving warning of danger, and thus adding to the safety of the herd."[20]

It appears that Smith did not accept the need for any enhancement of the social instincts in the later phases of human evolution. He held that our emotions and instincts were still essentially the same as those of the apes; the difference lies in the ability of our higher intelligence to control them.[21] For Smith, then, our emotional makeup had been shaped by the peaceful environment of the trees, and he became an active opponent of all efforts to postulate aggressive instincts shaped by our evolutionary past. It is this point which provides the link between his theory of human evolution and the other absorbing interest of his later career, the defense of an extreme form of diffusionism to explain the development of culture. He expounded the link explicitly in *The Evolution of Man* in 1924, and it became the central theme of later books such as *Human History* and *The Diffusion of Culture.*[22] Smith ridiculed Tylor's cultural evolutionism, with its assumption that the "psychic unity" of the race would guarantee the similarity of the inventions made by all societies.[23] He believed that civilization had been developed only once, in Egypt, and had been carried to the rest of the world by the deliberate expansion of the first civilized race.

It would be easy to see this hyperdiffusionism as a form of racism, based on the assumption that only one race had evolved highly enough to create civilization. In fact Smith's intention was very different. He certainly believed that each cultural and technological invention was the product of an isolated genius, who was often ignored at first by his less adventurous comrades.[24] But such inventions were not made more readily in one race than another. In prehistory, each refinement of stone toolmaking had spread very slowly over the world from its original starting point. The creation of the first highly organized society was an invention of the same kind, and could only have come about suddenly through an accidental combination of circumstances. Civilization "bears the impress of its wholly accidental origin: it is equally alien to the instinctive tendencies of human beings."[25] Warfare, slavery, and exploitation were not the natural expressions of human aggression, but highly artificial modes of behavior created by the greed of the first people to enjoy the fruits of the agricultural revolution. This revolution started when "some man of genius" first saw the possibility of harnessing the waters of the Nile for irrigation in a controlled system, inspired by the regular flooding of the river.[26] The rulers of this first civilization, the "Children of the Sun," had spread the myths and beliefs of Egypt over the whole world, including the Americas, by conquest. Smith urged modern statesmen to recognize that warfare was not the inevitable product of an instinct of pugnacity, but a highly artificial product of civilized society.[27] Before Egypt, there had been a "golden age" of peace over the whole earth.[28]

Although Smith does not refer to opponents of social Darwinism such as Drummond and Kropotkin, it seems likely that he would have been in sympathy with their efforts to portray nature in a more peaceful light. For all his references to natural selection, he did not believe that evolution required the creation of aggressive instincts. In the primates, at least, the growth of intelligence had resulted from the benevolent interaction of eye, hand, and brain within the safe environment of the forests. The plains represented a more dangerous habitat that had stimulated adaptive evolution in many of the other mammalian orders, but Smith shows no comprehension of the central thesis of Darwinism, which is that individuals must compete for resources in any environment, whether or not they are seriously threatened by predators.

F. Wood Jones was more explicit in his rejection of a Darwinian view of nature. In his *Arboreal Man,* Jones too adopted the metaphor of nature experimenting to reach its goal. In the formation of the mammals, "Nature has made several experiments in brain-building," leading to the parallel evolution of the marsupials and placentals.[29] The early primates had undergone an "arboreal apprenticeship" preparing them for the final steps toward humanity.[30] In the trees, life was much more peaceful than anything implied by the Darwinian struggle for existence.[31] Here the family group could originate in safety, with both parents helping to rear their helpless infants. "If higher ideals of conduct are to be acquired as an evolutionary process, it is in the family circle and in the society composed of families that these rudiments will be perfected."[32] The forests, at least, provided a tranquil but stimulating environment essential for the development of human mental and moral powers.

When he proposed his tarsioid theory of human origins, Jones was forced to admit that overspecialization for a forest environment had led the apes toward degeneration. His scorn for the Darwinian theory of an ape origin was now coupled explicitly with a rejection of natural selection: "Man is no new-begot child of the ape, bred of a struggle for existence upon brutish lines."[33] In the course of his career, Jones gradually moved toward Lamarckism, a mechanism that made convergence more plausible and also served to undermine the harsh image of nature created by Darwinism. In his *Design and Purpose* of 1942, he expressed open support for a philosophy of vitalism and teleology. Jones excused Darwin himself from wishing to promote the idea that nature was based on struggle, shifting the blame to Huxley and Haeckel. Darwin recognized the interdependence of all living things, a view that was at last becoming more popular with the rising interest in ecology. Kropotkin, "that neglected and tragic humanitarian," had tried to promote this more harmonious view of nature, but had been ignored by the over-enthusiastic Darwinians.[34] Convinced by the arguments of Huxley, even churchmen had now accepted that nature was amoral, and their acceptance had been responsible for much evil in the world.[35] Lamarckians such as Samuel Butler had resisted the trend, and had developed an alternative philosophy based on the assumption that all matter is in some sense alive.[36] If this was mysticism, it was a necessary product of our ignorance of the material causes underlying evolution. There might still be room for a spiritual element in nature, based on a recognition of the Creator as a Cosmic Mind rather than an anthropomorphic God, and of the spiritual perfectibility of the human mind.[37]

Jones kept these somewhat unconventional opinions largely out of sight until quite late in his career, but Robert Broom made no secret of his support for teleology. Broom had made his original reputation through his study of the mammal-like reptiles of South Africa, using these fossils to throw light on the origin of the mammals. The 1932 monograph in which he summed up his work in this field already contained a brief account of the ideas he would develop in his study of human evolution. Broom was convinced that something more than the natural selection of random variations was involved in progressive evolution, but at the same time he doubted that Lamarckism was the complete solution to the problem. He insisted that "we almost seem driven to assume that there is some controlling power which modifies the animal according to its needs and that the changes are inherited."[38] Noting that new initiatives in evolution always arose from unspecialized types, he offered the "daring suggestion" that nature had deliberately preserved a line of undifferentiated forms "as a sort of source from which all manner of what look a little like experiments might be made." The growth of the brain was only possible along this line, whereas the side branches were trapped by overspecialization. "In the side lines the evolutionary forces if we may personify them only seem to have thought of the immediate present; in the line of small generalized types, the evolutionary force is of a different type and seems to have foreseen the future."[39] Broom concluded by portraying mankind as the goal of this farsighted trend, all other living forms having been brought to an evolutionary halt by overspecialization.

> In fact we know of no invertebrates that could give rise to vertebrates. We know of no fishes that could again be the ancestors of amphibians; and no amphibians are now alive that could give rise to the primitive reptiles. And there are certainly no reptiles that could possibly evolve into mammals; nor existing mammals that could possibly evolve into apes. And it hardly seems possible that any living chimpanzee or gorilla or orang could be the ancestor of a new type of man.
>
> Apart from minor modifications evolution is finished. From which we may perhaps conclude that man is the final product; and that amid all the thousands of apparently useless types of animals that have been formed some intelligent controlling power has specially guided one line to result in man.[40]

One could hardly find a more explicit statement of the old idea that mankind is the main goal of evolution, or of the principle of theistic evolutionism once expounded by writers such as Mivart. Nor was this a casual speculation, since Broom went on to develop these ideas in his book entitled *The Coming of Man: Was It Accident or Design?* He still believed that some power had foreseen the appearance of mankind, but now suggested that it worked alongside a multitude of weaker forces whose creative efforts were more shortsighted. Intelligence had planned the various trends, but "curiously enough, the evolution has apparently not all been the result of one intelligence. We seem to see many agencies at work—some beneficient, some malignant; but amid it all some power has guided the main evolution to man. And man seems to be the end foreseen from the beginning."[41] Broom repeated his opposition to Darwinism and his halfhearted support for Lamarckism, but also brought in the evidence for nonadaptive orthogenesis. He repeated the metaphor of nature experimenting with the formation of mammals from generalized reptiles, but was forced to conclude that, since many of the experiments had ended in failure, "if there has been any intelligent force guiding evolution, that force only has been able to deal with the present and cannot plan for the future."[42] Clearly, this view was only meant to apply to the shortsighted forces in charge of adaptive radiation. The more farsighted power responsible for the main steps in evolution could certainly plan ahead. It had deliberately prevented the Therapsid reptiles from specializing so that they would later be free to evolve into mammals.[43] It had also produced the lobe-finned fishes, not because they were successful as fish, but because they were a necessary step toward the amphibians.[44]

Turning to the later stages of human evolution, Broom tried to clarify his position by developing the theme of a mulitplicity of guiding powers. Though many of the trends in primate evolution resembled the specializations found elsewhere in the animal kingdom, the growth of the brain resulted from an agency "apparently of a different order."[45] Presumably it was the same higher power that had preserved the line of generalized types as the basis of progressive evolution. Broom admitted that Lamarckian use-inheritance would promote brain development, but insisted that it was the increase of intelligence that had first permitted the use of tools—not the use of tools that stimulated the growth of intelligence.[46] In conclusion he argued that Lamarckism was right in its attempt to call in a psychic agency, but expressed his own belief that such an agency could not be limited to the minds of the individual animals. Evolution was directed from above by "some psychic or spiritual agency." The need felt by many biologists to call in vital forces to explain the functioning of the living body was an illustration of the same feeling on a smaller scale.[47] The ultimate goal of the most important trend "has not been merely the production of a large-brained erect walking ape, but the aim has been the production of human personalities, and the personality is evidently a new spiritual being that will probably survive the death of the body."[48] Now that biological evolution was finished, the spiritual agency was concentrating on the development of higher levels of human personality.

Broom was well aware of the embarrassment that his speculations caused to his scientific colleagues. He was lucky that his work on the fossil reptiles was of so high a standard that his reputation survived, although even so his career was a checkered one. His later discoveries of Autralopithecines brought him firmly back into the public arena, and many who read the reports of the new missing links may have been unaware of his unorthodox theoretical opinions. As late as 1951, Broom insisted on repeating his views in a popular account of his life's work.[49] It may seem strange that Broom—who had a reputation as a lady's man and whose honesty has been compared with that of a good poker player[50]—should have been so concerned to develop a spiritualized interpretation of nature. There is no doubt that he was sincere, however, and we must accept such apparent contradictions as part of the human situation. Although more outspoken than most, Broom gave expression to an anti-Darwinian feeling that was widespread among early twentieth-century paleontologists, and then remained true to his convictions as the climate of scientific opinion gradually changed.

Teilhard de Chardin was another scientist whose views on the spiritual quality of evolution were expressed in the postwar years. Active in paleoanthropology throughout the early part of the century, Teilhard had been convinced that evolution in general and human evolution in particular consisted of multiple parallel lines striving to reach an essentially spiritual goal. His work at the Piltdown and Peking sites was performed in the hope that fossil discoveries would confirm the parallel lines of human evolution predicted by the presapiens theory. Although expressed in a number of articles in the 1920s, Teilhard's theoretical position did not gain wide recognition until the posthumous publication of his complete philosophy of evolution in 1955 under the title *Le phenomène humain* (*The Phenomenon of Man*). The fact that such ideas could still attract wide attention at so late a date indicates the general public faith in a progressionist version of evolutionism. By the time Teilhard's views were published, however, they were out of touch with the revived Darwinism of the Modern Synthesis and received a mixed reaction from the scientists themselves. The concept of parallelism upon which Teilhard relied for his evidence that evolution is directed toward a goal was essentially a product of the earlier, non-Darwinian approach to evolution.

The ideas of Broom and Teilhard suggest that the teleological character of progressionism did not go unrecognized in the early twentieth century, even if many paleoanthropologists felt uncomfortable when it was pointed out too openly. Another illustration of this point can be seen in the work of a far more hardheaded individual: W. J. Sollas. In his *Ancent Hunters* of 1911, Sollas wrote of the development of the human brain as "that mystery of mysteries" and continued in a dismissive tone with the assertion: "Natural selection, that idol of the Victorian era, may accomplish much, but it creates nothing."[51] Opponents of Darwinism had always demanded to know the source of the higher types that succeeded in the struggle for existence, and had implied that they must be produced by a purposeful trend in variation. Sollas reflected on the power of the mind, especially in a state of inspiration, to transcend its apparent limits, and concluded "that the fundamental cause of the whole process of evolution is in reality an affair of the mind."[52] Although he never followed these speculations up in detail, and was aware of the offense they caused in some quarters, Sollas repeated them in the later editions of his book as a means of expressing continued support for those who saw evolution as the product of a more purposeful force than natural selection.[53]

Although willing to see the creation of the human mind as a central theme in evolution, Sollas did not follow Elliot Smith, Wood Jones, and Broom in their efforts to portray nature as a system of harmonious and benevolent interactions. For Sollas, as for many Lamarckians, progress was achieved only through mental effort. Those who made the effort advanced the cause of mental evolution, but any individual or any species that did not exert itself fully must expect to pay the price of failure: death or extinction. By treating the removal of these failures as an essential part of the process, Sollas thus presented a much harsher image of progressive evolution. His opinions bring us face to face with an apparent paradox in the history of evolutionism: that a progressionist and almost teleological view of nature could be used to stress either the benevolence or the harshness of the system's mode of operation.

The Image of Conflict

This paradox is by no means confined to theories dealing with the origin of mankind, but expresses a division that runs through the whole history of evolutionary thought.[54] For all that he has been labeled a social Darwinist, Herbert Spencer saw the inheritance of acquired characters as the main driving force of evolution. Spencer believed that *laissez-faire* individualism was the best way of encouraging the individual efforts essential for use-inheritance to function, and was prepared to accept that it would impose penalties on those who would not make the effort. Ernst Haeckel was also a Lamarckian, but his social Darwinism stressed racial conflict as the mechanism for eliminating the less progressive types. The American neo-Lamarckians were also quite prepared to dismiss the lower races as nature's failures, doomed to extinction. For every Lamarckian such as Samuel Butler or Kropotkin who viewed nature as a harmonious system designed to bring about moral as well as intellectual progress, there were several more who preferred to concentrate on the penalties that nature imposed on the failures. The claim that conflict was necessary for progress was thus by no means of an exclusively Darwinian origin. Most theories of evolution at the time were progressionist in tone, and few were free of the assumption that progress was intended to occur by nature or its Creator. Darwinism was but one manifestation of the desire to use progress as a means of justifying the elimination of those who, either through bad luck or laziness, had failed to keep up with the times.

It is a curious irony of history that even Lamarck himself noted the possibility that conflict might play a role in human evolution. He thought that the first race of apes to gain a prehuman level of intelligence would have driven all their rivals into remote corners of the earth and prevented their further advancement.[55] The race theorists of the later nineteenth century adopted a slightly different view. As the human species spread over the whole earth, those races that had moved into a less challenging environment would have gradually fallen behind. The highest form of mankind was, of course, the white race, formed in the stimulating environment of the northern regions. Once this race gained access to the territories occupied by the less developed forms of mankind, it would inevitably enslave or exterminate them.

Such views were still being expounded by Lamarckians in the early twentieth century, although the air of confidence had now begun to evaporate. We can see this approach in the work of the embryologist Ernest William MacBride, who is best known for his defense of Paul Kammerer in the "case of the midwife toad." MacBride developed a theory in which those organisms that responded in a constructive way to the challenges of the environment led the way in progressive evolution. It was he who declared at the end of the Zoological Society's debate on Wood Jones's tarsioid theory that "man did not evolve in response to an innate tendency lodged in a monkey's constitution, but in response to needs created by a change in the environment."[56] In his *Introduction to the Study of Heredity* of 1924, MacBride applied his theory to the race question, insisting that the Mediterranean race was inferior to the Nordic because it had evolved in a softer environment. This position led him to call for the compulsory sterilization of the Mediterranean element of the British population, which he believed had left Ireland for the industrial slums and now threatened to swamp the country with its overbreeding.[57]

MacBride took for granted the archaeologists' discovery that the racial composition of the European population had been disturbed by numerous migrations. It was this movement of races *after* their basic character had evolved that most interested Sollas. In a sense he wished to revive Lamarck's suggestion that migrations are caused by the appearance of higher races, which expand their territory and drive their inferiors before them. Racial character is not the product of the environment in which the race now lives, at least in the case of inferior races that have been forcibly driven to their present locations. In the preface to *Ancient Hunters,* Sollas declared: "I find little evidence of indigenous evolution, but much to suggest the influence of migrating races."[58] He implied that this view was considered heretical at the time, but he soon had the pleasure of announcing that his heresy had become the new orthodoxy.[59]

The purpose of Sollas's book was to locate the surviving remnants of the earlier stages in human evolution, now driven to Australia, the far north, and other marginal locations. The ancestors of the inferior modern races—or at least peoples with identical levels of culture—had originally occupied Europe, as revealed by the sequence of cultural epochs identified by the archaeologists. The implication was "that the surviving races which represent the vanished Palaeolithic hunters have succeeded one another over Europe in the order of their intelligence: each has yielded in turn to a more highly developed and more highly gifted form of man."[60] Nature was clearly a progressive system, presumably because of the role played by mind in evolution, but Sollas was now led to reflect on the moral implications of how the world dealt with those who did not keep up with the race.

> What part is to be assigned to justice in the government of human affairs? So far as the facts are clear they teach in no equivocal terms that there is no right which is not founded on might. Justice belongs to the strong, and has been meted out to each race according to its strength; each has received as much justice as it deserved. What perhaps is most impressive in each of the cases we have discussed is this, that the dispossession by a new-comer of a race already in occupation of the soil has marked an upward step in the intellectual progress of mankind. It is not priority of occupation, but the power to utilise, which establishes a claim to the land. Hence it is a duty which every race owes to itself, and to the human family as well, to cultivate by every possible means its own strength: directly it falls behind in the regard it pays

> to this duty, whether in art or science, in breeding or in organization for self-defence, it incurs a penalty which Natural Selection, the stern but beneficient tyrant of the organic world, will assuredly exact, and that speedily, to the full.[61]

It may seem odd that Sollas should now appeal to natural selection, but there were many opponents of Darwinism who were quite willing to call in that mechanism at the racial level in order to perform the purely negative task of eliminating the unfit. In fact, Sollas's words here echo those used by that enthusiastic Darwinist and social Darwinist Karl Pearson, showing how close the supporters and opponents of natural selection could come on the race question.[62]

Sollas's words reflect the "might is right" philosophy of imperialism quite explicitly, but his concept of racial competition was in many ways quite typical of the time. We have already seen in chapter 4 how the exponents of the presapiens theory tended to assume that the Neanderthals had been driven to extinction by invading representatives of modern *Homo sapiens*. Boule, Keith, and may others shared this assumption, which in turn seems to have been modeled on the anthropologists' belief that Europe had been populated by means of a series of invasions. The claim that superiority of technology gives one race the right to dispossess the "lower" forms of mankind constitutes the essential logic of imperialism. It can hardly have been a coincidence that such a theory should have become popular in paleoanthropology at the very time when empire-building had come to figure so prominently in European political thought.

It is significant that Elliot Smith, although he accepted that the brutal Neanderthals had died out, did not dwell at length on the means by which they had been replaced. He was interested in Sollas's ideas because of their implications for diffusionism, but was careful not to compromise his less violent image of human evolution by stressing the element of racial competition. Louis Leakey wrote even less that could be taken as support for struggle as the cause of racial extinction. He argued that different species of humanity had been able to live side by side, in effect leaving the extinction of the Neanderthals unexplained. In this case, Leakey's rapport with the black Africans among whom he was raised must have encouraged him to turn a blind eye to what most of his contemporaries saw as the obvious cause of the Neanderthal extinction.

Sollas did not believe that the Paleolithic races had evolved in Europe, but specified no other location as the center of dispersal, and may even have believed that each race originated in a different location. He favored the view that the climatic revolutions of the Pleistocene had stimulated the later phases of human evolution, but had no wish to limit the effects of these revolutions to any particular location.[63] The belief that the progressive steps in evolution were always triggered by changes in the environment was, of course, by no means confined to paleoanthropologists. The paleontologist Richard Swan Lull issued one of the most powerful statements of this view in the conclusion of his *Organic Evolution* of 1917.

> Thus time has wrought great changes in the earth and sea, and these changes, acting directly or through climate, have always found somewhere in the unending chain of living beings certain groups whose plasticity permitted their adaptation to the newly arising conditions. The great heart of nature beats, its throbbing stimulates the pulse of life, and not until that heart is stilled forever will the rhythmic tide of evolution cease to flow.[64]

The claim that evolution only responds to an external stimulus bears a superficial resemblance to one of the principles of Darwinism, but like many of the progressionists we have encountered, Lull was really interested more in the chosen few who would be able to respond to the challenge. Applied to human origins, Lull saw his position as lending support to the theory that located the focus of evolution not in the ice ages, but in the gradual drying-up of central Asia in Miocene and Pliocene times.[65]

The central Asia theory had been inspired by W. D. Matthew's interpretation of mammalian evolution, according to which the stimulating climate of this region throughout the Tertiary had made it the center of progress.[66] Matthew's whole thesis rested on the assumption that the modern representatives of lower types are to be found not at the center of evolution but in the marginal locations to which they have been driven by more successful forms appearing at the center. While stressing constant progress in central Asia, it also highlighted the waves of migration caused by the pressure of higher upon lower types. In effect, Matthew provided a focus for Sollas's more general belief that racial competition and migration were the inevitable byproducts of progressive evolution. As expressed by Davidson Black, the theory implied that the earliest forms of humanity "in turn were overwhelmed or forced to migrate elsewhere by the constant pressure of competition with more progressive generations within the broad extent of their original dispersal centre."[67]

If one considers that Black had studied for a time under Elliot Smith, it is clear that his support for the central Asia theory led him to deviate from Smith's more peaceful image of human evolution. Black was no Darwinian, and accepted that initiative was the character that distinguished the progressive types that faced up to the more stimulating environment. But for him the consequence was not a golden age of prehistory, but a series of conflicts between successive waves of racial types.

An entirely different argument for the evolutionary significance of racial competition came from Sir Arthur Keith. Sollas and Black had not needed to postulate an aggressive instinct in mankind, since they saw conflict and migration as unfortunate consequences of a progressive evolutionary process that was actually driven by entirely different forces. Keith denied the significance of migration and held that competition between tribal groups was the chief cause of evolution. In this sense, Keith was more of a Darwinist, and indeed he seems to have had a kind of love-hate relationship with the Darwinian theory throughout his career. In his *Darwinism and What It Implies* of 1928 he wrote: "Competition is not confined to human rivalries and struggles; it pervades the whole kingdom of life; it is the basis of Darwin's doctrine of evolution; it has been, and ever will be, the means of progressive evolution."[68] Yet only a few years later his *Darwinism and Its Critics* defended the general theory of evolution by insisting that far more than natural selection was involved.[69] Keith was really interested in competition only at the tribal rather than the individual level. In chapter 8 we have already seen that he preferred to explain adaptive evolution by appealing to the purposeful character of individual growth, thereby adopting the teleological viewpoint of Lamarckism without a direct appeal to use-inheritance.

Keith developed an early interest in the origin of races, although he did not share Sollas's view that racial differences could be explained in terms of progress or adaptation. He suspected that hormonal effects might be responsible for the creation of racial characters, but, as he later explained, this left him puzzled as to how a group of individuals with a particular hormonal balance could remain isolated long enough for their characters to become fixed in the race.[70] Geographical barriers might work, of course, but Keith doubted that they would be effective enough. Already during World War I he had begun to feel that tribal warfare played a role in evolution.[71] In 1916 he drew attention to the gregarious instinct in mankind and postulated that at a very early stage in our evolution, human beings had joined into competing tribal groups—of which the modern nations were the civilized equivalent.

"The evolutionary human unit in the past has been the primitive tribe; the world was covered with a mosaic of them; it is clear that in the future the unit is to be a gigantic one—one which will occupy a whole continent."[72] Throughout the rest of his career, Keith tried to fill in the details of his idea that the tribal instinct might explain the isolation and evolution of racial characters.

In 1931 Keith contributed an article on the origin of races to a collected volume entitled *Early Man.* He acknowledged that modern racial divisions made the work of the League of Nations more difficult, but felt that it was necessary to inquire why the sense of racial identity was so deeply rooted. The further we go back in the fossil record, he believed, the greater the number of racial types. This observation suggested that the division of the species into races was nature's way of trying to evolve higher forms of mankind through experimentation. Only in modern times have the early diverse types gradually consolidated themselves into four major groupings.[73] The early divisions could not have existed had primitive mankind wandered freely over the earth, so already each incipient race must have been bound to its own homeland.[74] What we now call patriotism is the modern manifestation of the "herd instinct" that had once been essential for evolution. Although transformed and no longer fulfilling its original role, this instinct, Keith felt, was still important. H. G. Wells and the supporters of a world state condemned patriotism as an emotion incompatible with peace, but in response Keith noted that the most enlightened races now governed those areas whose original inhabitants had been incapable of exploiting the land properly.[75] Elimination of the empire-building instinct would lead to the "utter domestication" of mankind, with disastrous consequences for future evolution.[76]

World War II, far from dampening Keith's enthusiasm, merely reinforced his belief that conflict was essential for evolution. In his *Essays on Human Evolution* and *A New Theory of Human Evolution* two years later, he expounded the final version of his thesis. Nature had conditioned human instincts to ensure the maintenance of barriers between tribes, so that each could develop its full potential in isolation.[77] Both Christianity and the campaign for a world state were out of touch with the realities of human nature.[78] Keith now openly criticized Elliot Smith's theory of a prehistoric golden age, insisting that the early tribes of mankind had always been in conflict with one another.[79] The *New Theory* tried to link this interpretation of racial evolution with more general questions on the origin of mankind. Keith now admitted that the whole issue would be greatly simplified if the Piltdown fragments could be ignored, and favored the Australopithecines as an important stage in human evolution.[80] He now held that these dark-skinned "Dartians" had spread over the world from their African cradle, diversifying into geographical races, each of which had evolved independently toward fully human status in its own area.[81] Tribal competition had been vital both to the original differentiation and to the later development of the races.

Keith's theory rested on the highly un-Darwinian assumption that human instincts have been developed, not to benefit the individual, but to allow the evolutionary process to work. But in the *New Theory* he also pointed to another possible source of human aggression, which could be accounted for in more realistic terms. This was the hunting instinct, the importance of which had been noted by a couple of earlier writers.[82] It was widely assumed that mankind had engaged in hunting at least since the invention of effective stone weapons, but there had been few efforts to explore the potential implications of this way of life for the evolution of the mind. Most authorities seem to have believed that hunting had commenced so late in our history that it could have had little effect on our instincts. They would probably have agreed with Charles Lyell, who had argued in 1872 that early mankind had lived by gathering vegetable food in a tropical environment, and had only begun to hunt after the expansion into the temperate zone.[83] We have seen that Elliot Smith was more worried about how our ancestors protected themselves from predators when they ventured out onto the plains than about the need for hunting to supply food. An American, Charles Morris, had suggested as early as 1890 that mankind rose to become head of the animal kingdom by struggling against all the other species and forcing them to fear him, but no one seems to have taken this idea seriously.[84]

In fact, Keith was able to cite only two writers who had explored the possibility that the early adoption of a hunting lifestyle might have played a role in human evolution. In 1913 the English physician, Harry Campbell, had suggested that the ancestors of the human race had abandoned the trees in a search for animal food.[85] At first limited to "vermin," they had gradually learned to trap and hunt larger game. By 1917 the stimulus of the war had led Campbell to inquire into the origin of what he took to be a natural combative instinct in mankind. The desires for food and for a mate were both relevant, of course, but he now argued that it was the hunting life that contributed most to the refinement of this instinct. Hunting may even have been responsible for the emergence of human intelligence.

> This agile, intelligent, hand-endowed precursor of man was compelled to rely upon his intelligence in hunting his prey; for blind instinct he had to substitute strategy; for natural weapons, weapons made by hand. Once the pre-human ape started his career of intelligent, as against instinctive, hunter, he began a struggle in which it was inevitable that a higher and ever higher grade of intelligence should continue to evolve by natural selection until sufficient mental capacity had been attained to render its possessor supreme in the chase.[86]

In other words, mankind had the instinct to hunt, but no instincts governing *how* to hunt, and the result had been the growth of intelligence. Although a supporter of natural selection, Campbell had no explanation of how the original hunting instinct originated. He suggested that the emergence of tribal life led to conflict, often over hunting grounds, and that the constant warfare had sharpened the fighting instinct that had its origin in hunting itself.

In an apparently independent development, the psychologist Carveth Read also began to speculate on the importance of hunting. Like Campbell, he accepted natural selection as the mechanism of evolution. He began to think about the cause of the transition from the animal to the human mind, and was not satisfied, he tells us, "to say year after year that hands and brains were plainly so useful that they must have been developed by Natural Selection."[87] The answer, he declared in *The Origin of Man and His Superstitions* of 1920, was the effect of the transition to a hunting way of life. He doubted that the search for a new source of food was prompted by a change in the climate. More probably a "spontaneous" variation had given some apes a taste for animal food.[88] Natural selection thus only came into play to back up an essentially purposeless genetic mutation. The adoption of an upright posture was necessary to facilitate hunting, as was the increase in social activity. Early mankind adopted a lifestyle resembling that of the wolves, becoming "in short, a sort of wolf-ape (*Lycopithecus*)."[89] Tribal warfare would also have been encouraged, and our ancestors would have killed off all rival carnivores, including "such experiments in human nature as Pithecanthropus, Eoanthropus and Neanderthalensis."[90] The instincts generated by the hunting lifestyle explain why modern people are so imperfectly sociable—like other carnivores, we tend to fight over the kill after we have cooperated to hunt it.[91]

No one seems to have paid much attention to Campbell and Read. Matt Cartmill notes that even Raymond Dart—later a strong exponent of the hunting hypothesis—was not aware of their work.[92] Cartmill suggests that paleoanthropologists were not yet ready for so harsh an image of human evolution. It was certainly possible to think of paleolithic hunting more as a source of social activity than as the product of an instinctive blood lust. Against this interpretation, however, must be set the willingness of Keith, Sollas, and others to see tribal warfare as an important component of human evolution. Reports that early mankind had engaged in cannibalism also proliferated, based especially on the discoveries at Krapina and later at Chou Kou Tien. Although Elliot Smith had led a move to create a more pacific image of early human nature, his position can hardly be said to have initiated a consensus.

Campbell and Read were ignored not because their theory was totally at variance with orthodox views of human nature, but, as Cartmill admits, because most paleoanthropologists at the time had failed to recognize the importance of specifying the advantages that resulted from hominization. It may be significant that both Campbell and Read were supporters of Darwinian natural selection, significant not because this position gave them an instinctive bias toward a vision of nature based on struggle, but because it allowed them to escape at least partially from the progressionist assumptions of their time. These assumptions, as Read plainly stated, encouraged most thinkers to imagine that the advantages of being human were so obvious that there was no need to specify them in detail. Read knew that those advantages would have to be realized within a particular way of life adopted by our early ancestors. Yet even he was not able to think of the advantages permitting a move into a new environment or a new ecological niche, since he fell back on the idea of a spontaneous variation giving a taste for flesh. Therefore his theory contained a fatal weakness, because the one thing that no one at the time was prepared to accept in a theory of human origins was an element of pure chance.

Raymond Dart assumed from the start that *Australopithecus* engaged in the hunting of at least small animals, although it was only later in his career that he came to see hunting as a major force in the shaping of human nature. In his announcement of the discovery at Taungs he stressed the stimulating environment of the African veldt, but seems to have been more concerned about the dangers offered by large predators. "Darwin has said 'no country in the world abounds in a greater degree with dangerous beasts than Southern Africa,' and, in my opinion, Southern Africa, by providing a vast open country with occasional wooded belts and a relative scarcity of water, together with a fierce and bitter mammalian competition, furnished a laboratory such as was essential to this penultimate phase of human evolution."[93] In a popular article, Dart concentrated more on the benefits that an intelligent creature could gain from the plains environment. The material in which the Taungs skull was found contained the remains of turtles, birds, and small mammals, almost certainly the food of *Australopithecus* itself. Dart believed that "we have in *Australopithecus* a troglodytic anthropoid which, in addition to, and probably because of its increased intelligence and its skill in using its hands as hands and its feet as feet, had become sufficiently weaned from its frugivorous tropical diet to vary its table with the fruits of the chase."[94] The hunting of small game was thus a factor that enabled the emerging hominids to live comfortably on the plains, although there is no hint of organized big-game hunts here. We can readily understand Dart's claim that such a way of life would be a great stimulus to the growth of intelligence, but as yet there was no suggestion that mankind was developing into a bloodthirsty monster.

As late as 1940, Dart was still emphasizing the positive aspects of *Australopithecus*'s predatory lifestyle, as the means by which an intelligent ape had enjoyed a new and rich environment. His language was, however, beginning to carry overtones that brought out the violent aspects of this way of life. *Australopithecus* was "a cave-dwelling, plains-frequenting, stream-searching, bird-nest-rifling and bone-cracking ape, who employed destructive implements in the chase and preparation of his carnivorous diet."[95] The Mapangsat site now seemed to reveal more evidence of tool-using by the Australopithecines, particularly the use of bones as weapons, leading Dart to coin the term "osteodontokeratic culture" to denote the use of bone, tooth, and horn tools. Large animals had been killed there, so large that Dart originally thought they were the prey of advanced humans, but now he was sure that the Australopithecines had already been hunting big game.[96] More disturbing was the jaw of a twelve-year-old who had apparently been killed by a blow to the chin. By 1948, this evidence had suggested to Dart that the Australopithecines had been a violent race, killing big game and their own fellows quite readily. He now argued that "the thug technique of bashing heads in with any handy bone or brick had a heritage of at least a million years."[97]

In his *Adventures with the Missing Link* of 1959, Dart still held that it was the initiative of the early Australopithecines that had led them onto the plains, but he also assumed that they had developed a preference for eating meat. "The ancestors of *Australopithecus* left their fellows in the trees of central Africa through a spirit of adventure and the more attractive fleshy food that lay in the vast savannas of the southern plains."[98] They began to use tools, not because of their larger brains, but because they had instinctively picked up sticks to hunt and defend themselves, thus forcing themselves into bipedalism. Dart now spoke openly of modern humanity's loathsome cruelty as a continuation of the blood lust of our carnivorous and cannibalistic ancestors.[99]

It is significant that Dart was encouraged to prepare this account of his work by the Hollywood scenario writer Robert Ardrey, who had become fascinated with the idea that evolution had programmed aggressive instincts into humanity.[100] In books such as *The Territorial Imperative* and *The Hunting Hypothesis,* Ardrey drove home this theme, becoming a leading exponent of what was sometimes known as the anthropology of aggression. Liberals and socialists alike condemned this approach as an attempt to portray capitalism and nationalism as the inevitable products of human instincts. Most social scientists were by now completely opposed to the claim that human beings have built-in instincts of a kind that may affect their social behavior. In these postwar years Dart had allied himself with a trend that ran counter to many aspects of academic orthodoxy, but was capable of arousing intense public interest. In some respects, the resulting debate foreshadowed the more recent controversy over the application of sociobiology to the human species.[101]

There is, however, one important difference between the two debates. Sociobiology is very much a product of modern Darwinism; it holds that natural selection will inevitably lead to organisms being programmed with instincts that will ensure reproductive success. Although Campbell and Read were Darwinists of a sort, neither used natural selection to explain the origin of the hunting instinct. Raymond Dart emerged from Elliot Smith's school, in which intelligence and initiative were seen as the driving forces of human evolution, and there is nothing to suggest that he was committed to natural selection. Keith, whose new theory also depended on the concept of instinctively violent behavior, did not appeal to natural selection to explain how the instinct was created. In both cases, the source of the violent instincts lay in a distinctly non-Darwinian view of evolution. If the climate of opinion in the 1950s and 1960s was receptive to the idea of aggressive instincts, the demand was satisfied by a view of human nature that had been developing alongside the more optimistic approach to evolution in the prewar years. The new Darwinism of the Modern Synthesis was slower in responding to the same demand—if we accept the critics' claim that sociobiology has now replaced the anthropology of aggression in the attempt to provide a scientific legitimation for conservative values.

Ardrey claimed that the brain-growth theory of human evolution had been promoted by anthropologists who wished to see our capacity for cultural development as something that cut us off completely from the forces of biological evolution. The rejection of instinct theory by the social sciences was thus, in effect, responsible for the success of the Piltdown fraud.[102] This claim cannot be accepted, if we take it to mean that paleoanthropologists such as Keith and Elliot Smith were influenced by the early twentieth-century revolution in sociology and cultural anthropology. Keith's growing belief in the importance of the tribal instinct clearly violated the assumptions of the new sociology, and Smith's diffusionism did not allow for the equal value of all cultural developments. However, there is a sense in which Smith's theory of human evolution did create a situation in which the social sciences were left with the freedom to cut themselves off from paleoanthropology. By paying comparatively little attention to the growth of social behavior, and by stressing the role of initiative, this theory offered few challenges to the claim that culture has no foundations in biological evolution. Smith shared the sociologists' distrust of the concept of instinctive aggression, but went even further in his determination to show that all human violence is unnatural. Diffusionism held that only one culture has ever become violent, and it was a result of a geographical accident. Smith's viewpoint had its roots in a pacific image of the natural world, but it was far less at variance with the new social sciences than the alternatives that appealed to the role of instinct in evolution.

In the 1940s, the gradual acceptance of the Australopithecines as ancestral to modern humanity began to undermine Smith's theory of brain growth. But the simultaneous reemergence of Darwinism as a component of the Modern Synthesis did not generate immediate support for a more violent image of human evolution. If the new Darwinism created a situation in which hunting could gradually come to be seen as more important, it did so not by stressing the violence of nature but by destroying the old progressionism and forcing everyone to think more carefully about the benefits that had been conferred by bipedalism and the move out onto the open plains. Individuals or groups must have benefited from the transition in their daily lives, and the additional source of food offered by hunting seemed to offer such a benefit. The paleoanthropologists of the 1960s thus became prepared to consider the possibility of our origin from a "killer ape," despite the suspicions of the social scientists. As Matt Cartmill suggests, though, the theory seems also to have responded to a general trend in postwar culture; not so much a self-congratulatory revival of conservative values as a more general fear that we may indeed be plagued by an uncontrollable blood lust.[103] The image of "man the hunter" served a mythical function, expressing a deep-seated feeling that may have been an inevitable product of the nuclear age.

Campbell's anticipation of the hunting hypothesis had been inspired by the horrors of World War I, but the war seems to have had surprisingly little effect in creating a climate of opinion receptive to the idea. The struggle in the trenches caused many intellectuals to lose faith in the progressive character of Western civilization, but most paleoanthropologists escaped this disillusionment. The concept of racial or tribal struggle had become popular before the war, and seems to have remained acceptable, although few went along with Keith's elaboration of the idea. But competition at this level could more easily be reconciled with progress, by assuming that some mechanism was needed to eliminate those who fell behind and threatened to hold up the general advance. Theories of human origin continued to stress that the growth of human intelligence was an inevitable product of nature's struggle to achieve perfection. Nothing could more clearly illustrate the limited impact of Darwin's theory based on the natural selection of individual variations by the environment, however effective his writings may have been in converting the world to a general belief in evolution. A progressionist, if not downright teleological view of nature continued to prevail in paleoanthropology, as indeed it did in several other areas of evolution theory. Only after World War II did the loss of faith in progress spread into this area of science, aided by the Modern Synthesis' rout of the anti-Darwinian mechanisms of evolution.

Modern theories have again challenged the image of the killer ape, and alternatives to the hunting hypothesis are freely, if somewhat inconclusively, debated. A more comforting account of our ancestors as peaceful gatherers has emerged once again, perhaps with the females playing a more significant role than the aggressive males. It would be easy to sneer at the inconclusive nature of the debates, and there is no shortage of modern writers who take up any excuse to belittle evolutionism. But at least the latest hypotheses look for realistic explanations of why our ancestors took the path toward becoming human. We no longer take it for granted that the human mind is the natural goal of the evolutionary process. If the progressionism of the earlier theories reflects a more sophisticated version of the teleology originating from the old belief in divine creation, the hunting hypothesis brought us back to the pessimistic outlook of the fall from grace. The latest theories try to avoid this last source of despair without slipping back into the old, facile optimism.

Epilogue

At several points in this study we have referred to the dramatic change in thinking on human evolution that took place in the 1940s. In the previous decade, ideas based on an earlier, anti-Darwinian climate of opinion still flourished. Parallel evolution, whether by Lamarckism or orthogenesis, was still taken for granted, even if the theories of Wood Jones and Osborn were seen as overenthusiastic applications of this principle. The various "genera" of fossil hominids such as *Eoanthropus, Pithecanthropus,* and *Sinanthropus* were still taken seriously as the products of independent lines of human evolution. Even the Neanderthal race was seen as a distinct species, perhaps a distinct genus. Many authorities believed in the extreme antiquity of a large-brained human form, making it difficult to present Dart's *Australopithecus* as the sort of ape-human intermediate to be expected on theoretical grounds. In the course of the 1940s, however, the situation changed completely, and paleoanthropology began to take on its modern form. This epilogue briefly surveys this transition, sketching in a little more detail the factors chiefly responsible for undermining the old climate of opinion.

The old order did not vanish overnight. Scientists who had matured within the earlier tradition found it difficult to throw off their old training and continued to publish as though nothing had happened. The 1949 edition of E. A. Hooton's *Up from the Ape* still had the white races originating from *Eoanthropus,* whereas the Australians came from a separate *Pithecanthropus* stock.[1] Louis Leakey reissued *Adam's Ancestors* in 1953 with a diagram indicating his continued acceptance of a separate subfamily of Palaeoanthropinae, including the Neanderthals.[2] Leakey now treated the Australopithecines as a third separate line, originating from the Palaeoanthropinae in the early Pliocene. Throughout his later career he sought evidence that the genus *Homo* had existed before the then-known Australopithecines, a search ultimately crowned with success by the discovery of *Homo habilis* in 1959.[3] But if Leakey was able to preserve a modernized image of multiple-branching hominid evolution, his dismissal of the Neanderthals as members of a distinct subfamily soon came to be regarded as outdated. Few authorities now countenanced the parallel and independent evolution of two forms as similar as modern *Homo sapiens* and the Neanderthals.

Perhaps the clearest indication of the change that had taken place during the 1940s was the consensus achieved at the 1950 Cold Spring Harbor Symposium on Quantitative Biology, devoted to "The Origin and Evolution of Man." Here S. L. Washburn, in summing up the new approach to primate evolution, concluded with the following observation:

> . . . it might be repeated that this is an appropriate time to reconsider the problems of the origin of man; for the traditional phylogenies have been upset by the discovery of new fossils; the old theories of orthogenesis, irreversibility and the supremacy of non-adaptive characters have been proved false; and because experimental procedures offer methods of raising some conclusions beyond the level of individual opinion.[4]

The new fossils to which Washburn referred included the discoveries from China and Java that had allowed Weidenreich and others to reemphasize the importance of *Pithecanthropus* as a stage in human evolution. Equally important were Broom's Australopithecines, which had now confirmed that small-brained yet upright hominids had occupied southern Africa during the early Pleistocene. By themselves, these discoveries might eventually have forced all paleoanthropologists to reconsider the interpretation of human evolution that had remained popular in the 1920s and 1930s. The Piltdown remains had been left in an increasingly isolated position, paving the way for acceptance of the view that bipedalism had preceded the great expansion of the human brain.

Note, however, that Washburn linked acceptance of these new fossils to the revolution in biological thinking that had now begun to sweep away the old theories of directed evolution. The revival of Darwinism in the form of the Modern Synthesis of natural selection and genetics helped to reveal the teleological character of the progressionist theory of brain development, and undermined the concept of parallel evolution which had been used to dismiss so many inconvenient fossils to side branches unconnected with the main line of human evolution. Without this theoretical revolution, the fossil discoveries might not have had the same impact, or their implications would at least have taken much longer to be appreciated. Neither Davidson Black nor Robert Broom had been forced to give up his belief in a progressionist version of evolution by his fossil discoveries. If the older theoretical climate had continued unchallenged, it is quite possible that most paleoanthropologists would have been able to accept the new fossils without abandoning the belief that brain development has been the driving force of human evolution. But the Modern Synthesis forced the next generation of scientists to think more carefully about the process of change, and made it easier to see the latest fossils as evidence that the achievement of an upright posture had preceeded the major expansion of the human brain.

The Modern Synthesis drew upon the techniques of population genetics pioneered in the 1930s by scientists such as R. A. Fisher, J.B.S. Haldane, and Sewall Wright.[5] The resulting genetical theory of natural selection was translated into terms more familiar to the field naturalists by Theodosius Dobzhansky's *Genetics and the Origin of Species* of 1937. Books such as Julian Huxley's *Evolution: The Modern Synthesis* and Ernst Mayr's *Systematics and the Origin of Species* (both of 1942) helped to promote the new Darwinism. Perhaps most important in the present context was George Gaylord Simpson's *Tempo and Mode in Evolution* of 1944, which showed that the fossil record was compatible with the new evolutionary mechanism. The evidence for orthogenesis was weighed and found wanting: wherever sufficient fossils were available to reconstruct a phylogeny in detail, the course of evolution turned out to be a haphazard branching process rather than a linear ascent toward a fixed goal. Simpson himself contributed to the 1950 Symposium, emphasizing the limited extent of parallelism and the adaptive character of all major evolutionary developments.[6]

As Washburn indicated, the new theoretical approach undermined the plausibility of the assumptions upon which the belief in extensive parallelism within primate evolution had rested. Washburn himself stressed the adaptive character of the major steps leading toward mankind. He appealed to new fossil evidence suggesting that the brains of the earliest mammals were not advanced beyond those of the reptiles, thus discrediting the theory that the brain led the way throughout mammalian evolution. It now appears that the brain developed only in response to external pressures. Our primate ancestors went through three major locomotive adaptations and one major reorganization of the senses, the brain doubling in size after each event. Washburn concluded: "Viewed in this way, the remarkable size of the human brain is due to the number of times the organ had to adjust to a new way of life."[7] The final step, unique to mankind, was the rapid expansion of the brain after the acquisition of an upright posture.

Washburn argued that attention had now shifted from the classification of fossil hominids to the evolutionary mechanism itself, thereby exposing the lack of any real justification for the plethora of specific and generic names assigned to the fossils. The desire to be identified with a new type of early hominid had certainly led the discoverers of the fossils to exaggerate their significance. But in implying that the range of species and genera had been created out of indifference to evolution theory, Washburn may have oversimplified the situation. The multiplication of taxa had not taken place in violation of the older evolutionary ideas, since these had allowed for parallel and independent lines of human evolution. It was the Darwinian character of the new theory that encouraged a dramatic simplification of hominid taxonomy. The interbreeding of modern human races now made them members of the same species by definition. Although there were still a few efforts to treat the races as distinct species, the majority of anthropologists and archaeologists were already beginning to turn against this approach, possibly out of growing dismay at the use to which racist doctrines had been put in Nazi Germany. Once the races were accepted as belonging to a single species, it was clear that modern *Homo sapiens* exhibits a considerable degree of variability. It thus became obvious that too much weight had been given to the comparatively trivial morphological differences upon which the earlier classification of fossil hominids had been based.

Ernst Mayr took this theme up directly at the 1950 Symposium, advocating the inclusion of all the fossils within the genus *Homo.*[8] *Pithecanthropus* and *Sinanthropus* would become merely subspecies of *Homo erectus,* now widely assumed to be the species ancestral to *Homo sapiens.* Most contributors to the symposium agreed that the Neanderthal race would have to be included as a subspecies of *Homo sapiens,* although it was still widely accepted that the rapid disappearance of this race from Europe meant that our own ancestors had constituted a different racial group. Although the Neanderthals show some resemblance to *Homo erectus,* they were regarded as a divergent and now extinct racial branch of *Homo sapiens,* rather than the original form of our species. In the 1960s, C. Loring Brace revived the "Neanderthal phase of man" theory (see chapter 4) and the status of the Neanderthals has remained controversial ever since.

Mayr suspected that even the Australopithecines ought to be included within the genus *Homo,* although the subsequent consensus accepted a separate genus, *Australopithecus,* containing a number of distinct species to accommodate the later discoveries of Broom and others. The second element in the reinterpretation of human origins was the recognition of this genus as the best indication of the ancestral form from which the earliest forms of *Homo* had evolved. The uniting of the Java and Peking specimens into a single species, *Homo erectus,* made a far more convincing ancestor for *Homo sapiens* than the increasingly problematic Piltdown type, thus confirming that a completely upright posture had been achieved before the full development of the brain. In these circumstances, something resembling *Australopithecus* became the most plausible link in a chain that would lead from an ape to *Homo erectus*. The South African finds that confirmed the upright stature of the Australopithecines thus filled in what was now, in Washburn's words, an "anticipated stage in human evolution."[9]

Few were prepared to accept Weinert's close linkage of mankind and the chimpanzee, but Gregory's view of a fairly recent emergence of the human line from a more generalized ape ancestor now seemed to have been vindicated. Gregory himself had begun to see the Australopithecines as a possible link in his scheme even before the war, and was now fully behind the new interpretations.[10] The new fossil discoveries confirmed that the Australopithecines were human in certain important respects and indicated, in Washburn's opinion, that the first step in the emergence of the human form had been the acquisition of an upright posture.[11] The same view was now being advocated in Britain by Le Gros Clark. In 1946 and again in 1950, Clark argued that Broom's discoveries in South Africa filled in what appeared to be the first step in human evolution from the apes.[12] Although once a leading supporter of orthogenesis in primate evolution, Clark now accepted the Darwinism of the Modern Synthesis. In his *History of the Primates* of 1949 (designed to replace Smith Woodward's guide to the fossil hominids in the British Museum) he argued that the similarities between Australopithecines and humans were too precise for them to have originated independently by parallel evolution.[13] This point suggests that, by themselves, the new fossil discoveries were not enough to generate a new and simpler image of human evolution. In the old climate of opinion, convincing evidence that the Australopithecines were bipedal could still have been accommodated by treating them as another parallel line, participating in some but not all of the trends essential for hominization. In a 1955 survey of the fossils, Clark followed Simpson in decisively rejecting orthogenesis in favor of adaptive evolution.[14]

Like several other contributors to the 1950 Symposium, Washburn mentioned the possibility that the acquisition of an upright posture might have taken place by what Simpson called quantum evolution.[15]

This was a process by which a small, isolated population might evolve a new adaptive structure rapidly, and by passing through an intermediate, nonadaptive stage (something like the rapid speciation events in the modern theory of punctuated equilibria). Several supporters of the Modern Synthesis were at first willing to compromise in this limited way with earlier views of nonadaptive evolution. It may be that such an idea made it easier for paleoanthropologists to accept the upright posture as the key step in human evolution, since it freed them from having to worry too much about the transition process itself. As Stephen Gould has shown, however, Simpson himself soon abandoned quantum evolution, and the Modern Synthesis as a whole hardened into a more rigid adaptationism.[16] Even in the early stages, it was still necessary to specify the adaptive benefits of the final stage in the process (a fully upright posture), and the emphasis on the adaptive character of human evolution could only increase thereafter.

The geological position of the Australopithecine fossils meant that they might be too late to serve as the ancestors of modern humanity. Even so, they were almost certainly representative of a stage through which the human line had passed somewhat earlier. Clark used a diagram to show how the Australopithecines could have survived for some time alongside the earliest members of the genus *Homo* that had evolved from them.[17] In his 1950 paper, Ernst Mayr suggested that only one hominid species could have existed at any one time, given the wide ecological diversity allowed by increased intelligence.[18] It soon became apparent, however, that the Australopithecines had indeed survived in Africa long after a larger-brained form had appeared. The first step in the process was the recognition of the coexistence of the robust and gracile forms of *Australopithecus*. The robust species was clearly not ancestral to mankind, and represented a side branch of hominid evolution leading to extinction. At first, though, it still seemed possible that an earlier form of the more lightly built *Australopithecus africanus* might be the direct ancestor of the genus *Homo*.

We have already seen that Louis Leakey remained convinced that none of the Australopithecines were ancestral to mankind. In 1953—the date of the last edition of Leakey's *Adam's Ancestors*—the fraudulent character of the Piltdown remains was finally confirmed.[19] *Eoanthropus* could therefore no longer be used as a non-Australopithecine ancestor for mankind. Yet Leakey was committed to the view that some more human form had existed alongside the Australopithecines. His faith was so strong that when a robust Australopithecine skull was unearthed along with stone tools in 1959, he named it as a new genus, *Zinjanthropus*, and claimed that it represented the true ancestor of humanity. When evidence for a more advanced form at an even earlier date turned up shortly afterward, Leakey soon abandoned this position to proclaim the new form as the true ancestor under the name *Homo habilis*.[20] Despite an ongoing controversy, the majority of paleoanthropologists have accepted *Homo habilis* and have assumed that this species was ancestral to *H. erectus* and thus eventually to *H. sapiens*. The discovery of a well-preserved skull by Louis's son Richard in 1972 considerably strengthened the case for *Homo habilis*. With this view, even the gracile *Australopithecus africanus* could not be a human ancestor. Along with its more robust counterparts, it was another side branch leading to a dead end.

The possibility of an Australopithecine ancestry for mankind has not disappeared, however, since further discoveries have revealed more primitive members of the genus at an even earlier date. In 1974, Donald Johanson's "Lucy"—the remains of a new species named *Australopithecus afarensis*—extended the genus back to an age of 3.5 million years. In Johanson's view, *Australopithecus afarensis* is the ancestor of both the later Australopithecines and the genus *Homo*.[21] In a sense this new picture of human evolution is a compromise between the more extreme positions advocated earlier. The old view in which even *Homo erectus* and the Neanderthals were dismissed as side branches having nothing to do with the origin of modern humans has broken down, yet so has the image of a straight-line ascent from ape to human. The branching is still there in the modern view, but on a much smaller scale than was possible when a large degree of evolutionary parallelism seemed plausible. Branching itself is compatible with Darwinism, and indeed is to be expected on the basis of a theory in which the opportunistic exploitation of new environments and ecological niches is the driving force of evolution. What is no longer plausible is the claim that, once separated, two or more lines could independently evolve fully human characters.

As the split between *Homo* and *Australopithecus* has been pushed farther back in time by later discoveries, there have also been dramatic changes in our thinking on the more basic question of the point at which the ape and hominid lines of evolution diverged. The postwar years saw a refinement of dating techniques that finally made nonsense out of the dates assigned to the geological periods by earlier paleoanthropologists such as Sollas and Keith. In the 1930s, Keith had placed the beginning of the Miocene at less than a million years before the present. Many geologists would already have regarded such a date as far too recent, and in the

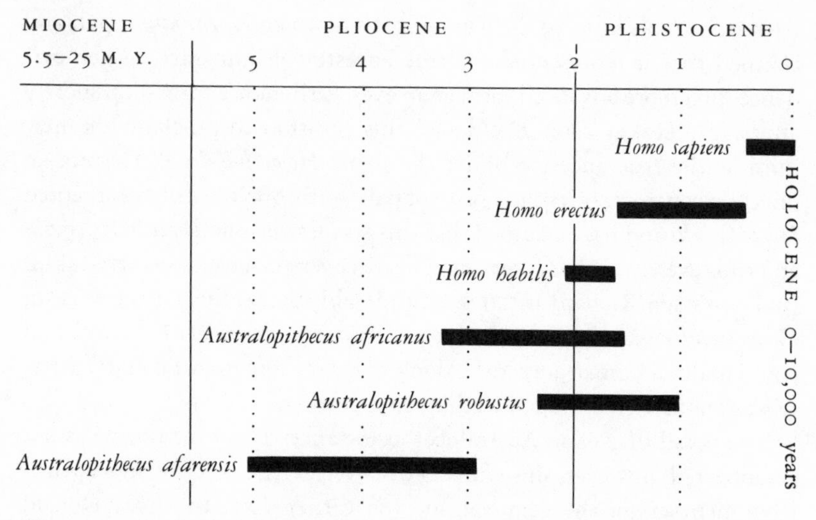

Figure 17. Geological distribution of hominids according to modern evidence. The horizontal scale indicates millions of years (M.Y.) before the present.

1950s better techniques of radioactive dating confirmed that the time scale against which human evolution must be measured would have to be extended considerably. The start of the Miocene is now dated at twenty million years before present, and the length of the Pleistocene was officially doubled in 1961 from one to two million years. On this time scale, the split between *Homo* and *Australopithecus* can be placed sometime after the dating of "Lucy" (3.5 million years), and yet the ape-hominid divergence could still be placed in the early Miocene, fifteen million years or more ago. This dating was still popular among paleoanthropologists in the early 1970s, when it began to be challenged by new evidence from molecular biology. This evidence confirmed that humans are genetically much more closely related to the African than to the Asian apes. When set against the "molecular clock" established on the basis of evolutionary developments for which there is more fossil evidence, the close genetic similarity between humans and chimpanzees seemed to suggest a split less than five million years ago.[22] The modern consensus now places the divergence of the gorilla and chimpanzee from the human stock in the period between ten and six million years ago.[23]

The fossil evidence (see fig. 17) now confirms that the initial step in the differentiation of humans from apes was the acquisition of bipedalism and an upright posture, probably as an adaptation to a foraging lifestyle on the expanding savannas of Africa. The various Australopithecine species are presumed to have coexisted, eventually along with *Homo habilis,* because of some form of ecological diversity. Only with the emergence of *Homo,* approximately two million years ago, were the first tools made and a significant increase in brain size begun. But what were the causes of these crucial steps in human evolution: what advantage did bipedalism offer the early Australopithecines, and why did the dramatic increase in brain size begin at a later date? Modern paleoanthropology is a more complex discipline, in which the techniques of the anatomist and paleontologist are supplemented by a whole range of studies designed to extract from the fossils the maximum amount of information about the behavior and environment of the early hominids. The fossil record is so fragmentary that there is still plenty of scope for disagreement. So crucial are these questions now thought to be, however, that there has been no shortage of hypotheses to be debated. In fact, a number of different ideas have enjoyed some degree of popularity over the last few decades.

During the 1960s, there was much discussion of the claim that early humans developed their unique character as an adaptation to a lifestyle based on hunting. Raymond Dart's evidence for the use of weapons by the Australopithecines was discredited, but Dart's views on the bloodthirsty nature of our early ancestors were picked up and popularized by Robert Ardrey and others (chapter 9). The so-called anthropology of aggression supposed that modern humans are endowed with violent instincts inherited from these hunting ancestors. This interpretation of human nature was roundly condemned by sociologists and social anthropologists, who were now committed to the view that our behavior is not controlled by instincts of any kind. Paleoanthropologists, though, tended to adopt a less sensationalized version of the theory, in which the ability to hunt with weapons was seen as the prime advantage conferred by bipedalism. Even this approach, however, tended to emphasize the extent to which the human species has become separated from the rest of the animal kingdom through its adoption of organized violence. More recently, attention has switched to the hypothesis that food-gathering rather than actual hunting is the true benefit derived from bipedalism. This move was in part encouraged by a desire to get away from the male-oriented atmosphere of the hunting hypothesis, in which the male plays the active role while the female stays at home. More generally, the image of food-gathering provides an alternative to the atmosphere of violence that seems to pervade the hunting hypothesis. In recent developments of the idea anthropologists such as Glynn Isaac present food-sharing as an explanation of growing sociability and intelligence.[24]

Some paleoanthropologists are now prepared to admit that there may be a link between theories of human evolution and changing cultural values.[25] The image of man the hunter became popular at the time of the cold war, and is still supported by those who wish to see a competitive element as an inherent part of human nature. To evaluate such cultural links in the postwar decades is beyond the scope of the present study. We have seen that the scarcity of hard evidence left the scientists of an earlier period free to construct theories that accorded with their own preconceptions about humanity. Many of these earlier theories have now been decisively refuted, no doubt at least in part because of the fossil discoveries of the last few decades, and the wide range of scientific disciplines now brought to bear on them, although the fossils still leave considerable room for speculation. But the real transition to the modern approach to paleoanthropology involves a complex interaction between scientific and social forces. As late as the 1930s, theories of human evolution still took for granted an element of progressionism that was becoming increasingly out of touch with the cultural values of the time. The destruction of this last vestige of Victorian optimism was catalyzed, if not caused, by the application of the renewed Darwinism of the Modern Synthesis to the area of human evolution. At the same time, the new theory discredited the concept of parallel evolution so freely employed by the earlier generation of paleontologists.

The adoption of a broadly Darwinian framework for theorizing about human origins has focused attention onto a different series of issues. It is no longer possible to assume that an inherent tendency toward the growth of intelligence could have produced an enlargement of the brain independently in several different lines of primate or hominid evolution. Since Darwinism treats each step in evolution as a new adaptive initiative, in response to either a change in the environment or the exploration of a new ecological opportunity, the "escape from the trees" cannot be seen as the culmination of a longstanding progressive trend. The elimination of progressionism created the climate in which paleoanthropologists could accept the fossil evidence for bipedalism as the breakthrough responsible for the separation from the apes. The fossils have now defined the key steps: first the adoption of bipedalism, and then the enlargement of the brain that created the genus *Homo*. Attention must now concentrate on an evaluation of the adaptive and ecological initiatives that made these transitions possible. Bipedalism may represent an adaptation to a foraging lifestyle in a savanna environment, but we need to know precisely in what way it was advantageous before we can accept that natural selection would have promoted such a step. Similarly, we need to ask why food-sharing and the subsequent increase in sociability encouraged the growth of intelligence. The earlier generation of paleoanthropologists simply did not see these issues as being quite so crucial. They not only took the progressive development of intelligence for granted, but also saw the advantages of bipedalism as being so obvious that there was no need to investigate them in detail. In a sense, the concentration of effort on certain key questions that has helped to create the modern discipline of paleoanthropology is a product of the dramatic change in our thinking on the mechanism of biological evolution, a change that has forced us to think more carefully about the adaptive benefits of the key steps in the origin of mankind.

The destruction of progressionism thus represents far more than the rejection of a somewhat overoptimistic expectation that nature tends to move in the direction we would prefer. It entirely alters the style of explanation that is considered suitable in the search for human origins. The stress placed on opportunistic adaptation rather than necessary progress forces us to construct a genuinely historical account of the events that shaped our past evolution. Since there is no inbuilt trend, we must try to define the adaptive steps that separated the various lines of evolution. By looking to the contingency of a historical explanation, rather than the necessary unfolding of a progressive trend, we emphasize an entirely different style of explanation. In effect, we can now construct a narrative or story, based on the unique events that were experienced by our ancestors, but not by those of the other primates. The fascination of modern anthropologists with the narrative style is a genuine reflection of the structure of explanation now required in the field. Whether our modern stories can be evaluated by scientific testing is an entirely different question.

Earlier theories of human origins, however, show that different assumptions about the nature of evolution itself require different styles of explanation. The narrative style played a much smaller role when so much could be left to the automatic unfolding of evolutionary trends. We must by all means try to understand the structure of modern explanations and their nonscientific implications. Since those explanations do adopt a narrative style, a comparison with folk tales and creation myths may well help to illustrate how the theories express their cultural messages. But we should not assume that the style of explanation preferred today must always have been the accepted mode of tackling the issues. In a pre-Darwinian world (which in paleoanthropology, at least, extended into the 1930s), the directionalist concept of evolution encouraged a different approach to the explanation of human origins. The historian must be aware of such changes in the way we conceive a problem, and accept that the current style of explanation is not always the best model upon which to base an analysis of earlier theories.

Notes

The complete reference for every item cited in the notes may be found in the bibliography.

Introduction

1. For an account of Lyell's fears, see Bartholomew, "Lyell and Evolution."

2. For discussions of the *Vestiges* debate see Gillispie, *Genesis and Geology,* and Millhauser, *Just before Darwin.*

3. For a sample of the vast literature on the Darwinian revolution, see Eiseley, *Darwin's Century;* Greene, *The Death of Adam;* Moore, *The Post-Darwinian Controversies;* Ruse, *The Darwinian Revolution,* and Young, "The Historical and Historiographical Contexts of the Nineteenth-Century Debate on Man's Place in Nature." For further references, see Bowler, *Evolution,* chaps. 6–8.

4. The more important literatures includes Hofstadter, *Social Darwinism in American Thought;* Bannister, *Social Darwinism;* and Greta Jones, *Social Darwinism and English Thought.* For further references see Bowler, *Evolution,* chap. 10.

5. For more details, see chapter 3.

6. Moore, *Man, Time, and Fossils;* Wendt, *In Search of Adam* and *From Ape to Adam;* Reader, *Missing Links.*

7. Brace, "The Fate of the 'Classic' Neanderthals."

8. Spencer, ed., *A History of American Physical Anthropology.*

9. Cartmill, "Four Legs Good, Two Legs Bad."

10. Landau, "Human Evolution as Narrative." Landau's dissertation, "The Anthropogenic," contains a more detailed study of some of the theories mentioned in this article.

11. Kuhn, *The Structure of Scientific Revolutions.* There is, of course, an immense literature on Kuhn's thesis. For a brief discussion in the context of evolution theories, see Bowler, *Evolution,* chap. 1.

12. Holtzman, "On Brace's Notion of 'Hominid Catastrophism.'" See also the "Reply" by Brace. The term "presentism" is derived from Stocking's *Race, Culture, and Evolution,* p. 3.

13. Cartmill, "Basic Primatology and Prosimian Evolution." p. 171.

14. Isaac, "Aspects of Human Evolution." p. 515. For an outline of the general discussion of whether evolutionary hypotheses are "scientific," see Bowler, *Evolution,* chap. 12.

15. There is now a substantial literature on the sociological approach to the history of science; for a recent survey see Shapin, "History of Science and Its Sociological Reconstruction."

16. Blavatsky, *The Secret Doctrine,* vol. 2, esp. bk. 2, pt. 3.
17. Beer, *Darwin's Plots.*
18. Landau, "Human Evolution as Narrative," p. 262.
19. For discussions of late-nineteenth-century efforts to reconstruct the course of evolution, see Rudwick, *The Meaning of Fossils;* Bowler, *Fossils and Progress;* Gould, *Ontogeny and Phylogeny;* and Desmond, *Archetypes and Ancestors.*
20. Hull, "Historical Entities and Historical Narratives."
21. McCown and Kennedy, *Climbing Man's Family Tree.*

1. The Evidence of Human Antiquity

1. Daniel's *A Hundred and Fifty Years of Archaeology* accepts a positive role for evolution, pp. 63–66, but this view is repudiated by Gruber, "Brixham Cave and the Antiquity of Man"; Grayson, *The Establishment of Human Antiquity,* p. 211; and Freeman, "Evolutionism and Arch(a)eology."
2. Grayson, *The Establishment of Human Antiquity.*
3. On the history of paleontology, see Green, *The Death of Adam;* Rudwick, *The Meaning of Fossils;* and Bowler, *Fossils and Progress.*
4. For a guide to the immense literature on the uniformitarian-catastrophist debate see Bowler, *Evolution,* chap. 5. The standard biography of Lyell is Wilson, *Charles Lyell: The Years to 1841;* see also Rudwick, "The Strategy of Lyell's *Principles of Geology*" and "Uniformity and Progression."
5. Geikie, *The Great Ice Age;* see also his *Prehistoric Europe* and *The Antiquity of Man in Europe.* See Hamlin, "James Geikie, James Croll, and the Eventful Ice Age."
6. See Burchfield, *Lord Kelvin and the Age of the Earth.*
7. Keith, *The Antiquity of Man,* pp. 307–8 and for an even shorter estimate, 2d ed., 2:304. See also Geikie, *The Antiquity of Man in Europe,* p. 302.
8. See Grayson, *The Establishment of Human Antiquity;* A. Laming-Emperaire, *Origines de l'archéologie préhistorique en France;* and Lyon, "The Search for Fossil Man."
9. See Grayson, *The Establishment of Human Antiquity* and Gruber, "Brixham Cave and the Antiquity of Man." Many professionals regarded Lyell's work in this field as derivative, despite the popular success of his writings; see Bynum, "Charles Lyell's *Antiquity of Man* and Its Critics."
10. The translation of Worsaae's *Primal Antiquities of Denmark* appeared in 1849. On the origins of the three-age system, see Daniel, *A Hundred and Fifty Years of Archaeology,* pp. 38–54 and Daniel, ed., *Towards a History of Archaeology.*
11. Lubbock, *Prehistoric Times,* pp. 1–2, although later in the book Lubbock also uses the term "archaeolithic" for paleolithic.
12. Lartet, "Nouvelles recherches sur la coexistence de l'homme et des grandes mammifères fossiles," p. 231. Lartet's fourth period, the age of the aurochs or wild cattle, is the neolithic.
13. Lartet and Christy, *Reliquiae Aquitanicae,* pp. 9–10.
14. De Mortillet, *Le Préhistorique,* pp. 482–83.
15. Ibid., p. 374.
16. De Mortillet, "Promenades préhistoriques à l'Exposition universelle," p. 368.
17. De Mortillet, *Le Préhistorique,* pp. 103–5.

18. Ibid., pp. 248–49.

19. Hammond, "Anthropology as a Weapon of Social Combat" and "The Expulsion of the Neanderthals from Human Ancestry."

20. De Mortillet, *Le Préhistorique,* pp. 85–97. See Daniel, *A Hundred and Fifty Years of Archaeology,* pp. 97–99, 230–31.

21. Boule, *Les hommes fossiles,* chap. 5. Boule did not deny the possibility of Tertiary humans, but found the eoliths unconvincing.

22. Moir, *The Earliest Men,* pp. 1–2. For an earlier summary of Moir's discoveries see his *Pre-Palaeolithic Man.*

23. Lankester, "On the Discovery of a New Type of Flint Implement."

24. Dawkins, *Cave Hunting,* p. 353.

25. See Daniel, *A Hundred and Fifty Years of Archaeology,* pp. 236–51.

26. Louis Leakey, *Adam's Ancestors,* pp. 133–34, 140–43.

27. See for instance Leakey and Goodall, *Unveiling Man's Origins;* Moore, *Man, Time, and Fossils;* and Reader, *Missing Links* (the last an excellent and well-illustrated modern account). For a survey of the current fossil evidence see Smith and Spencer, eds., *The Origins of Modern Humans.*

28. Lartet, "Note sur un grand singe fossile."

29. See Lartet and Christy, *Reliquiae Aquitanicae,* pp. 82–115. On the Cro Magnon race, see de Quatrefages and Hamy, *Crania Ethnica,* pt. 1, chap. 2, and de Quatrefages, *The Human Species,* chap. 27.

30. See Sollas, *Ancient Hunters,* chap. 12.

31. Verneau, "Anthropologie," in Villeneuve, ed., *Les Grottes de Grimaldi,* vol. 1, fasc. 1.

32. Boule, *Fossil Men,* p. 272.

33. Keith, *Antiquity of Man,* chap. 10 (2d ed., chap. 12).

34. Leakey, *The Stone Age Races of Kenya,* chap. 2, also *Adam's Ancestors,* pp. 206–7. Although Leakey named the type *Homo kanamensis,* he insisted that his new species was close to *H. sapiens.* The jaw is now attributed to *Homo habilis,* the earliest member of the genus *Homo.*

35. See Marston, "The Swanscombe Skull." The full report on the skull is in the *Journal of the Royal Anthropological Institute* 68 (1938):17–98.

36. Schaafhausen, "On the Crania of the Most Ancient Races of Man."

37. Huxley, *Man's Place in Nature,* chap. 3.

38. King, "The Reputed Fossil Man of the Neanderthal."

39. I have used Fraipont's shorter report of 1889, "Les Hommes de Spy."

40. Boule, "L'homme fossile de La Chapelle-aux-Saints."

41. Dubois's original monograph is *Pithecanthropus erectus: Ein Menschenaehnliche Uebergangsform aus Java.* His articles contributing to the debate are listed in the bibliography.

42. See Dubois, "*Pithecanthropus erectus*—a Form from the Ancestral Stock of Mankind," p. 454.

43. Dubois, "Remarks upon the Brain-cast of *Pithecanthropus erectus,*" p. 95.

44. Dubois, "On the Fossil Human Skulls Recently Discovered in Java," pp. 5–6.

45. See for instance Reader, *Missing Links,* chap. 3 and Millar, *The Piltdown Men.* Hammond argues that the concept of branching evolution made the fraud more plausible; see "A Framework of Plausibility for an Anthropological Forgery."

46. Dawson and Woodward, "On the Discovery of a Palaeolithic Human Skull and Mandible."

47. For Keith's opposition to the apelike character of the reconstruction, see the discussion reported in the *Quarterly Journal of the Geological Society of London* 69 (1912):148. On the larger capacity of the skull, see ibid., vol. 70 (1914):98. In more detail see Keith, "The Piltdown Skull and Brain Cast" and *The Antiquity of Man,* chaps. 18–22 (2d ed., chaps. 26–30, 33). On the discovery of the canine tooth, see Dawson and Woodward, "Supplementary Note on the Discovery of a Palaeolithic Human Skull and Mandible." More human remains were later discovered nearby; see Woodward, "Fourth Note on the Piltdown Gravel."

48. Keith, "The Reconstruction of Fossil Human Skulls."

49. Boule, *Fossil Men,* pp. 170–71, 471–72.

50. Miller, "The Jaw of Piltdown Man" and "The Piltdown Jaw." In the latter paper Miller records the support of noted American authorities such as W. D. Matthew and W. K. Gregory. One British anatomist, at least, shared this view; see Waterston, "The Piltdown Mandible."

51. Miller, "The Jaw of Piltdown Man," p. 1.

52. Marston, "The Swanscombe Skull," p. 394.

53. Oakley and Hoskins, "New Evidence on the Antiquity of Piltdown Man."

54. On the exposure of the fraud, see Weiner, *The Piltdown Forgery.*

55. Grafton Elliot Smith is the candidate suggested in Millar, *The Piltdown Men;* Sollas was implicated by his successor at Oxford, J. A. Douglas, as reported by Halstead, "New Light on the Piltdown Hoax?" Teilhard de Chardin is suspected by Stephen J. Gould, *The Panda's Thumb,* pp. 108–24 and *Hen's Teeth and Horse's Toes,* pp. 201–50. The case against Sir Arthur Conan Doyle is set out by Winslow, "The Perpetrator of Piltdown" and refuted by Langham, "Sherlock Holmes, Circumstantial Evidence, and Piltdown Man." Dr. Langham's tragic death in 1983 prevented him from completing his book on the affair, but his conclusions may be published posthumously.

56. William P. Pyecraft et al., *Rhodesian Man,* see esp. pp. 46–51.

57. Black's most complete account is his "On the Discovery, Morphology, and Environment of *Sinanthropus pekinensis.*" See also Hood, *Davidson Black.*

58. Von Koenigswald and Weidenreich, "The Relationship between *Pithecanthropus* and *Sinanthropus.*"

59. Weidenreich, "The Skull of *Sinanthropus pekinensis*" and "Giant Early Man from Java and South China."

60. Dart, "*Australopithecus africanus:* The Man-ape of South Africa." See Dart, *Adventures with the Missing Link* and Clark, *Man-apes or Ape-men?*

61. Keith, "The Fossil Anthropoid Ape from Taungs" and "The Taungs Skull." See also *New Discoveries Relating to the Antiquity of Man,* chaps. 4, 5. For Dart's reply, see his "The Taungs Skull."

62. Dart and Shellshear, "The Origin of the Motor Neuroblasts of the Anterior Cornu of the Neural Tube"; Dart, "The Misuse of the Term 'Visceral,'" and "The Anterior End of the Neural Tube and the Anterior End of the Body." Le Gros Clark suggested that this early work may have soured the reception of Dart's discovery; see *Man-apes or Ape-men?* p. 19. I am grateful to Professor Dart himself for supplying me with the references to these papers.

63. See for instance Sollas, "The Taungs Skull" and "On a Sagittal Section of the Skull of *Australopithecus africanus*."

64. Broom, "Some Notes on the Taungs Skull," "Note on the Milk Dentition of *Australopithecus*," and "The Age of *Australopithecus*." On Broom's career, see his *Finding the Missing Link* and Findlay, *Dr. Robert Broom, F.R.S.*

65. Broom, "A New Fossil Anthropoid Skull from South Africa" and "The Pleistocene Anthropoid Apes of South Africa."

66. Broom, "An Ankle Bone of the Ape-man, *Paranthropus robustus*."

67. Broom and Schepers, *The South-African Fossil Ape-Men.*

68. Le Gros Clark, "Significance of the Australopithecinae" and *The Fossil Evidence for Human Evolution.*

69. See Ashton and Zuckerman, "Some Quantitative Dental Characters of Fossil Anthropoids" and "Statistical Methods in Anthropology." Zuckerman reiterates these views in his autobiography, *From Apes to Warlords*, pp. 61–62.

70. See Johanson, *Lucy.*

2. The Framework of Debate

1. Classic accounts of the Darwinian revolution include Eiseley, *Darwin's Century;* Himmelfarb, *Darwin and the Darwinian Revolution;* and more recently Ruse, *The Darwinian Revolution.* For more general accounts of the history of evolutionism see Greene, *The Death of Adam* and Bowler, *Evolution.*

2. Note that the translation of Lamarck's *Zoological Philosophy* did not appear until 1914. For a biography and account of Lamarck's theory, see Burkhardt, *The Spirit of System.*

3. On progressionism, see Bowler, *Fossils and Progress,* and on the embryological parallel, see Gould, *Ontogeny and Phylogeny* chap. 3.

4. On Chambers's theory see Hodge, "The Universal Gestation of Nature." On the reception of *Vestiges,* see Millhauser, *Just before Darwin* and Gillispie, *Genesis and Geology.*

5. See Ellegård, *Darwin and the General Reader.* A more cautious approach to the rapidity of the takeover can be found in David Hull et al., "Planck's Principle," but this study also shows that about 75 percent of scientists had converted to evolution by the early 1870s.

6. On the reaction to Darwinism in different countries, see Glick, ed., *The Comparative Reception of Darwinism.* On France, see Conry, *L'introduction du Darwinisme en France.*

7. For an outline of the role played by anti-Darwinian theories in the late nineteenth and early twentieth centuries, see Bowler, *The Eclipse of Darwinism.*

8. On Huxley's theory of the "persistence of type," see Desmond, *Archetypes and Ancestors,* chap. 3.

9. On phylogeny-building in the late nineteenth century, see Rudwick, *The Meaning of Fossils,* chap. 5; Gould, *Ontogeny and Phylogeny,* chap. 4; Bowler, *Fossils and Progress,* chap. 6; and Desmond, *Archetypes and Ancestors,* chaps. 4, 5.

10. On the concept of designed evolution, see Bowler, *The Eclipse of Darwinism,* chap. 3.

11. On the growth of Lamarckism see Bowler, *The Eclipse of Darwinism,* chaps. 4–6 and on orthogenesis, chap. 7. On the American school see also Gould, *Ontogeny and Phylogeny,* chap. 4. Works by Butler, Cope, and Osborn are cited in the bibliography.

12. See Haas and Simpson, "Analysis of some Phylogenetic Terms." This article notes that early works such as Arthur Willey's *Convergence in Evolution* of 1911 used "convergence" to mean independent adaptation to similar ends, but rejects this definition for modern usage. Parallelism is attributed to an internal mechanism—inherited restriction on variation—but since the authors reject orthogenesis, they are forced to argue that parallelism is normally adaptive. I suggest that in earlier decades "parallelism" was generally used for trends that were not adaptive.

13. On the mutation theory see Bowler, *The Eclipse of Darwinism,* chap. 8.

14. De Vries, *The Mutation Theory,* 1:156–57.

15. Casey, "The Mutation Theory."

16. Thomson, *What Is Man?,* pp. 28–29.

17. See Dubois, "On the Fossil Human Skulls Recently Discovered in Java," p. 6.

18. See Provine, *The Origins of Theoretical Population Genetics;* and Mayr and Provine, eds., *The Evolutionary Synthesis.*

19. See Lubbock, *On the Origin of Civilisation,* chap. 3; McLennan, *Studies in Ancient History* (a reissue of his *Primitive Marriage* of 1865); and Morgan, *Ancient Society,* pt. 3. On the debate see Harris, *The Rise of Anthropological Theory,* pp. 189–97. Other works on the history of anthropology include Burrow, *Evolution and Society;* Hatch, *Theories of Man and Culture;* and Stocking, *Race, Culture, and Evolution.*

20. Maine, *Dissertations on Early Law and Custom,* pp. 206–7, notes the support expressed in Darwin's *Descent of Man,* pp. 589–90.

21. Westermark, *The History of Human Marriage,* vol. 2, chap. 20. The first edition of this work appeared in 1891.

22. McDougall appealed to natural selection to explain how the instinct to accept the patriarch's authority was developed; see *Introduction to Social Psychology,* pp. 282–86, although he later became a Lamarckian. For other examples of the instinct theory, see Shand, *The Foundations of Character;* Briffault, *The Mothers;* and Alverdes, *The Psychology of Animals.* Even the founder of functionalism, Bronislaw Malinowski, accepted a role for instinct in his *Sex and Repression in Savage Society;* see Harris, *The Rise of Anthropological Theory,* pp. 545–67 and Hatch, *Theories of Man and Culture,* chap. 6. In general, the social sciences of the early twentieth century repudiated the link with evolution, and zoologists too began to criticize the oversimplified use of animal studies to throw light on human behavior; see Zuckerman, *The Social Life of Monkeys and Apes.*

23. See esp. Lubbock's *On the Origin of Civilisation.*

24. Tylor, *Researches on the Early History of Mankind,* p. 13. The first edition was published in 1865.

25. Ibid., chap. 7. Note also Tylor's later interest in the link between the Australian aborigines and stone-age cultures: "On the Survival of Palaeolithic Conditions in Tasmania and Australia."

26. Morgan, *Ancient Society,* p. 11.

27. Ibid., p. 59.

28. See Burrow, *Evolution and Society.*

29. Tylor, *Anthropology,* pp. 60, 73–74.

30. Morgan, *Ancient Society,* pp. 38–40, 59.

31. See Hatch, *Theories of Man and Culture,* p. 32.

32. McGee, "The Trend of Human Progress," p. 416. W. J. McGee was John Wesley Powell's protégé at the Bureau of American Ethnology and followed his teacher's view of cultural evolution, see Stocking, *Race, Culture, and Evolution,* chap. 10.

33. Tylor, *Anthropology,* p. 54.

34. See Gillespie, "The Duke of Argyll, Evolutionary Anthropology, and the Art of Scientific Controversy."

35. Broca, "Sur le volume et la forme du cerveau," and on the expansion of the brains in Europeans, "Sur la capacité des crânes de parisiens des diverses époques." On craniometry, see Gould, *The Mismeasure of Man,* chap. 3.

36. Cope's anthropological papers are reprinted in his *On the Origin of the Fittest;* for a detailed account of his ideas see Haller, *Outcasts from Evolution,* pp. 187–202. Similar views are expressed by McGee in "The Trend of Human Progress" and by Brinton, "The Aims of Anthropology." See Stocking, *Race, Culture, and Evolution,* chap. 6. For other discussion of the history of race theory see Barzun, *Race;* Greene, *The Death of Adam,* chap. 8; Montague, *Man's Most Dangerous Myth;* Stanton, *The Leopard's Spots;* and Stepan, *The Idea of Race in Science.*

37. See for instance Berry and Robinson, "The Place in Nature of the Tasmanian Aboriginal," and Cross, "On a Numerical Determination of the Relative Positions of Certain Biological Types on the Evolutionary Scale."

38. Spencer, *Principles of Psychology,* 1:581–82; see also *Principles of Sociology,* 1:23–29.

39. Spencer, *Principles of Sociology,* 1:104.

40. Spencer, *Principles of Psychology,* 1:422–23.

41. Spencer, *Principles of Sociology,* 1:61. Spencer thought that modern primitives have often degenerated from a higher state; see 1:106.

42. Ibid., 3:325.

43. Ratzel, *The History of Mankind,* 1:15–25. This is a translation of Ratzel's *Volkerkunde,* first published 1885–88. On diffusionism, see Harris, *The Rise of Anthropological Theory,* chap. 14.

44. Rivers, "President's Address." See Slobodin, *W.H.R. Rivers,* and Langham, *The Building of British Social Anthropology.*

45. On Boas and his influence see Harris, *The Rise of Anthropological Theory,* chap. 9, 10; Hatch, *Theories of Man and Culture,* chap. 2; and Cravens, *The Triumph of Evolution,* chap. 3. Boas certainly realized that his approach had severed the link between physical and cultural anthropology; see his 1936 essay on this topic reprinted in Boas, *Race, Language, and Culture,* pp. 172–75.

46. See Prichard's *Researches into the Physical History of Man,* which first appeared in 1813.

47. See Knox, *The Races of Man* (a reissue of the 1850 edition); Morton, *Crania Americana* and *Crania Aegyptica;* and Nott and Gliddon, *Indigenous Races of the Earth.*

48. On the Paris society see Hammond, "Anthropology as a Weapon of Social Combat," and Harvey, "Evolutionism Transformed."

49. Broca, "L'ordre des primates."

50. Ibid., pp. 397–98.

51. Broca, "Sur le transformisme." On human origins see pp. 191–93, p. 234.

52. Broca, *On the Phenomenon of Hybridity.* He claimed crosses between whites and Australian aborigines to be infertile; see pp. 45–49. See also Pouchet, *The Plurality of the Human Race.*

53. See Stocking, "What's in a Name?"

54. Ripley criticized Deniker for failing to recognize what were mere local variations; see *The Races of Europe,* pp. 598–601.

55. Stocking, *Race, Culture, and Evolution,* chap. 3; and Brace, "The Roots of the Race Concept in American Physical Anthropology."

56. Topinard, *Elements d'anthropologie générale,* pp. 202–3.

57. Deniker, *The Races of Man,* pp. 3–7.

58. Müller, *Lectures on the Science of Language,* p. 199.

59. See Taylor, *The Aryan Controversy* and T. H. Huxley, "The Aryan Question and Prehistoric Man," reprinted in Huxley, *Man's Place in Nature* (1894 ed.) pp. 271–328.

60. These ancient human types are discussed in chapter 2. On the cranial index, see Gould, *The Mismeasure of Man,* chap. 3.

61. See Sergi, *The Mediterranean Race,* and Ripley, *The Races of Europe.*

62. See Urry, "Englishmen, Celts, and Iberians."

63. De Quatrefages, *The Prussian Race,* and Boule, "La Guerre."

64. For a detailed discussion of this question see Poliakov, *The Aryan Myth.*

65. See for instance Vacher de Lapouge, *L'Aryen,* and Grant, *The Passing of the Great Race.*

66. Wallace's paper, "The Development of Human Races under the Law of Natural Selection" is reprinted in his *Natural Selection and Tropical Nature,* pp. 167–85.

67. Haeckel, *The History of Creation,* 2:303–5.

68. Ibid., pp. 307–10. Note that Haeckel saw the Australian aborigines as relics of the ancestral form of the more progressive Lissotrichi, a view later taken up by Hermann Klaatsch.

69. Vogt, *Lectures on Man.* The polytypic theory of human origins is discussed in Chapter 6.

3. Up from the Ape

1. A valuable collection of writings on human phylogeny is *Climbing Man's Family Tree,* edited by McCown and Kennedy. A useful guide to the early twentiety-century controversies is Keith, *The Construction of Man's Family Tree.* The title of this chapter is derived from E. A. Hooton's survey, *Up from the Ape.*

2. Lamarck, *Zoological Philosophy,* pp. 171–72. On early ideas about the links between apes and humans see Greene, *The Death of Adam,* chap. 6. On Lamarck see Burkhardt, *The Spirit of System.*

3. On Lyell's reaction to transmutation see Bartholomew, "Lyell and Evolution," and Corsi, "The Importance of French Transformist Ideas." On Owen see Desmond, "Richard Owen's Reaction to Transmutation in the 1830s," esp. pp. 40–43.

4. Chambers, *Vestiges of the Natural History of Creation* (1844), pp. 267–68. On the reaction to *Vestiges* see Millhauser, *Just before Darwin,* which makes few references to the ape link.

5. Chambers, *Vestiges* (1844), pp. 277–323, esp. p. 307. See also 5th ed., p. 322.

6. See Haller, *Outcasts from Evolution,* pp. 187–202.

7. Lankester, *Diversions of a Naturalist,* pp. 275–76.

8. Darwin, *Origin of Species,* p. 488.

9. Huxley, "On the Zoological Relations of Man with the Lower Animals." See di Gregorio, *T. H. Huxley's Place in Natural Science,* chap. 4.

10. The three sections of the 1863 edition of *Man's Place in Nature* appears as the first three chapters of the volume in Huxley's *Collected Essays* under the same title. Citations below are to the latter, which is far more readily available.

11. *Man's Place in Nature,* pp. 125–26.

12. Ibid., pp. 116–17, 140, 144.

13. Ibid., p. 97.

14. Broca, "L'ordre des Primates," p. 397.

15. Huxley, *Man's Place in Nature,* pp. 107–8.

16. Ibid., p. 119.

17. Ibid., pp. 204–5. A footnote added to this edition notes Huxley's subsequent opposition to the idea of giving the Neanderthal race the status of a distinct species, see pp. 207–8.

18. Ibid., p. 208.

19. Darwin, *Descent of Man,* pp. 152–54. Note that it is the second edition cited here.

20. The diagram is reproduced in Gruber, *Darwin on Man,* p. 197. It implies a slightly more distant relationship between humans and apes than the *Descent,* since it shows us as being no more closely related to the chimpanzee and gorilla than we are to the orangutan or the gibbon.

21. Darwin, *Descent of Man,* pp. 155–61.

22. Wallace, *Darwinism,* pp. 455–56. On Wallace's view of mental evolution, see Kottler, "Alfred Russel Wallace, the Origin of Man, and Spiritualism."

23. Darwin, *Descent of Man,* p. 3. Haeckel's *Natürliche Schopfungsgeschichte,* later translated as *The History of Creation,* appeared in 1873.

24. Haeckel, *History of Creation,* 2:293–94.

25. Ibid., pp. 326–28. See also pp. 307–10. It should be noted that in his first survey of evolution, the *Generelle Morphologie* of 1866, Haeckel had implied a closer link with the African apes; see 2:426–29 and table 8.

26. Keith, *The Construction of Man's Family Tree,* p. 10.

27. See the diagram in Dubois, *Pithecanthropus erectus: Eine menschenaehnliche Uebergangsform aus Java,* p. 8; "*Pithecanthropus erectus*—a Form from the Ancestral Stock of Mankind," p. 449; and "The Place of *Pithecanthropus* on the Genealogical Tree," p. 245.

28. See Dubois, "On *Pithecanthropus erectus:* A Transitional Form between Man and Apes," p. 17, and "The Place of *Pithecanthropus* on the Genealogical Tree," p. 245.

29. Schwalbe, "The Descent of Man," p. 129.

30. See Weinert, *Ursprung der Menschheit.*

31. Gregory, *Man's Place among the Anthropoids*, p. 214.

32. Gregory predicted that our true ancestors were Miocene ground-apes closely related to *Pithecanthropus;* see *The Origin and Evolution of Human Dentition,* p. 413. *Pithecanthropus* itself was a stagnant remnant of this ancestral form; see p. 360.

33. See Keith, "The Extent to Which the Posterior Segments of the Body Have Been Transmuted." See also the account in Keith, *The Construction of Man's Family Tree,* pp. 13–16.

34. Gregory, *The Origin and Evolution of Human Dentition,* pp. 411–12. Guy Pilgrim, the discoverer of *Sivapithecus,* regarded it as the first stage in the separation of the human from the gibbon line, *Dryopithecus* being already committed to the great ape branch; see his "New Sewalik Primates."

35. See Nuttall, *Blood Immunity and Blood Relationships.*

36. See for instance Gregory, "The Upright Posture of Man."

37. Miller, "Conflicting Views on the Problem of Man's Ancestry."

38. Dawson and Woodward, "On the Discovery of a Palaeolithic Human Skull," p. 139. The more human appearance of the young ape, and its implications in terms of the recapitulation theory, had already been pointed out in Germany by J. Ranke, "Ueber die individuellen Variationen im Schadelbau des Menschen" and by J. Kollmann, "Neue Gedanken über das alte Problem von der Abstammung des Menschen." See also Hill-Tout, "The Phylogeny of Man from a New Angle." These ideas are discussed more fully in chapter 5. On the general link between the recapitulation theory and ideas on human evolution, see Gould, *Ontogeny and Phylogeny,* chap. 10.

39. Le Gros Clark, *Early Forerunners of Man,* p. 15.

40. Morton, "Human Origins," pp. 200–201. See also his "Evolution of the Human Foot."

4. Neanderthals and Presapiens

1. Vallois, "Neanderthals and Praesapiens" and "La grotte de Fontéchevade."

2. Brace, "The Fate of the 'Classic' Neanderthals: A Consideration of Hominid Catastrophism."

3. Brace, "Tales of the Phylogenetic Woods." In 1973, Stepehn F. Holtzman accused Brace of distorting history to fit his own theory; see "On Brace's Notion of 'Hominid Catastrophism.'" Brace responded in his "Reply." My own sympathies lie with Holtzman, since whatever the validity of Brace's argument on the link between Cuvier and Gaudry, the origins of the Neanderthal-phase theory seem to lie in a linear and hence very un-Darwinian notion of human evolution.

4. Hammond, "The Expulsion of the Neanderthals from Human Ancestry."

5. Watson, *Palaeontology and the Evolution of Man.*

6. Huxley, *American Addresses,* pp. 55–60. The first edition of this work appeared in 1877.

7. See Hammond, "Anthropology as a Weapon of Social Combat."

8. See Gasman, *Ernst Haeckel and the Monist League.*

9. See chapter 2, and Burrow, *Evolution and Society.*

10. See chapter 2.

11. Cleland, "Terminal Forms of Life."

12. Fraipont, "Les hommes de Spy," p. 348.

13. Cope, "The Genealogy of Man," p. 328.

14. Ibid., p. 331. Cope did not accept Dubois's interpretation of *Pithecanthropus* and suspected that the Java specimen was a small-brained type of Neanderthal; see *The Primary Factors of Organic Evolution,* pp. 169–70.

15. Cope, "The Genealogy of Man," p. 335.

16. See for instance Cross, "On a Numerical Determination of the Relative Positions of Certain Biological Types in the Evolutionary Scale."

17. Berry and Robertson, "The Place in Nature of the Tasmanian Aboriginal."

18. Sollas, *Ancient Hunters,* chap. 4.

19. Ibid., p. 170. Note how reluctant Sollas was to abandon this identification later; see the 3d ed., p. 258.

20. Grant, *The Passing of the Great Race,* p. 108.

21. G. de Mortillet, *Le Préhistorique,* p. 104. On Gaudry's suggestion, see p. 125.

22. Ibid., pp. 248–49.

23. G. and A. de Mortillet, *Le Préhistorique* (1903 ed.), pp. 122–27.

24. Ibid., pp. 301–2.

25. Haeckel, *The Last Link,* p. 25.

26. Dubois, "On *Pithecanthropus erectus:* A Transitional Form between Men and Apes," p. 13.

27. Haeckel, *Last Words on Evolution,* p. 76.

28. Hrdlička, "The Neanderthal Phase of Man," p. 250.

29. Brace, "The Fate of the 'Classic' Neanderthals."

30. Schwalbe, *Studien zur Vorgeschichte des Menschen,* p. 14. See also "Kritische Besprechung von Boule's Werk," p. 601, but note that the diagram here is modified to make *Pithecanthropus* a direct ancestor in both cases.

31. Schwalbe, *Studien zur Vorgeschichte des Menschen,* p. 14.

32. Schwalbe, "Kritische Besprechung von Boule's Werk," p. 602.

33. Schwalbe, "The Descent of Man," p. 129.

34. Sollas, "On the Cranial and Facial Characters of the Neanderthal Race," p. 336.

35. Ibid., p. 337.

36. Keith, *Ancient Types of Man,* p. 78.

37. Ibid., p. 93.

38. Ibid., pp. 78–79.

39. Ibid., p. 118.

40. Ibid., p. 119.

41. Ibid., pp. 121, 131. See also Keith, "The Early History of the Gibraltar Cranium."

42. Keith, *Ancient Types of Man,* p. 133.

43. Sollas, *Ancient Hunters* (1st ed.), pp. 50–51.

44. Brace, "The Fate of the 'Classic' Neanderthals." See Holtzman, "On Brace's Notion of 'Hominid Catastrophism' " and Hammond, "The Expulsion of the Neanderthals from Human Ancestry."

45. Boule, "L'homme fossile de La Chapelle-aux-Saints," pp. 213–14. Note that this monograph, although published in several parts as listed in the bibliography, was also paginated as a unit. This internal pagination is used here.

46. Ibid., p. 253.
47. Ibid., p. 248.
48. Ibid., p. 247.
49. Ibid., p. 265. See also Boule, *Fossil Men,* p. 212.
50. Boule, *Fossil Men,* p. 243. See *Les hommes fossiles,* p. 242. Boule goes on to admit that some Neanderthal traits may be physiological specializations, and thus not truly primitive, but suggests that it would be difficult to distinguish between the two modes of origin.
51. Boule, "L'homme fossile de La Chapelle-aux-Saints," p. 270.
52. Boule, Review of *Le Préhistorique.* See also *Fossil Men,* p. 251.
53. See, for instance, the discussion of European races in Boule, *Fossil Men,* pp. 316–48.
54. Boule, "L'homme fossile de La Chapelle-aux-Saints," p. 264. See also *Fossil Men,* p. 110.
55. Boule, *Fossil Men,* pp. 453–54. *Les hommes fossiles,* pp. 450–51.
56. Ibid.
57. Boule, *Fossil Men,* p. 106. *Les hommes fossiles,* p. 106.
58. Boule, *Fossil Men,* e.g., pp. 460–63. *Les hommes fossiles,* pp. 459–60.
59. Boule, *Fossil Men,* p. 172. *Les hommes fossiles,* p. 172.
60. Boule, *Fossil Men,* p. 174. *Les hommes fossiles,* p. 174.
61. Boule, "Le Sinanthrope," p. 20.
62. Ibid., p. 22.
63. Boule, *Fossil Men,* p. 463. *Les hommes fossiles,* p. 460.
64. Brace, "Tales of the Phylogenetic Woods."
65. Keith, *An Autobiography,* p. 319. It may be added that Keith was unlikely to have been influenced by T. H. Huxley's views on the extreme antiquity of mankind, since he rejoiced in the fact that he was not trained in the Huxley tradition; see ibid., p. 73.
66. Keith, *The Antiquity of Man,* pp. 148–50.
67. Ibid., p. 115.
68. Keith, *An Autobiography,* p. 318. See also *The Antiquity of Man,* pp. 178, 183.
69. Keith, *The Antiquity of Man,* pp. 209–10. See also pp. 498–99.
70. Keith, "Modern Problems Relating to the Antiquity of Man."
71. Keith, *The Antiquity of Man,* p. 178.
72. Ibid., pp. 269–70.
73. Keith stressed that a form could advance in one part of its structure but not in another; see ibid., pp. 431–32. For his overall views on Piltdown see ibid., pp. 504–5 and the evolutionary tree as frontispiece. Hammond, "A Framework of Plausibility for an Anthropological Forgery" argues that the branching image of evolution helped the Piltdown fraud to gain acceptance.
74. Keith, *The Antiquity of Man,* p. 136.
75. Keith, *An Autobiography,* p. 340. See also "A New Theory of the Descent of Man."
76. Keith, *New Discoveries Relating to the Antiquity of Man,* p. 171.
77. Keith, *An Autobiography,* pp. 118–19.
78. Keith, *The Antiquity of Man,* p. 500.

79. See for instance Keith, "Modern Problems Relating to the Antiquity of Man," p. 756; *An Autobiography,* pp. 331–36 and *The Antiquity of Man,* p. 466.

80. Keith, *The Antiquity of Man,* p. 144.

81. Ibid., p. 148.

82. Ibid., p. 151.

83. Ibid., p. 136.

84. Keith, "On Certain Factors Concerned in the Evolution of Human Races" and, later on, his *Essays on Human Evolution* and *A New Theory of Human Evolution.*

85. On racial extermination see Pearson, *The Grammar of Science,* p. 369. Pearson was a leading exponent of eugenics: controlled breeding to improve the British race so that it could meet the challenge of rival powers. For a survey of the literature on this movement, see Bowler, *Evolution,* chap. 10.

86. Hammond, "The Expulsion of the Neanderthals from Human Ancestry."

87. Sollas, *Ancient Hunters* (3d ed.), pp. 599–600.

88. Ibid., pp. 256–57.

89. Ibid., p. 171, referring to a footnote on p. 120 of the first edition.

90. Ibid., p. ix (reprinting the preface to the 2d ed.).

91. First published in *Storyteller Magazine* for 1921 and reprinted in Wells, *The Short Stories of H. G. Wells.*

92. Elliot Smith, *The Evolution of Man,* pp. 3–6.

93. See Dawson and Woodward, "On the Discovery of a Palaeolithic Human Skull," p. 139.

94. Spurrell, *Modern Man and His Forerunners,* p. 44.

95. Keith, *An Autobiography,* p. 324.

96. Osborn, *Men of the Old Stone Age,* pp. 144, 491.

97. On Leakey's career see the autobiographical *White African* and *By the Evidence.* See also Cole, *Leakey's Luck.*

98. See Leakey, *Stone Age Africa;* also, *Adam's Ancestors,* pp. 104–5.

99. See Leakey, *The Stone Age Races of Kenya.*

100. See, for instance, Leakey, *Adam's Ancestors,* p. 202.

101. Ibid., pp. 79, 105, 133.

102. See ibid., p. 204 and Leakey, *The Stone Age Races of Kenya,* pp. 122–23.

103. Leakey, *Adam's Ancestors,* p. 176.

104. Ibid., pp. 165–70, 200–201.

105. Ibid., pp. 186, 228.

106. Ibid., pp. 228–29. See Zuckerman, "Sinanthropus and Other Fossil Men." In his autobiography, *From Apes to Warlords,* p. 62, Zuckerman defends this distinction as being equivalent to the modern one between *Homo erectus* and *H. sapiens.* He forgets that the Neanderthals are now placed with *H. sapiens,* though he had put them with the Palaeoanthropidae.

107. Leakey, *Adam's Ancestors,* pp. 219–20.

108. Ibid. (1953 ed.), p. 186 and diagram facing p. 112.

109. Ibid. (1934 ed.), p. 176.

110. Ibid., (1953 ed.), p. 161.

111. Gates, *Human Ancestry.*

112. Keith, *The Antiquity of Man* (2d ed.), pp. xi–xii.

113. Ibid., p. xii.

114. Ibid., pp. 416–17.

115. Keith and McCown, *The Stone Age of Mount Carmel,* 2:16–17.

116. Brace, "The Fate of the 'Classic' Neanderthals," p. 11.

117. Gregory, *The Origin and Evolution of Human Dentition,* pp. 458–60.

118. Black, "On the Discovery, Morphology, and Environment of *Sinanthropus pekinensis,*" pp. 74, 93–95, 113.

119. Hrdlička, "The Peopling of the Earth." See also the earlier "The Peopling of Asia."

120. Hrdlička, "The Most Ancient Skeletal Remains of Man," p. 533.

121. Hrdlička, "The Neanderthal Phase of Man," p. 250. See Brace, "Tales of the Phylogenetic Woods," p. 417. Hrdlička repeated his views in his "The Skeletal Remains of Early Man," pp. 335–36, 344–49.

122. Hrdlička, "The Neanderthal Phase of Man," pp. 253, 258.

123. Ibid., p. 260.

124. Ibid., p. 271.

125. Ibid., p. 270.

126. Verneau, "La race de Néanderthal et la race de Grimaldi."

127. Weinert, "*Pithecanthropus erectus,*" p. 542. On the Neanderthals see *Ursprung der Menschheit,* chap. 7.

128. Weidenreich, "Entwicklungs und Wassentypen des *Homo primigenius,*" p. 59.

129. Ibid., p. 60.

130. Weidenreich, "The Skull of *Sinanthropus pekinensis,*" pp. 232–33. For brief outlines of Weidenreich's later views see his "Some Problems Dealing with Ancient Man" and "The 'Neanderthal Man' and the Ancestors of *Homo sapiens.*"

131. Weidenreich, *Apes, Giants, and Man,* p. 30.

132. Weidenreich, "The Skull of *Sinanthropus pekinensis,*" p. 250. On racial variations see p. 248.

133. Weidenreich, *Apes, Giants, and Man,* pp. 106–7.

134. See the diagram ibid., p. 25, and for a full discussion of Australopithecus see Weidenreich, "The Skull of *Sinanthropus pekinensis,*" pp. 266–73.

135. Coon, *The Races of Europe,* pp. 50–51.

136. Coon, *The Origin of Races,* pp. 28–29.

137. Straus and Cave, "Pathology and the Posture of Neanderthal Man." This interpretation of Boule's work has been challenged, however, by Boaz, "American Research on Australopithecines and Early *Homo.*"

138. Brace, "The Fate of the 'Classic" Neanderthals." Note that pp. 19–32 of the article as cited consist of repsonses to Brace's position by a wide range of authors.

5. The Tarsioid Theory

1. Darwin, *Descent of Man,* p. 152.

2. Dawson and Woodward, "On the Discovery of a Palaeolithic Human Skull," p. 139.

3. Cope, "The Genealogy of Man," p. 324.

4. On Mivart's role in the opposition to Darwinism see Bowler, *The Eclipse of Darwinism,* chap. 3.

5. Mivart, *Man and Apes,* p. 176.

6. Ibid., p. 180.

7. Ibid., p. 185.

8. Ibid., p. 190. Mivart was convinced that our moral sense makes a complete link with the animal kingdom impossible; see his review of Darwin's *Descent of Man,* reprinted in Mivart, *Essays and Criticism,* 2:1–59.

9. Munro, *Prehistoric Problems,* p. 180.

10. Klaatsch, "Entstehung und Entwickelung des Menschengeschlechtes," pp. 183–206.

11. Ibid., p. 162.

12. Ibid., p. 185.

13. Ibid., pp. 176–80.

14. Haeckel, *Last Words on Evolution,* p. 71.

15. Ranke, "Ueber die individuellen Variationen im Schädelbau des Menschen."

16. Kollmann, "Neue Gedanken über das alte Problem von der Abstammung des Menschen." For Schwalbe's criticisms, see his *Studien zur Vorgeschichte des Menschen,* pt. 1. On Kollmann's theory and more generally on the recapitulation theory applied to human evolution, see Gould, *Ontogeny and Phylogeny,* chap. 10.

17. Kollmann, "Neue Gedanken," pp. 18–19.

18. Weidenreich, *Giant Early Man from Java and South China,* pp. 11–12.

19. Hill-Tout, "The Phylogeny of Man from a New Angle," pp. 48, 62. Hill-Tout was rebutted by W. K. Gregory; see the latter's "The Biogenetic Law and the Skull Form of Primitive Man."

20. Hill-Tout, "The Phylogeny of Man," pp. 63–65.

21. Ibid., pp. 67, 77.

22. Ibid., p. 80.

23. Adloff, *Das Gebiss des Menschen und Anthropomorphen;* see especially the diagram p. 131.

24. Sergi, *L'origine umane,* pp. 77–79.

25. Ibid., pp. 101, 106.

26. Hubrecht, *The Descent of the Primates,* pp. 19–21.

27. Ibid., p. 23.

28. Ibid., pp. 24, 26.

29. Ibid., pp. 27–28.

30. Ibid., p. 31. On p. 38 Hubrecht notes that Klaatsch supported a similar position.

31. Sollas, "Presidential Address," p. lxxx, quoting p. 275 of the 4th edition of Wiedersheim's *Der Bau des Menschen.* I have not seen this edition of the work; the speculation does not occur in the 2d edition or in the English translation.

32. Wood Jones, *Arboreal Man,* p. 4.

33. Ibid., pp. 12–13, 23.

34. Ibid., pp. 32–33, 44.

35. Ibid., p. 73.

36. Ibid., pp. 120–21, 153–54.

37. Ibid., p. 46.

38. I have used the version of "The Origin of Man" reprinted in Dendy, (ed.), *Animal Life and Human Progress.*

39. Ibid., p. 105.

40. Ibid., pp. 106–7.

41. Ibid., pp. 108–10.

42. Ibid., p. 110.

43. Ibid., pp. 112–15.

44. Ibid., p. 122.

45. Ibid., pp. 126–28.

46. Ibid., p. 131.

47. Smith Woodward et al., "Discussion on the Zoological Position and Affinities of Tarsius," p. 471. Grafton Elliot Smith's contribution to the discussion occupies pp. 465–75.

48. Ibid., p. 472.

49. Wood Jones, *Man's Place among the Mammals,* pp. 4, 10–11.

50. Ibid., chap. 7.

51. Ibid., chap. 8.

52. Ibid., p. 62.

53. Ibid., p. 155.

54. Ibid., p. 174.

55. Ibid., pp. 213–14.

56. Ibid., p. 328.

57. Ibid., p. 355.

58. Ibid., p. 329.

59. Ibid., p. 349.

60. Ibid., p. 353.

61. Wood Jones, *Hallmarks of Mankind,* p. 28.

62. Ibid., pp. 44–45.

63. Gregory, *Man's Place among the Anthropoids,* p. 29.

64. Ibid., p. 109.

65. Hooton, *Up from the Ape* (1931 ed.), p. 105 n.

66. Osborn, "Hesperopithecus, the Anthropoid Primate of Western Nebraska."

67. For example, Osborn, "Why Central Asia?" We shall return to the central Asian theory in chapter 7.

68. Citations below are to the most accessible version of Osborn's paper, "Recent Discoveries Relating to the Origin and Antiquity of Man," in the *Proceedings of the American Philosophical Society.* The paper also appeared in *Palaeobiologica* (1927), 1:189–202, and this version, which appears to contain a few extra passages, is reprinted in McCown and Kennedy, *Climbing Man's Family Tree,* pp. 285–301.

69. Osborn, "Recent Discoveries," p. 383. In *Man Rises to Parnassus,* Osborn argued that the arboreal stage was neither extensive nor important; see p. 214.

70. Osborn, "Recent Discoveries," e.g., pp. 376, 382.

71. Ibid., p. 377.

72. Osborn, *Man Rises to Parnassus,* pp. 83–84.

73. Osborn, "Recent Discoveries," p. 380. For the full story of Osborn's views on *Eoanthropus,* see *Man Rises to Parnassus,* pp. 51–68.

74. Osborn, "Recent Discoveries," p. 388.

75. Ibid., p. 381. See also Osborn, *Man Rises to Parnassus,* p. 83.

76. Osborn, "Recent Discoveries," p. 387; *Man Rises to Parnassus,* pp. 89–90.

77. Osborn, *Man Rises to Parnassus,* p. 199.

78. Ibid., p. 201. The "soft" environment of the tropics was held to have restricted the Negroid race to an "arrested brain development," p. 206.

79. Ibid., p. 199.

80. See ibid., pp. 220–21.

81. Osborn, "Recent Discoveries," p. 379.

82. On Osborn's general theory of evolution, see Bowler, *The Eclipse of Darwinism,* pp. 131–32, 174–76.

83. Osborn, *Evolution and Religion in Education,* a collection of Osborn's addresses and articles published in the course of the debate from 1922 to 1926.

84. Osborn, *Man Rises to Parnassus,* p. 84.

85. Hooton, *Up from the Ape* (1931 ed.), p. 105 n.

86. Gregory, "A Critique of Professor Osborn's Theory of Human Origins," pp. 133–34. Gregory's other articles on this theme are: "The Origin of Man from the Anthropoid Stem—When and Where?" and "How Near Is the Relationship of Man to the Chimpanzee-Gorilla Stock?" The arguments presented are similar to those used later against Frederic Wood Jones in Gregory's *Man's Place among the Anthropoids.*

87. Boule, *Fossil Men,* pp. 453–54; *Les hommes fossiles,* pp. 450–51.

88. Leakey, *Adam's Ancestors* (1934 ed.), p. 177; see also the diagram p. 227.

89. Broom and Schepers, *The South African Fossil Ape-Men,* p. 132.

90. Ibid., p. 257.

91. Ibid., pp. 261–63.

92. Straus, "The Riddle of Man's Ancestry."

6. Polytypic Theories

1. See Broca, *On the Phenomonon of Hybridity in the Genus Homo,* and Pouchet, *The Plurality of the Human Race.* The background to the monogenist-polygenist debate is outlined in chapter 2.

2. Vogt, *Lectures on Man,* p. 172.

3. Ibid., pp. 194, 299–307.

4. Ibid., p. 436.

5. Ibid., pp. 445–46.

6. Ibid., pp. 464–65.

7. Ibid., p. 466. Vogt did not specify in detail how the racial types were related to the different apes, but he did suggest that the dolichocephalic races came from the chimpanzee and gorilla, and the brachycephalics came from the orangutan.

8. Ibid., p. 468.

9. Wallace, "The Development of the Human Races under the Law of Natural Selection," reprinted in Wallace, *Natural Selection and Tropical Nature,* pp. 167–85.

10. Darwin to Wallace, 28 May 1864, in Marchant, ed., *Alfred Russel Wallace: Letters and Reminiscences,* pp. 126–29. See Wallace's reply, pp. 129–30.

11. Huxley, "On the Methods and Results of Ethnology," reprinted from the *Fortnightly Review* of 1865 in Huxley, *Man's Place in Nature* (1894 ed.), pp. 209–52; see esp. pp. 251–52.

12. Darwin, *Descent of Man,* pp. 171–72.

13. Ibid., p. 174.

14. Ibid., pp. 177–78.

15. Haeckel, *History of Creation,* 2:294. On Haeckel's views about race, see chapter 2.

16. Biographical information on Klaatsch is derived from the introduction to his *The Evolution and Progress of Mankind,* pp. 15–29.

17. Klaatsch, "Entstehung und Entwickelung des Menschengeschlechtes," discussed in chapter 5.

18. Klaatsch, "Die Aurignac-Rasse und ihre Stellung im Stammbaum der Menschheit"; summarized in Klaatsch, "Menschenrassen und menschenaffen," and in Wegner, "A New Theory of the Descent of Man."

19. Klaatsch, "Die Aurignac-Rasse," pp. 530–35.

20. Ibid., p. 567.

21. Wegner, "A New Theory of the Descent of Man," pp. 120–21.

22. Klaatsch, *The Evolution and Progress of Mankind,* pp. 99, 103.

23. Ibid., p. 99.

24. Klaatsch, "Die Aurignac-Rasse," p. 566.

25. Klaatsch, *The Evolution and Progress of Mankind,* p. 270.

26. Ibid.

27. Ibid., p. 107.

28. Ibid., p. 27.

29. Keith, "A New Theory of the Descent of Man." In his *Autobiography,* p. 340, Keith insisted that the theory aroused little interest in Britain.

30. Von Bonin, "Klaatsch's Theory of the Descent of Man."

31. Keith, "Klaatsch's Theory of the Descent of Man," p. 510.

32. Von Buttel-Reepen, *Man and His Forerunners,* pp. 75–76.

33. Gray, "The Differences and Affinities of Palaeolithic Man and the Anthropoid Apes," p. 119.

34. Ibid., p. 120.

35. Duckworth, *Prehistoric Man,* p. 139.

36. Sergi, *Le origine umane,* pp. 150–52.

37. Ibid. Even Keith was scornful of Ameghino's work; see *The Antiquity of Man,* chap. 17 (2d ed., chap. 25). Boule was rather less critical: see *Les hommes fossiles,* pp. 426–34.

38. Sergi, *Le origine umane,* p. 153.

39. Ibid., pp. 15–16.

40. Montandon, "L'ologénism ou ologenèse humaine."

41. Vallois, "Les preuves anatomiques de l'origine monophyletique de l'homme."

42. Crookshank, *The Mongol in Our Midst,* pp. 7–10.

43. Ibid., pp. 16–19 and more especially the 3d ed., pp. 13–21, 32, 38, 41.

44. Ibid. (1st ed.), pp. 22–23, 31.

45. Ibid., pp. 44–46.

46. Ibid., pp. 107, 115; 3d ed., pp. 380–83.

47. Ibid. (3d ed.), p. xii and esp. chap. 4.

48. Hooton, "The Asymmetrical Character of Human Evolution," pp. 138–40.

49. Hooton, *Up from the Ape,* p. 386. The "French anthropologist" who proposed human evolution on a global scale is clearly Montandon, although no citation is given.

50. Ibid., p. 390.

51. Ibid., p. 395.

52. Keith, "Origins of the Modern Races of Mankind."

53. Ibid., p. 194. Keith thus disagreed with some aspects of Leakey's interpretation of the African discoveries, see Keith, *New Discoveries Relating to the Antiquity of Man,* pp. 171–73.

54. Weidenreich, "The Skull of *Sinanthropus Pekinensis,*" p. 246. On Weidenreich's views on racial evolution, see chapter 4.

55. For Keith's views on Peking man see his *New Discoveries,* chap. 18.

56. Gates, *Human Ancestry from a Genetical Point of View,* preface.

57. Ibid., chap. 2.

58. Ibid., pp. 90, 114.

59. Ibid., pp. 237–49.

60. For Gates's complete scheme of human evolution, see the diagram, ibid., p. 161.

61. Ibid., p. 249.

62. See Coon, *The Origin of Races,* and the 1949 edition of Hooton's *Up from the Ape.*

63. See Croizat, *Panbiogeography,* pp. 564–91.

7. Brain, Posture, and Environment

1. Gould, *Ever Since Darwin,* chap. 26. Gould is referring to Engel's essay *Der Anteil der Arbeit an der Menschwerdung der Affen,* published posthumously in 1896. See also Meyer, *Nature, Human Nature, and Society,* chap. 9.

2. The development of the progressionist viewpoint is surveyed in Bowler, *Fossils and Progress;* on Agassiz and Chambers see chap. 3, on Dana p. 94 and on Marsh p. 137. For Cleland's views see his "Terminal Forms of Life."

3. On the teleological character of much late-nineteenth-century evolutionism, see Bowler, *The Eclipse of Darwinism* and Moore, *The Post-Darwinian Controversies.*

4. On de Mortillet, see Hammond, "Anthropology as a Weapon of Social Combat," and on Vogt see chapter 6.

5. Tylor, *Anthropology,* p. 113.

6. See chapter 2.

7. Tylor, *Anthropology,* p. 54. On Brinton's position see his "The Aims of Anthropology."

8. Spencer, *Principles of Sociology,* 1:104.

9. Ibid., p. 630 n. It is significant that even here Spencer's interest in the question is confined to a footnote.

10. Fiske, *Outlines of Cosmic Philosophy,* 2:342–44.
11. Ibid., pp. 294–95.
12. Romanes, *Mental Evolution in Man,* pp. 154–55.
13. Ibid., pp. 370–76.
14. Ward, "Relation of Sociology to Anthropology," p. 243.
15. Ibid., pp. 243–44.
16. Darwin, *Descent of Man,* pp. 62–63. Darwin opposed the concept of the "promiscuous horde," pp. 589–90.
17. Ibid., p. 98.
18. Ibid., pp. 49–51.
19. Ibid., p. 53.
20. Ibid.
21. Ibid., p. 64.
22. Ibid. This view was shared by Charles Lyell, *Principles of Geology* (11th ed.), 2:471.
23. Darwin, *Descent of Man,* pp. 155–56.
24. Haeckel, *History of Creation,* 2:299–300.
25. Ibid., p. 293.
26. Ibid., p. 327.
27. E.g., Haeckel, *The Last Link.*
28. Wallace, "The Development of Human Races under the Law of Natural Selection," reprinted in Wallace, *Natural Selection and Tropical Nature,* pp. 167–85.
29. Wallace, "The Limits of Natural Selection as Applied to Man," reprinted in *Natural Selection and Tropical Nature,* pp. 186–214, see p. 198.
30. Wallace, *Darwinism,* p. 458.
31. Lankester, "From Ape to Man," reprinted in Lankester, *Diversions of a Naturalist,* pp. 236–44; see esp. p. 243.
32. Lankester, "The Skeleton of Apes and of Man," reprinted ibid., pp. 245–52; see esp. p. 246.
33. Ibid., p. 248.
34. Morris, "The Making of Man," pp. 498–99. See also his "From Brute to Man."
35. Munro, "President's Address"; also reprinted as chap. 2 of Munro, *Prehistoric Problems;* see the latter, pp. 90–93.
36. See the letter from Huxley quoted in Munro, *Prehistoric Problems,* pp. 94–95.
37. Ibid., pp. 107–8.
38. Ibid., pp. 96, 115–16.
39. Wallace, *Darwinism,* p. 459.
40. Ibid., p. 460.
41. Landau, "Paradise Lost: The Theme of Terrestriality in Human Evolution."
42. Dubois also referred to Munro's discussion of the erect posture; see his *Pithecanthropus erectus; Eine Menschenaehnliche Uebergangsform aus Java,* p. 33 n.
43. Sollas, "Presidential Address" (1910), p. lxv. On Klaatsch see p. lxxx.
44. Ibid., p. lxxxiii.
45. Ibid., p. lxxxii.
46. Ibid., p. lxxxiii.

47. Ibid., p. lxxxiv. Sollas quotes pp. 204 and 206 of Klaatsch's "Entstehung und Entwickelung des Menschengeschlechtes."

48. Ibid., p. lxxxv.

49. Sollas, *Ancient Hunters* (3d ed.), pp. 190–91.

50. Elliot Smith, "The Piltdown Skull." See also his *The Evolution of Man,* p. 62.

51. Keith, *The Antiquity of Man,* p. 434.

52. Smith Woodward, *A Guide to the Fossil Remains of Man,* p. 25.

53. This view is hinted at ibid., p. 3, and will be discussed more fully in chapter 8.

54. Elliot Smith, "Discussion on the Origin of Mammals." See Smith, *The Evolution of Man,* p. 21. Citations to the 1912 "President's Address" below are to the version reprinted in *The Evolution of Man,* chap. 1, except where otherwise stated.

55. Smith, *The Evolution of Man,* pp. 22, 36. The criticism of Sollas in the original "President's Address," p. 590, was removed from the reprinted version.

56. See Keith, "The Extent to Which the Posterior Segments of the Body Have Been Transmuted and Suppressed in the Evolution of Man and Allied Primates," and *The Human Body,* chap. 5.

57. See for instance Keith, *The Antiquity of Man,* p. 434.

58. Keith, "Man's Posture: Its Evolution and Disorders," p. 451. Landau, "Paradise Lost," p. 15, cites Keith's remarks as evidence of his deep interest in the transition to a terrestrial life, yet it is significant that the relevant passage occurs in Keith's description of Lamarck's theory—nowhere else in this series of lectures does he address the question directly.

59. Smith, *The Evolution of Man,* p. 36.

60. Ibid.

61. Ibid., p. 39.

62. Davison, *Men of the Dawn,* p. 18. Cf. Smith, *The Evolution of Man,* pp. 30–31.

63. Smith, *The Evolution of Man,* p. 40.

64. Smith, "President's Address" (1912), p. 594.

65. Smith, *The Evolution of Man,* p. 28.

66. Ibid., p. 13.

67. Ibid., pp. 10–11. Smith also regarded the loss of pigment as a general trend that had gone furthest in the Nordics.

68. Manouvrier, "On *Pithecanthropus erectus,*" p. 228.

69. Matthew's "Climate and Evolution" appeared originally in the *Annals* of the New York Academy of Sciences, and was reissued posthumously in book form in 1939. Citations below are to the latter.

70. Matthew, *Climate and Evolution,* pp. 10–11.

71. Ibid., pp. 40–45.

72. Ibid., pp. 7–8.

73. Lull, *Organic Evolution,* p. 672.

74. Barrell, "Probable Relations of Climatic Change to the Origin of the Tertiary Ape-Man." On p. 17 Barrell acknowledges his debt to Matthew and Lull.

75. Lull, *Organic Evolution,* p. 691.

76. Osborn's 1900 article "The Geological and Faunal Relations of Europe and America during the Tertiary Period" is cited as the inspiration for the expeditions by

Roy Chapman Andrews, *The New Conquest of Central Asia,* pp. 3–4. See also Osborn's foreword to Andrews's *On the Trail of Ancient Man,* pp. vii–xi.

77. Osborn, "Why Central Asia?" pp. 264–65.

78. Ibid., pp. 266–67.

79. Osborn, "The Plateau Habitat of the Pro-Dawn Man."

80. Osborn, "Recent Discoveries Relating to the Origin and Antiquity of Man," e.g., p. 382.

81. Osborn, *Man Rises to Parnassus,* p. 215.

82. Osborn's 1896 paper "A Mode of Evolution Requiring neither Natural Selection nor the Inheritance of Acquired Characters (Organic Selection)" is reprinted in Baldwin, *Development and Evolution,* pp. 335–52. See Bowler, *The Eclipse of Darwinism,* pp. 81, 131–32. Organic selection is often known as the "Baldwin effect."

83. Osborn, *Man Rises to Parnassus,* pp. 214–18.

84. Black, "Asia and the Dispersal of Primates," pp. 134–35, 141. For details of Black's career see the biography by Dora Hood.

85. Black, "Asia and the Dispersal of Primates," p. 158.

86. Ibid., p. 175.

87. Gregory, *Origin and Evolution of Human Dentition,* p. 413.

88. See for instance Gregory, *Our Face from Fish to Man,* pp. 69–70.

89. Smith Woodward, "Recent Progress in the Study of Early Man," p. 133.

90. Boule, *Les hommes fossiles,* p. 454; *Fossil Men,* p. 458.

91. Dart, "*Australopithecus africanus:* The Man-ape of South Africa," p. 197.

92. Ibid., pp. 198–99.

93. Ibid., p. 198. See also Dart, "Taungs and Its Significance," p. 317.

94. Elliot Smith, *The Evolution of Man* (2d ed.), p. 10.

95. Ibid., pp. 62–65.

96. Ibid., pp. 61, 66–67.

97. Sollas, "The Taungs Skull."

98. Discussion following Sollas, "On a Sagittal Section of the Skull of *Australopithecus africanus,*" p. 11.

99. Wells, *The Outline of History,* pp. 38, 42.

100. H. G. Wells, Julian Huxley, and G. P. Wells, *The Science of Life,* p. 484.

101. Miller, "Conflicting Views on the Problem of Man's Ancestry," p. 240.

8. Trends in Human Evolution

1. On the development of Lamarckism and orthogenesis, see Bowler, *The Eclipse of Darwinism.* The role of orthogenesis in theories of human evolution is stressed in Fleagle and Jungers, "Fifty Years of Higher Primate Phylogeny."

2. See chapter 2.

3. Wood Jones, "The Origin of Man," p. 119.

4. Jones, *Man's Place among the Mammals,* p. 361.

5. Ibid., p. 26. On Kammerer see Koestler, *The Case of the Midwife Toad,* and Bowler, *The Eclipse of Darwinism,* pp. 92–100.

6. Jones, *Man's Place among the Mammals,* p. 29.

7. Ibid., pp. 32–36.

8. E.g., ibid., p. 42, on which Jones mentions the nonadaptive trends described by Watson.

9. Jones, *Trends of Life,* p. 41.

10. Broom, *The Coming of Man,* p. 19.

11. Ibid., pp. 21–22.

12. Ibid., pp. 46–64.

13. Ibid., p. 191.

14. Ibid., p. 196; see also p. 206.

15. Broom and Schepers, *The South African Fossil Ape-Men,* p. 261.

16. Broom, *Finding The Missing Link,* pp. 95–101.

17. Broom and Schepers, *The South African Fossil Ape-Men,* pp. 132, 260.

18. Osborn, "Recent Discoveries Relating to the Origin and Antiquity of Man," p. 376.

19. Osborn, "Is the Ape-Man a Myth?" p. 8.

20. Boule, *Les hommes fossiles,* p. 450; *Fossil Men,* pp. 453–54.

21. Boule, *Fossil Men,* p. 110; *Les hommes fossiles,* p. 110.

22. Leakey, *Adam's Ancestors* (1934 ed.), e.g., the diagram on p. 227.

23. Vogt, *Lectures on Man,* p. 466.

24. Klaatsch, "Entstehung und Entwickelung des Menschengeschlechtes," pp. 332–33.

25. Ibid., p. 194.

26. See Wegner's account of Klaatsch's views, "A New Theory of the Descent of Man," pp. 120–21; see also chapter 6.

27. Klaatsch, *Evolution and Progress of Mankind,* p. 96.

28. Ibid., p. 99.

29. Ibid., p. 103.

30. Gray, "The Differences and Affinities of Palaeolithic Man and the Anthropoid Apes," p. 119.

31. Sergi, *Le origine umane,* pp. 77–79.

32. Smith Woodward, "Missing Links among Extinct Animals," p. 785.

33. Ibid.

34. Ibid., p. 787. See also Dawson and Woodward, "On the Discovery of a Palaeolithic Human Skull and Mandible," p. 139.

35. Woodward, *A Guide to the Fossil Remains of Man,* p. 3.

36. Woodward, *The Earliest Englishman,* p. 76.

37. Woodward, "President's Address," p. 468.

38. Woodward, *The Earliest Englishman,* p. 114.

39. Elliot Smith in Smith Woodward et al., "Discussion on the Zoological Position and Affinities of Tarsius," p. 471.

40. Le Gros Clark *Early Forerunners of Man,* p. 286.

41. Ibid., p. 6; see also p. 287. Clark returned to the theme of parallelism in his "Evolutionary Parallelism and Human Phylogeny."

42. Clark, *Early Forerunners of Man,* p. 6; see also p. 166.

43. Ibid., p. 22.

44. Ibid., pp. 282–84.

45. Ibid., p. 79.

46. Ibid., p. 131.

47. Ibid., p. 139.

48. Ibid., p. 23.

49. Ibid., p. 288.

50. Le Gros Clark, *The Fossil Evidence for Human Evolution,* pp. 17–18.

51. Hooton, *Up from the Ape,* p. 115.

52. Hooton, "Doubts and Suspicions Concerning Certain Functional Theories of Primate Evolution," p. 232.

53. Ibid., p. 224.

54. Hooton, *Up from the Ape* (1931 ed.), p. 141.

55. Ibid., p. 142.

56. Ibid., p. 390. See also Hooton's "The Asymmetrical Character of Human Evolution."

57. Kollmann, "Neue Gedanken über das alte Problem von der Abstammung des Menschen."

58. For details of Bolk's theory see Gould, *Ontogeny and Phylogeny,* chap. 10.

59. Bolk, "On the Problem of Anthropogenesis," p. 469.

60. Ibid.

61. Ibid., p. 471.

62. See Keith, "The Adaptational Machinery Concerned in the Evolution of Man's Body," p. 267 and *Concerning Man's Origin,* p. 23.

63. Keith, *Ancient Types of Man,* p. 120.

64. Keith, *Autobiography,* pp. 339, 394.

65. Keith, "The Adaptational Machinery," p. 264.

66. See Keith's reflections on this issue in his "Fifty Years Ago," p. 265.

67. Keith, "The Adaptational Machinery," p. 265. On Cunningham's theory see Bowler, *Eclipse of Darwinism,* p. 102.

68. Keith, "The Adaptational Machinery," pp. 257–58 and *Concerning Man's Origin,* pp. 26–27.

69. Keith, *Antiquity of Man* (2d ed.), 2:725.

70. Ibid., p. 727.

71. Ibid., p. 456.

72. Keith, "Origins of the Modern Races of Mankind."

73. Weidenreich, *The Skull of Sinanthropus pekinensis,* p. 253.

74. Ibid., p. 258. In his *Apes, Giants, and Man,* p. 111, Weidenreich presents the trend from round to long skulls in modern races as a continuing adjustment to the erect posture.

75. Weidenreich, "Lamarckismus."

76. Weidenreich, *Giant Early Man,* pp. 113–18; *Apes, Giants, and Man,* pp. 57–62.

77. Weidenreich, *Giant Early Man,* pp. 11–12.

78. Gates, *Human Ancestry from a Genetical Point of View,* e.g., pp. 4, 13.

79. See Cartmill, "Basic Primatology and Prosimian Evolution." Simpson contributed a general statement of the principle of polytypic evolution based on adaptive trends to the 1950 Cold Spring Harbor Symposium on the Origin and Evolution of Man; see his "Some Principles of Historical Biology Bearing on Human Origins."

9. Accident or Design?

1. Darwin, *Descent of Man,* pp. 62–63, 132. Darwin also postulated a Lamarckian effect in the growth of social behavior.

2. See Drummond, *The Ascent of Man,* and Kropotkin, *Mutual Aid.* On the background to these developments see Bowler, *Evolution,* chap. 10.

3. Cleland, "Terminal Forms of Life," p. 359.

4. See Boule, *Les hommes fossiles,* p. 110; see also, chapter 8.

5. Wegner, "A New Theory of the Descent of Man," p. 120. See also Klaatsch, *The Evolution and Progress of Mankind,* p. 71.

6. Gould, "Paleontology," p. 154.

7. Gregory stressed the importance of transitions to a new mode of life as a means of rebutting the claim that the whole of human evolution was the unfolding of a single trend. See his "The Role of Undeviating Evolution and Transformation in the Origin of Man." For hints at a Lamarckian explanation of the effects of new habits, see, for example, *Man's Place among the Anthropoids,* p. 217.

8. Smith Woodward, "Recent Progress in the Study of Early Man," p. 136.

9. Hooton, "The Asymmetrical Character of Human Evolution," p. 136.

10. Hooton, *Up from the Ape* (1931 ed.), p. 390.

11. Elliot Smith, *The Evolution of Man,* p. 20.

12. Ibid., p. 27.

13. Ibid., p. 77.

14. Ibid., p. vi.

15. Ibid., pp. 19–20.

16. Ibid., p. 31.

17. Ibid., p. 42.

18. Ibid., p. 34.

19. Smith, "President's Address," p. 595. This passage was not included in the reprinted version in *The Evolution of Man.*

20. Ibid.

21. Smith, *The Evolution of Man,* p. 64.

22. On Smith's diffusionism see Daniel, "Elliot Smith, Egypt, and Diffusionism" and the biography by Warren R. Dawson.

23. Smith, *The Evolution of Man,* p. 92.

24. Ibid., p. 103.

25. Ibid., p. 122.

26. Smith, *Human History,* p. 272. See also *The Evolution of Man,* pp. 125–26.

27. Smith, *Human History,* pp. 11, 191.

28. Smith, *The Evolution of Man,* p. 131.

29. Wood Jones, *Arboreal Man,* p. 196.

30. Ibid., p. 221.

31. Ibid., p. 182.

32. Ibid., p. 189.

33. Jones, "The Origin of Man," p. 131.

34. Jones, *Design and Purpose,* p. 46.

35. Ibid., p. 71.

36. Ibid., pp. 72–73.

37. Ibid., pp. 76, 81.
38. Broom, *The Mammal-like Reptiles of South Africa,* p. 313.
39. Ibid., p. 332.
40. Ibid., p. 333.
41. Broom, *The Coming of Man,* pp. 11–12.
42. Ibid., pp. 66–67.
43. Ibid.
44. Ibid., p. 89.
45. Ibid., p. 117.
46. Ibid., p. 191.
47. Ibid., p. 197.
48. Ibid., p. 221.
49. Broom, *Finding the Missing Link,* pp. 95–101.
50. Findlay, *Dr. Robert Broom, F.R.S.,* p. 101.
51. Sollas, *Ancient Hunters,* p. 405. The passage is repeated in the 3d ed., p. 666.
52. Ibid.
53. Ibid., 3d ed., p. ix.
54. For more detailed expositions of the views expressed in this paragraph see Bowler, *The Eclipse of Darwinism,* chaps. 1, 4, and *Evolution,* chaps. 9, 10.
55. Lamarck, *Zoological Philosophy,* pp. 170–71.
56. MacBride in Smith Woodward et al., "Discussion on the Zoological Position and Affinities of Tarsius," p. 498. For details of MacBride's views see Bowler, "E. W. MacBride's Lamarckian Eugenics."
57. MacBride, *An Introduction to the Study of Heredity,* chap. 9.
58. Sollas, *Ancient Hunters* (1911 ed.), p. viii.
59. Ibid., preface to 1915 ed., reprinted in 3d ed., p. ix.
60. Ibid. (1911 ed.), p. 382; see also 3d ed., p. 599.
61. Ibid. (1911 ed.), p. 383; 3d ed., pp. 599–600.
62. See Pearson, *The Grammar of Science,* p. 369.
63. Sollas, *Ancient Hunters* (1911 ed.), pp. 50–51.
64. Lull, *Organic Evolution,* p. 691.
65. Ibid., pp. 671–72.
66. Matthew, *Climate and Evolution;* see chapter 7.
67. Black, "Asia and the Dispersal of Primates," p. 175.
68. Keith, *Darwinism and What It Implies,* pp. 18–19.
69. Keith, *Darwinism and Its Critics,* pp. 3–5.
70. Keith, *A New Theory of Human Evolution,* p. 3.
71. Keith, "War as a Factor in the Evolution of Races." I have not seen the original of this article, which appeared in the *St. Thomas' Hospital Gazette* for 1915. The version cited in the bibliography below is a French translation by Marcellin Boule, who was himself interested in the anthropological implications of the war.
72. Keith, "On Certain Factors Concerned in the Evolution of Human Races," p. 33.
73. Keith, "The Evolution of Human Races, Past and Present," pp. 48–49.
74. Ibid., p. 50.
75. Ibid., pp. 52, 54.

76. Ibid., p. 55.
77. Keith, *Essays on Human Evolution,* essay 6.
78. Ibid., essays 12–18.
79. Ibid., pp. 176–77.
80. Keith, *A New Theory of Human Evolution,* p. 229.
81. Ibid., pp. 247, 258.
82. Ibid., p. 254.
83. Lyell, *Principles of Geology* (11th ed.), 2:470–71.
84. Morris, "From Brute to Man."
85. Campbell, "Man's Mental Evolution, Past and Future," p. 1260.
86. Campbell, "The Biological Aspects of Warfare," p. 435.
87. Read, *The Origin of Man and His Superstitions,* p. v.
88. Ibid., p. 3.
89. Ibid., p. 8.
90. Ibid., p. 27.
91. Ibid., pp. 35, 46.
92. Cartmill, "Four Legs Good, Two Legs Bad," p. 69.
93. Dart, "*Australopithecus africanus:* The Man-Ape of South Africa," p. 199.
94. Dart, "Taungs and Its Significance," p. 321.
95. Dart, "The Status of *Australopithecus,*" p. 178.
96. Dart, "Cultural Status of the South African Man-Apes," p. 325.
97. Ibid., p. 327.
98. Dart and Craig, *Adventures with the Missing Link,* p. 195.
99. Ibid., p. 201.
100. Ibid., preface.
101. It would be inappropriate to discuss the details of the debate over sociobiology here; for further information see the references in Bowler, *Evolution,* chap. 11.
102. Ardrey, *The Territorial Imperative,* pp. 15–16.
103. Cartmill, "Four Legs Good, Two Legs Bad."

Epilogue

1. See the diagram in Hooton, *Up from the Ape* (1949 ed.), p. 413.
2. Leakey, *Adam's Ancestors* (1953 ed.), diagram facing p. 213.
3. For the story of Leakey's later work see for instance Reader, *Missing Links.*
4. Washburn, "The Analysis of Primate Evolution with Particular Reference to the Origin of Man," p. 76.
5. On the emergence of the Modern Synthesis see Provine, *Origins of Theoretical Population Genetics;* Mayr and Provine, eds., *The Evolutionary Synthesis;* and Grene, ed., *Dimensions of Darwinism.* For a brief survey, see Bowler, *Evolution,* chap. 11.
6. Simpson, "Some Principles of Historical Biology Bearing upon Human Origins."
7. Washburn, "The Analysis of Primate Evolution," p. 75.
8. Mayr, "Taxonomic Categories in Fossil Hominids."

9. Washburn, "The Analysis of Primate Evolution," p. 67.

10. Gregory, "Fossil Man-Apes of South Africa" and "The Bearing of the Australopithecinae upon the Problem of Man's Place in Nature."

11. Washburn, "The Analysis of Primate Evolution," p. 70.

12. Wilfrid Le Gros Clark, "Significance of the Australopithecinae," and "New Palaeontological Evidence Bearing on the Evolution of the Hominoidea."

13. Clark, *History of the Primates,* pp. 68–69.

14. Clark, *The Fossil Evidence for Human Evolution,* pp. 17–18.

15. Washburn, "The Analysis of Primate Evolution," p. 70. Quantum evolution is discussed in the conclusion to Simpson's *Tempo and Mode in Evolution.*

16. See Gould, "Paleontology."

17. Clark, *The Fossil Evidence for Human Evolution,* p. 8.

18. Mayr, "Taxonomic Categories in Fossil Hominids," p. 116.

19. See Weiner, *The Piltdown Forgery.*

20. See Reader, *Missing Links,* chaps. 7, 9.

21. See Johanson and Edey, *Lucy.*

22. See Gribbin and Cherfas, *The Monkey Puzzle.*

23. See Pilbeam, "The Descent of Hominoids and Hominids."

24. See Isaac, "Aspects of Human Evolution."

25. See for instance Cartmill, "Four Legs Good, Two Legs Bad" and also Lewin, *Human Evolution,* pp. 26–27. The latter is an excellent survey of modern ideas on human origins.

Bibliography

Adloff, P. *Das Gebiss des Menschen und der Anthropomorphen. Vergleichenden-anatomische Untersuchungen. Zugleich ein Beitrag zur menschlichen Stammesgeschichte.* Berlin: Julius Springer, 1908.

Agassiz, Louis. *An Essay on Classification,* ed. Edward Lurie. Cambridge, Mass.: Harvard University Press, 1962.

Allen, Garland E. *Life Science in the Twentieth Century.* New York: Wiley, 1975.

Alverdes, F. *The Psychology of Animals in Relation to Human Psychology.* London: Kegan Paul, Trench, Trubner & Co., 1932.

Andersson, J. Gunnar. *Children of the Yellow Earth: Studies in Prehistoric China,* trans. E. Classen. London: Kegan Paul, 1934.

Andrews, Roy Chapman. *On the Trail of Ancient Man: A Narrative of the Field Work of the Central Asiatic Expeditions.* New York: G. P. Putnam's Sons, 1926.

———, ed. *The New Conquest of Central Asia: A Narrative of the Explorations of the Central Asiatic Expedition in Mongolia and China, 1921–1930.* New York: American Museum of Natural History, 1932.

Ardrey, Robert. *The Territorial Imperative: A Personal Inquiry into the Animal Origins of Property and Nations.* New York: Delta Books, 1966.

———. *The Hunting Hypothesis: A Personal Conclusion Concerning the Evolutionary Nature of Man.* London: Collins, 1976.

Argyll, George Douglas Campbell, 8th Duke of. *The Reign of Law.* London: Alexander Strahan, 1867. 5th ed., 1868.

———. *Primeval Man: An Examination of Some Recent Speculations.* London: Alexander Strahan, 1869.

———. *Organic Evolution Cross-Examined.* London: Murray, 1898.

Aston, E. H., and Zuckerman, Solly. "Some Quantitative Dental Characters of Fossil Anthropoids." *Phil. Trans. Roy. Soc. Lond.* 234B (1950): 485–520.

———. "Statistical Methods in Anthropology." *Nature* 168 (1951): 1117–18.

Baitsell, George Alfred, ed. *The Evolution of Man.* New Haven: Yale University Press, 1922.

Baldwin, James Mark. *Development and Evolution: Including Psychophysical Evolution, Evolution by Orthoplasy, and the Theory of Genetic Modes.* New York: Macmillan, 1902.

Bannister, Robert C. *Social Darwinism: Science and Myth in Anglo-American Social Thought.* Philadelphia: Temple University Press, 1979.

Barrell, Joseph. "Probable Relations of Climatic Change to the Tertiary Ape-Man." *Scientific Monthly* 4 (1917): 16–26.

Bartholomew, Michael. "Lyell and Evolution: An Account of Lyell's Response to the Prospect of an Evolutionary Ancestry for Man." *Brit. J. Hist. Sci.* 6 (1973):261–303.

———. "Huxley's Defense of Darwin." *Annals of Science* 32 (1975):525–35.

Barzun, Jacques, *Race: A Study in Modern Superstition.* London: Methuen, 1938.

Beer, Gillian. *Darwin's Plots: Evolutionary Narrative in Darwin, George Eliot, and Nineteenth-Century Fiction.* London: Routledge & Kegan Paul, 1983.

Bergson, Henri. *Creative Evolution,* trans. Arthur Mitchell. New York: Henry Holt, 1911.

Berry, Richard, J. A., and Robertson, A.W.D. "The Place in Nature of the Tasmanian Aboriginal as Deduced from a Study of His Calvarium." *Proc. Roy. Soc. Edinburgh* 31 (1910):41–69 and 34 (1914):144–89.

Black, Davidson. "Asia and the Dispersal of Primates." *Bull. Geol. Soc. China* 4 (1925):133–83.

———. "Tertiary Man in Asia: The Chou Kou Tien Discovery." *Nature* 118 (1927):733–34.

———. "Further Hominid Remains of Lower Quaternary Age from the Chou Kou Tien Deposits." *Nature* 120 (1927):954.

———. "The Croonian Lecture—On the Discovery, Morphology, and Environment of *Sinanthropus pekinensis.*" *Phil. Trans. Roy. Soc. Lond.* 123 (1934):57–120.

Blavatsky, H. P. *The Secret Doctrine: The Synthesis of Science, Religion, and Philosophy.* 3d ed. Point Loma, Calif.: Aryan Theosophical Press, 1925. 2 vols.

Boas, Franz. *Race, Language, and Culture.* New York: Macmillan, 1940.

———, ed. *General Anthropology.* 1938. Reprint. New York: Johnson Reprint Crop., 1965.

Boaz, Noel T. "American Research on Australopithecines and Early *Homo.*" In *A History of American Physical Anthropology,* ed. Frank Spencer, 239–60. New York: Academic Press, 1982.

Bolk, Louis. *Die Entstehung des Menschenkinnes: Ein Beitrag zur Entwickelungsgeschichte des Unterkiefers.* Amsterdam: Koninklijke Akademie van Wetenschappen, 1924.

———. "The Chin Problem." *Proc. Section Sciences, Koninklijke Akademie van Wetenschappen, Amsterdam* 27 (1924):329–44.

———. *Das Problem der Menschwerdung.* Jena: Gustav Fischer, 1926.

———. "On the Problem of Anthropogenesis." *Proc. Section Sciences, Koniklijke Akademie van Wetenschappen, Amsterdam* 29 (1926):465–75.

———. "Origin of Racial Characteristics in Man." *Am. J. Phys. Anthrop.* 13 (1929):1–28.

Boule, Marcellin. "Description de l'hyaena brevirostris du Pliocene de Sainzelles près le Puy (Haute Loire)." *Annales des sciences naturelles,* 7th ser. 15 (1893):85–97.

———. Review of G. and A. de Mortillet, *Le préhistorique. L'Anthropologie* 12 (1901):427–31.

———. "L'homme fossile de La Chapelle-aux-Saints." *L'Anthropologie,* 19 (1908):519–25.

———. "L'homme fossile de La Chapelle-aux-Saints." *Annales de paléontologie* 6 (1909):109–72; 7 (1910):18–56, 85–192; 8 (1911):1–71.

———. "L'homme fossile de Piltdown, Sussex (Angleterre)." *L'Anthropologie* 23 (1912):742–44.

———. "La guerre." *L'Anthropologie* 25 (1914):575–80.

———. "La paléontologie humaine en Angleterre." *L'Anthropologie* 26 (1915):1–67.

———. *Les hommes fossiles: Elements de paléontologie humaine.* Paris: Masson, 1921.

———. *Fossil Men: Elements of Human Palaeontology,* trans. Jessie Elliot Ritchie and James Ritchie. Edinburgh: Oliver & Boyd, 1923.

———. "Le Sinanthrope." *L'Anthropologie* 47 (1937):1–22.

———, and Anthony, Raoul. "L'encéphale de l'homme fossile de La Chapelle-aux-Saints." *L'Anthropologie* 22 (1911):129–96.

Bowler, Peter J. *Fossils and Progress: Paleontology and the Idea of Progressive Evolution in the Nineteenth Century.* New York: Science History Publications, 1976.

———. *The Eclipse of Darwinism: Anti-Darwinian Evolution Theories in the Decades around 1900.* Baltimore: Johns Hopkins University Press, 1983.

———. *Evolution: The History of an Idea.* Berkeley and Los Angeles: University of California Press, 1984.

———. "E. W. MacBride's Lamarckian Eugenics and Its Implications for the Social Construction of Scientific Knowledge." *Annals of Science* 41 (1984): 245–60.

Brace, C. Loring. "The Fate of the 'Classic' Neanderthals: A Consideration of Hominid Catastrophism." *Current Anthropology* 5 (1964): 3–43.

———. "Reply" [to Stephan F. Holtzman]. *Current Anthropology* 14 (1973):308–9.

———. "Tales of the Phylogenetic Woods: The Evolution and Significance of Evolutionary Trees." *Am. J. Phys. Anthrop.* 56 (1981): 411–29.

———. "The Roots of the Race Concept in American Physical Anthropology." In *A History of American Physical Anthropology,* ed. Frank Spencer, 11–29. New York: Academic Press, 1982.

Breuil, Henri. "Notes de voyage paléolithique en Europe centrale. I. Les industries paléolithiques en Hongrie." *L'Anthropologie* 33 (1923):323–46.

———. "Palaeolithic Industries from the Beginning of the Rissian to the Beginning of the Wurm Glaciation." *Man* 26 (1926):176–79.

Briffault, Robert. *The Mothers: A Study of the Origins of Sentiments and Institutions.* 2 vols. London: Allen & Unwin, 1927.

Brinton, Daniel G. "The Aims of Anthropology: President's Address." *Proceedings of the American Association for the Advancement of Science,* 1895, pp. 1–17.

Broca, Paul. "Sur le volume et la forme du cerveau suivant les individues et suivant les races." *Bulletin de la Société d'Anthropologie de Paris* 2 (1861):139–207, 301–31, and 441–46.

———. "Sur la capacité des crânes parisiens des diverses époques." *Bulletin de la Société d'Anthropologie de Paris* 3 (1862):102–16.

———. *On the Phenomena of Hybridity in the Genus Homo,* ed. C. Carter Blake. London: For the Anthropological Society, Longmans, Green, Longman & Roberts, 1864.

———. "L'ordre des primates. Parallele anatomique de l'homme et des singes." *Bulletin de la Société d'Anthropologie de Paris,* 2d ser. 4 (1869):228–401.

———. "Sur le transformisme." *Bulletin de la Société d'Anthropologie de Paris,* 2d ser. 5 (1870):168–239.

Broom, Robert. "On the Organ of Jacobson and Its Relations in the 'Insectivora.'" *Proc. Zool. Soc. Lond.* 1915:157–62, 347–54.

———. "Some Notes on the Taungs Skull." *Nature* 115 (1925):560–71.

———. "Note on the Milk Dentition of *Australopithecus.*" *Poc. Zool. Soc. Lond.* 1929:85–88.

———. "The Age of *Australopithecus.*" *Nature* 125 (1930):814.

———. *The Mammal-like Reptiles of South Africa and the Origin of Mammals.* London: H. F. & G. Witherby, 1932.

———. *The Coming of Man: Was It Accident or Design?* London: H. F. & G. Witherby, 1933.

———. "A New Fossil Anthropoid Skull from South Africa." *Nature* 138 (1936):486–88.

———. "More Discoveries of *Australopithecus.*" *Nature* 141 (1938):828–29.

———. "The Pleistocene Anthropoid Apes of South Africa." *Nature* 142 (1938):377–79.

———. *Finding the Missing Link.* London: Watts & Co., 1950.

———, and Schepers, G.W.H. *The South-African Fossil Ape-Men: The Australopithecinae.* Pretoria: Transvaal Museum, 1946.

Burchfield, Joe D. *Lord Kelvin and the Age of the Earth.* New York: Science History Publications, 1975.

Burkhardt, Richard W., Jr. *The Spirit of System: Lamarck and Evolutionary Biology.* Cambridge, Mass.: Harvard University Press, 1977.

Burkitt, M. C. *Prehistory: A Study of Early Cultures in Europe and the Mediterranean Basin.* Cambridge: Cambridge University Press, 1921.

Burrow, J. W. *Evolution and Society: a Study in Victorian Social Thought.* Cambridge: Cambridge University Press, 1966.

Butler, Samuel. *Evolution, Old and New: Or the Theories of Buffon, Dr. Erasmus Darwin, and Lamarck, as Compared with that of Mr. Charles Darwin.* London: Hardwick & Bogue, 1879.

———. *Life and Habit.* New ed. London: A. C. Fifield, 1916.
———. *Unconscious Memory.* 3d ed. London: A. C. Fifield, 1920.
———. *Luck or Cunning as the Main Means of Organic Modification.* 2d ed. London: A. C. Fifield, 1920.
Bynum, William F. "Charles Lyell's *Antiquity of Man* and Its Critics." *J. Hist. Biology* 17 (1984):153–87.
Campbell, Harry. "Man's Mental Evolution, Past and Future." *Lancet* 1913: 1260–62, 1333–35, 1408–10, 1473–76.
———. "The Biological Aspects of Warfare." *Lancet* 1917:433–35, 469–71, 505–8.
———. "Man's Evolution from the Anthropoid." *Lancet* 1921:629.
Campion, George G., and Smith, G. Elliot. *The Neural Basis of Thought.* London: Kegan Paul, Trench, Trubner & Co., 1934.
Carpenter, William Benjamin. *Nature and Man: Essays Scientific and Philosophical.* New York: Appleton, 1889.
Cartailhac, Emile. *La France préhistorique d'après les sépultures et les monuments.* Paris: Alcan, 1889.
Cartmill, Matt. "Basic Primatology and Prosimian Evolution." In *A History of American Physical Anthropology,* ed. Frank Spencer, 147–86. New York: Academic Press, 1982.
———. "Four Legs Good, Two Legs Bad: Man's Place (if any) in Nature." *Natural History* 92 (1983):64–79.
Casey, Thomas L. "The Mutation Theory." *Science* n.s. 21 (1905):307–9.
Chambers, Robert. *Vestiges of the Natural History of Creation.* London: Churchill, 1844. 5th ed. London: Churchill, 1846; 11th ed. London: Churchill, 1860. Reprint introd. Sir Gavin de Beer. Leicester: Leicester University Press, 1969.
———. *Explanations: A Sequel to the Vestiges of the Natural History of Creation.* 2d ed. London: Churchill, 1846.
Clark, Sir Wilfrid E. Le Gros. *Early Forerunners of Man: A Morphological Study of the Evolutionary Origin of the Primates.* London: Baillière, Tindall & Cox, 1934.
———. "Evolutionary Parallelism and Human Phylogeny." *Man* 36 (1936):4–8.
———. "The Relationship between *Pithecanthropus* and *Sinanthropus.*" *Nature* 145 (1940): 70–71.
———. "Significance of the Australopithecinae." *Nature* 157 (1946):863–65.
———. *History of the Primates: An Introduction to the Study of Fossil Man.* London: British Museum, 1949.
———. "New Palaeontological Evidence Bearing on the Evolution of the Hominoidea." *Quart. J. Geol. Soc. Lond.* 105 (1950):225–64.
———. *The Fossil Evidence for Human Evolution.* Chicago: University of Chicago Press, 1955.

———. *Man-Apes or Ape-Men? The Story of Discoveries in Africa.* New York: Holt, Rinehart & Winston, 1967.

———, and Leakey, L.S.B. *The Miocene Hominoidea of East Africa.* London: British Museum, 1951.

Cleland, John. "Terminal Forms of Life." *J. Anat. and Physiol.* 18 (1884):345–62.

Cole, Sonia. *Leakey's Luck: The Life of Louis Seymour Bazett Leakey, 1903–1972.* London: Collins, 1975.

Coleman, William. *Biology in the Nineteenth Century: Problems of Form, Function, and Transformation.* New York: Wiley, 1971.

Conry, Yvette. *L'introduction du Darwinisme en France au XIXe siècle.* Paris: Vrin, 1974.

Coon, Carlton Stevens. *The Races of Europe.* New York: Macmillan, 1939.

———. *The Origin of Races.* London: Jonathan Cape, 1963.

Cope, Edward Drinker. "A Review of the Modern Doctrine of Evolution." *Am. Naturalist* 14 (1880): 166–79, 260–71.

———. "On Lemurine Reversion in Human Dentition." *Am. Naturalist* 20 (1886):941–47.

———. *The Origin of the Fittest: Essays on Evolution.* London: Macmillan, 1887.

———. "The Genealogy of Man." *Am. Naturalist* 27 (1893):321–35.

———. *The Primary Factors of Organic Evolution.* Chicago: Open Court, 1896. Reprinted with *The Origin of the Fittest,* New York: Arno, 1974.

Corsi, Pietro. "The Importance of French Transformist Ideas for the Second Volume of Lyell's *Principles of Geology.*" *Brit. J. Hist. Sci.* 11 (1978):221–44.

Cravens, Hamilton. *The Triumph of Evolution: American Scientists and the Heredity-Environment Controversy, 1900–1941.* Philadelphia: Univeristy of Pennsylvania Press, 1978.

Croizat, Leon. *Panbiogeography: Or an Introductory Synthesis of Zoogeography, Phytogeography, and Geology, with Notes on Evolution, Systematics, Ecology, Anthropology, etc.* Caracas: The Author, 1958.

Crookshank, F. G. *The Mongol in Our Midst: A Study of Man and His Three Faces.* London: Kegan Paul, Trench, Trubner & Co., 1924., 3d ed., 1931.

Cross, K. Stuart. "On a Numerical Determination of the Relative Positions of Certain Biological Types in the Evolutionary Scale." *Proc. Roy. Soc. Edinburgh* 31 (1910):70–84.

Cunningham, D. J. "President's Address, Anthropology Section." *Report of the British Association for the Advancement of Science,* 1901 meeting, reprinted *Nature* 64 (1901):539–45.

Cuvier, Georges. *Essay on the Theory of the Earth, with Geological Illustrations by Professor Jameson.* Edinburgh, 1813.

Dana, James Dwight. "Agassiz's Contributions to the Natural History of the United States." *Am. J. Sci.* 2d ser. 25 (1858):201–16, 321–41.

Daniel, Glyn. "Grafton Elliot Smith: Egypt and Diffusionism." In "The Concepts of Human Evolution," ed. S. Zuckerman. *Symposia of the Zoological Society of London* 33 (1973):407–47.

———. *A Hundred and Fifty Years of Archaeology.* London: Duckworth, 1975.

———, ed. *Towards a History of Archaeology.* London: Thames & Hudson, 1981.

Dart, Raymond A. "The Misuse of the Term 'Visceral.'" *J. Anatomy & Physiology* 56 (1921–22):177–88.

———. "Boskop Remains from the South-east African Coast." *Nature* 112 (1923):623–25.

———. "The Anterior End of the Neural Tube and the Anterior End of the Body." *J. Anatomy & Physiology* 58 (1923–24):181–205.

———. "*Australopithecus africanus:* The Man-Ape of South Africa." *Nature* 115 (1925):195–99.

———. "The Taungs Skull." *Nature* 116 (1925):462.

———. "Taungs and Its Significance." *Natural History* 26 (1926):315–27.

———. "The Status of *Australopithecus.*" *Am. J. Phys. Anthrop.* 26 (1940): 167–86.

———. "Cultural Status of the South African Man-Apes." *Annual Report, Smithsonian Institution,* 1955, pp. 317–38.

———, and Craig, Dennis. *Adventures with the Missing Link.* New York: Harper & Row, 1959.

———, and Shellshear, Joseph L. "The Origin of the Motor Neuroblasts of the Anterior Cornu of the Neural Tube." *J. Anatomy & Physiology* 56 (1921–22):77–95.

Darwin, Charles Robert. *On the Origin of Species by Means of Natural Selection: Or the Preservation of Favoured Races in the Struggle for Life.* London: Murray, 1859; 6th ed. London: Murray, 1872. Reprint introd. Ernst Mayr. Cambridge, Mass.: Harvard University Press, 1964.

———. *The Descent of Man and Selection in Relation to Sex.* 2d ed., revised. London: Murray, 1885.

Darwin, Francis, ed. *The Life and Letters of Charles Darwin.* 3 vols. London: Murray, 1887.

———. *More Letters of Charles Darwin.* 2 vols. New York: Appleton, 1903.

Davison, Dorothy. *Men of the Dawn: The Story of Man's Evolution to the End of the Old Stone Age.* London: Watts, 1934.

Dawkins, W. Boyd. *Cave Hunting: Researches on the Evidence of Caves Respecting the Early Inhabitants of Europe.* London: Macmillan, 1874.

Dawson, Charles, and Woodward, Arthur Smith. "On the Discovery of a Palaeolithic Human Skull and Mandible in a Flint-Bearing Gravel Overlying the Wealden (Hastings Beds) at Piltdown, Fletching (Sussex)." *Quart. J. Geol. Soc. Lond.* 69 (1912):117–44.

———. "Supplementary Note on the Discovery of a Palaeolithic Human Skull and Mandible at Piltdown (Sussex)." *Quart. J. Geol. Soc. Lond.* 70 (1914):82–93.

Dawson, Sir John William. *Fossil Men and Their Modern Representatives.* 2d ed. London: Hodder & Stoughton, 1883.

Dawson, Warren R., ed. *Sir Grafton Elliot Smith: A Biographical Record by His Colleagues.* London: Jonathan Cape, 1938.

De Beer, Sir Gavin. *Charles Darwin.* London: Nelson, 1963.

De Mortillet, Gabriel. "Promenades préhistoriques à l'Exposition universelle." *Matériaux pour l'histoire positive et philosophique de l'homme* 3 (1867):181–368.

———. *Le Préhistorique: Antiquité de l'homme.* Paris: C. Reinwald, 1883.

———, and de Mortillet, Adrien. *Le Préhistorique: Origine et antiquité de l'homme.* 3d ed. Paris: Schleicher, 1900.

Deniker, J. *The Races of Man: An Outline of Anthropology and Ethnography.* London: Walter Scott, 1900.

De Quatrefages, Jean-Louis Armand. *The Prussian Race Ethnologically Considered,* trans. Isabella Innes. London: Virtue & Co., 1872.

———. *The Human Species.* 2d ed. London: C. Kegan Paul, 1879.

———. *Hommes fossiles et hommes sauvages: Études d'anthropologie.* Paris: Baillière, 1884.

———, and Hamy, Ernest T. *Crania Ethnica: Les crânes des races humaines.* Paris: Baillière, 1882.

Desmond, Adrian. *Archetypes and Ancestors: Palaeontology in Victorian London, 1850–1875.* London: Blond & Briggs, 1982.

———. "Richard Owen's Reaction to Transmutation in the 1830s." *Brit. J. Hist. Sci.* 18 (1985):25–50.

De Vries, Hugo. *The Mutation Theory,* trans. J. B. Farmer and A. D. Darbyshire. 2 vols. London: Kegan Paul, Trench, Trubner & Co., 1910.

Di Gregorio, Mario A. *T. H. Huxley's Place in Natural Science.* New Haven: Yale University Press, 1984.

Dobzhansky, Theodosius. *Genetics and the Origin of Species.* New York: Columbia University Press, 1937.

Drummond, Henry. *The Ascent of Man.* 1894; 13th ed. New York: James Pott, 1904.

Dubois, Eugene. *Pithecanthropus erectus: Eine Menschenaenliche Uebergangsform aus Java.* Batavia: Landsdruckerei, 1894.

———. "The Place of '*Pithecanthropus*' in the Genealogical Tree." *Nature* 53 (1895–96): 245–46.

———. "On *Pithecanthropus erectus:* A Transitional Form between Man and the Apes." *Scientific Transactions of the Royal Dublin Society* 6 (1898):1–18.

———. "*Pithecanthropus erectus*—a Form from the Ancestral Stock of Mankind." *Annual Report, Smithsonian Institution,* 1898, pp. 445–59.

———. "Remarks upon the Brain-cast of *Pithecanthropus erectus.*" In *Proceedings of the 4th International Congress of Zoology,* ed. Adam Sedgwick, 78–95. Cambridge: Cambridge University Press, 1898.

———. "On the Fossil Human Skulls Recently Discovered in Java and *Pithecanthropus erectus.*" *Man* 37 (1937):1–7.

———. "Early Man in Java and *Pithecanthropus erectus.*" In *Early Man,* ed. G. G. MacCurdy, 315–22. Philadelphia and New York: Lippincott, 1937.

Duckworth, W.L.H. *Prehistoric Man.* Cambridge: Cambridge University Press, 1912.

Duncan, David. *The Life and Letters of Herbert Spencer.* Reissue. London: Methuen, 1911.

Dupont, E. "Classement des âges de la pierre en Belgique." *Congrès International d'Anthropologie et d'Archéologie Préhistorique, Comte Rendu* 6 (Brussels, 1872):459–79.

Durant, J. R. "The Beast in Man: An Historical Perspective on the Biology of Human Aggression." In *The Biology of Aggression,* ed. Paul F. Brain and David Benton, 17–46. Alphen aan den Rijn, Netherlands: Sijthoff & Noordhoff, 1981,

Eimer, Theodor. *Organic Evolution as the Result of the Inheritance of Acquired Characters According to the Laws of Organic Growth,* trans. J. T. Cunningham. London: Macmillan, 1890.

———. *On Orthogenesis and the Impotence of Natural Selection in Species Formation,* trans. T. J. McCormack. Chicago: Open Court, 1898.

Eiseley, Loren. *Darwin's Century: Evolution and the Men Who Discovered It.* New York: Doubleday, 1958.

Eldredge, Niles, and Tattersall, Ian. *The Myths of Human Evolution.* New York: Columbia University Press, 1982.

Ellegård, Alvar. *Darwin and the General Reader: The Reception of Darwin's Theory of Evolution in the British Periodical Press, 1859–72.* Göteborg: Acta Universitatis Göthenburgensis, 1958.

Engels, Friedrich. "Der Anteil der Arbeit an der Menschwerdung des Affen." In Karl Marx and Friedrich Engels, *Werke,* XX, pp. 444–55. Berlin: Dietz Verlag, 1968. 39 vols.

Evans, John. *The Ancient Stone Implements, Weapons, and Ornaments of Great Britain.* London: Longmans Green, Reader & Dyer, 1872; 2d ed., revised, 1897.

Findlay, G. H. *Dr. Robert Broom, F.R.S.: Palaeontologist and Physician, 1866–1951.* Cape Town: A. A. Balkema, 1972.

Fisher, Ronald Aylmer. *The Genetical Theory of Natural Selection.* Oxford: Clarendon Press, 1930.

Fiske, John. *Outlines of Cosmic Philosophy based on the Doctrine of Evolution, with Criticisms on the Positive Philosophy.* 2 vols. London: Macmillan, 1874.

Fothergill, Philip G. *Historical Aspects of Organic Evolution.* London: Hollis & Carter, 1952.

Fraipont, Julien. "Les hommes de Spy (La race de Canstadt ou de Néanderthal en Belgique)." *Congrès International d'Anthropologie et d'Archéologie Préhistorique, Comte Rendu.* 10 (Paris, 1888), 321–48.

Freeland, Guy. "Evolution and Arch(a)eology." In *The Wider Domain of Evolutionary Thought,* ed. David Oldroyd and Ian Langham, 175–219. Dordrecht: D. Reidel, 1983.

Galton, Francis. *Hereditary Genius: An Inquiry into its Laws and Consequences.* London: Macmillan, 1869.

Gasman, Daniel. *The Scientific Origins of National Socialism: Social Darwinism in Ernst Haeckel and the Monist League.* New York: American Elsevier, 1971.

Gates, R. Ruggles. "Parallel Mutations in *Oenothera biennis.*" *Nature* 89 (1912):659–60.

———. *The Mutation Factor in Evolution: With Particular Reference to Oenothera.* London: Macmillan, 1915.

———. *Human Ancestry from a Genetical Point of View.* Cambridge, Mass.: Harvard University Press, 1948.

Gaudry, Albert. *Les enchainements du monde animal dans les temps géologiques: mammifères tertiaires.* Paris: Savy, 1878.

———. "Contribution à l'histoire des hommes fossiles." *L'Anthropologie* 14 (1903):1–14.

Geikie, James. *The Great Ice Age and Its Relation to the Antiquity of Man.* 2d ed. London: Edward Stanford, 1877.

———. *Prehistoric Europe: A Geological Sketch.* London: Edward Stanford, 1881.

———. *The Antiquity of Man in Europe.* Edinburgh: Oliver & Boyd, 1914.

George, Wilma. *Biologist Philosopher: A Study of the Life and Writings of Alfred Russel Wallace.* New York: Abelard-Schuman, 1964.

Giddings, Franklin Henry. *The Principles of Sociology: An Analysis of the Phenomena of Association and of Social Organization.* New York: Macmillan, 1902.

Gillespie, Neal C. "The Duke of Argyll, Evolutionary Anthropology, and the Art of Scientific Controversy." *Isis* 68 (1977):40–54.

———. *Charles Darwin and the Problem of Creation.* Chicago: University of Chicago Press, 1979.

Gillispie, Charles C. *Genesis and Geology: A Study in the Relations of Scientific Thought, Natural Theology, and Social Opinion in Great Britain, 1790–1850.* Reprint. New York: Harper & Row, 1959.

Glick, Thomas F., ed. *The Comparative Reception of Darwinism.* Austin: University of Texas Press, 1972.

Gould, Stephen Jay. *Ontogeny and Phylogeny.* Cambridge, Mass.: Harvard University Press, 1977.

———. *The Panda's Thumb: More Reflections on Natural History.* New York: Norton, 1980.

———. *The Mismeasure of Man*. New York: Norton, 1981.

———. "Paleontology." In *The Evolutionary Synthesis,* ed. Ernst Mayr and W. B. Provine, 153–72. Cambridge, Mass.: Harvard University Press, 1982.

———. *Hen's Teeth and Horse's Toes: Further Reflections on Natural History.* New York: Norton, 1983.

Grant, Madison. *The Passing of the Great Race: Or the Racial Basis of European History.* 4th ed., revised, introd. Henry Fairfield Osborn. London: George Bell & Sons, 1921.

Gray, Asa. *Darwiniana: Essays and Reviews Pertaining to Darwinism.* New York: Appleton, 1876. Reprint ed. A. Hunter Dupree. Cambridge, Mass.: Harvard University Press, 1963.

Gray, John. "The Differences and Affinities of Palaeolithic Man and the Anthropoid Apes." *Man* 11 (1911): 117–20.

Grayson, Donald K. *The Establishment of Human Antiquity.* New York: Academic Press, 1983.

Greene, John C. *The Death of Adam: Evolution and Its Impact on Western Thought.* Ames: Iowa State University Press, 1959.

———. *Darwin and the Modern World View.* Baton Rouge: Louisiana State University Press, 1961.

———. *Science, Ideology, and World View: Essays in the History of Evolutionary Ideas.* Berkeley and Los Angeles: University of California Press, 1981.

Gregory, William King. "Studies on the Evolution of the Primates." *Bull. Am. Museum Nat. Hist.* 35 (1916):239–355.

———. *The Origin and Evolution of the Human Dentition.* Baltimore: Williams & Wilkins, 1922.

———. "The Biogenetic Law and the Skull Form of Primitive Man." *Am. J. Phys. Anthrop.* 8 (1925):373–78.

———. "How Near Is the Relationship of Man to the Chimpanzee-Gorilla Stock?" *Quart. Rev. Biol.* 2 (1927):549–60.

———. "The Origin of Man from the Anthropoid Stem—When and Where?" *Proc. Am. Philosophical Soc.* 66 (1927): 439–63.

———. "The Upright Posture of Man: A Review of Its Origin and Evolution." *Proc. Am. Philosophical Soc.* 67 (1928):339–76.

———. *Our Face from Fish to Man: A Portrait Gallery of Our Ancient Ancestors and Kinsfolk, Together with a Concise History of Our Best Features.* New York: G. P. Putnam's Sons, 1929.

———. "A Critique of Professor Osborn's Theory of Human Origin." *Am. J. Phys. Anthrop.* 14 (1930):133–61.

———. *Man's Place among the Anthropoids: Three Lectures on the Evolution of Man from the Lower Vertebrates.* Oxford: Clarendon Press, 1934.

———. "The Roles of Undeviating Evolution and Transformation in the Origin of Man." *Am. Naturalist* 69 (1935):385–404.

———. "Fossil Man-Apes of South Africa." *Nature* 143 (1939):25–26.

———. "The Bearing of the Australopithecinae upon the Problem of Man's Place in Nature." *Am. J. Phys. Anthrop.* n.s. 7 (1949):485–512.

———, and Hellman, Milo. "The Evidence of the Dentition on the Origins of Man." In *Early Man,* ed. G. G. MacCurdy, 185–92. Philadelphia and New York: Lippincott, 1937.

Grene, Marjorie, ed. *Dimensions of Darwinism: Themes and Counter-Themes in Twentieth-Century Evolutionary Thought.* Cambridge: Cambridge University Press, 1983.

Gribben, John, and Cherfas, Jeremy. *The Monkey-Puzzle: A Family Tree.* London: Bodley Head, 1982.

Gruber, Howard E. *Darwin on Man: A Psychological Study of Scientific Creativity.* London: Wildwood House, 1974.

Gruber, Jacob W. *A Conscience in Conflict: The Life of St. George Jackson Mivart.* New York: Columbia University Press, 1960.

———. "Brixham Cave and the Antiquity of Man." In *Context and Meaning in Cultural Anthropology,* ed. Milford E. Spiro, 373–402. New York: Free Press, London: Collier-Macmillan, 1964.

Haas, Otto, and Simpson, George Gaylord. "Analysis of Some Phylogenetic Terms, with Attempts and a Redefinition." *Proc. Am. Philosophical Soc.* 90 (1946):319–49.

Haeckel, Ernst. *Generelle Morphologie der Organismen: Allgemene Grundzüge der organischen Formen Wissenschaft, mechanische begründet durch die von Charles Darwin reformiste Descendenz-theorie.* 2 vols. Berlin: Georg Reimer, 1866.

———. *The History of Creation: Or the Development of the Earth and Its Inhabitants by the Action of Natural Causes. A Popular Exposition of the Doctrine of Evolution in General, and of That of Darwin, Goethe, and Lamarck in Particular,* trans. E. Ray Lankester. 2 vols. New York: Appleton, 1883.

———. *The Evolution of Man: A Popular Exposition of Human Ontogeny and Phylogeny.* 2 vols. London: Kegan Paul, Trench, Trubner & Co., 1883.

———. *The Last Link: Our Present Knowledge of the Descent of Man.* London: A. & C. Black, 1898.

———. "On Our Present Knowledge of the Origin of Man." *Annual Report, Smithsonian Institution,* 1898, pp. 461–80.

———. *Last Words on Evolution: A Popular Retrospect and Summary,* trans. Joseph McCabe. London: A. Owen, 1906.

———. *The Evolution of Man: A Popular Scientific Study,* trans. from the 5th ed., Joseph McCabe. London: Watts, 1907.

Haldane, J.B.S. *The Causes of Evolution.* London: Longmans Green, 1932.

Haller, John S., Jr. *Outcasts from Evolution: Scientific Attitudes of Racial Inferiority, 1859–1900.* Urbana: University of Illinois Press, 1971.

Halstead, L. B. "New Light on the Piltdown Hoax?" *Nature* 276 (1978):11–13.

Hamlin, Christopher. "James Geikie, James Croll, and the Eventful Ice Age." *Annals of Science* 39 (1982):565–83.

Hammond, Michael. "A Framework of Plausibility for an Anthropological Forgery: The Piltdown Case." *Anthropology* 3 (1979):47–58.

———. "Anthropology as a Weapon of Social Combat in Late Nineteenth-Century France." *J. Hist. Behavioral Sci.* 16 (1980):118–32.

———. "The Expulsion of the Neanderthals from Human Ancestry: Marcellin Boule and the Social Context of Scientific Research." *Social Studies of Science* 12 (1982):1–36.

Hamy, E. T. *Précis de paléontologie humaine.* Paris: Baillière, 1870.

Harris, Marvin. *The Rise of Anthropological Theory: A History of Theories of Culture.* London: Routledge & Kegan Paul, 1968.

Harvey, Joy. "Evolutionism Transformed: Positivists and Materialists in the *Société d'Anthropologie de Paris* from Second Empire to Third Republic." In *The Wider Domain of Evolutionary Thought,* ed. David Oldroyd and Ian Langham, 289–310. Dordrecht: D. Reidel. 1983.

Hatch, Elvin. *Theories of Man and Culture.* New York: Columbia University Press, 1973.

Heyer, Paul. *Nature, Human Nature, and Society: Marx, Darwin and the Human Sciences.* Westport, Conn.: Greenwood Press, 1982.

Hill-Tout, Charles. "The Phylogeny of Man from a New Angle." *Trans. Roy. Soc. Canada* 15 (1921):47–82.

Himmelfarb, Gertrude. *Darwin and the Darwinian Revolution.* London: Chatto & Windus, 1959.

Hodge, M. J. S. "The Universal Gestation of Nature: Chambers' *Vestiges* and *Explanations.*" *J. Hist. Biology* 5 (1972):127–52.

Hofstadter, Richard. *Social Darwinism in American Thought.* Revised ed. Boston: Beacon Press, 1955.

Holtzman, Stephen F. "On Brace's Notion of 'Hominid Catastrophism.'" *Current Anthropology* 14 (1973): 306–8.

Hood, Dora. *Davidson Black: A Biography.* Toronto: University of Toronto Press, 1964.

Hooton, Earnest Albert. "The Asymmetrical Character of Human Evolution." *Am. J. Phys. Anthrop.* 8 (1925):125–41.

———. "Doubts and Suspicions Concerning Certain Functional Theories of Primate Evolution." *Human Biology* 2 (1930):223–49.

———. *Up from the Ape.* London: Allen & Unwin, 1931. Revised ed. New York: Macmillan, 1949.

Howell, F. Clark. "The Evolutionary Significance of Variation and Varieties of 'Neanderthal Man.'" *Quart. Rev. Biol.* 32 (1957):330–47.

Howells, W. W. "Fossil Man and the Origin of Races." *Am. Anthropologist* 42 (1944):182–93.

Hrdlička, Ales. "The Most Ancient Skeletal Remains of Man." *Annual Report, Smithsonian Institution,* 1913, 491–552.
———. "The Peopling of Asia." *Proc. Am. Philosophical Soc.* 60 (1921):535–45.
———. "The Piltdown Jaw." *Am. J. Phys. Anthrop.* 5 (1922):337–47.
———. "The Taungs Ape." *Am. J. Phys. Anthrop.* 8 (1925):379–92.
———. "The Peopling of the Earth." *Proc. Am. Philosophical Soc.* 65 (1926):150–56.
———. "The Neanderthal Phase of Man." *J. Roy. Anthrop. Inst.* 56 (1927):249–74.
———. "The Skeletal Remains of Early Man." *Smithsonian Miscellaneous Collections* 83 (1930). (Whole vol.)
Hubrecht, A.A.W. *The Descent of the Primates.* New York: Charles Scribner's Sons, 1897.
Hull, David L. *Darwin and His Critics: The Reception of Darwin's Theory of Evolution by the Scientific Community.* Cambridge, Mass.: Harvard University Press, 1973.
———. "Historical Entities and Historical Narratives." In *Minds, Machines, and Evolution: Philosophical Studies,* ed. Christopher Hookway, 17–41. Cambridge: Cambridge University Press, 1984.
———, Tessner, Peter D., and Diamond, Arthur M.. "Planck's Principle: Do Younger Scientists Accept New Scientific Ideas with Greater Alacrity than Older Scientists?" *Science* 202 (1978):717–23.
Huxley, Julian S. *Evolution: The Modern Synthesis.* London: Allen & Unwin, 1942.
Huxley, Leonard, ed. *Life and Letters of Thomas Henry Huxley.* 3 vols. 2d ed. London: Macmillan, 1903.
Huxley, Thomas Henry. "On the Zoological Relations of Man with the Lower Animals." *Natural History Review* 8 (1861):67–84.
———. *Evidence as to Man's Place in Nature.* London: Williams & Norgate, 1863.
———. *American Addresses: With a Lecture on the Study of Biology.* New York: Appleton, 1888.
———. *Man's Place in Nature.* (Huxley, *Collected Essays,* vol. 7.) London: Macmillan, 1894.
Hyatt, Alpheus. *Genesis of the Arietidae.* Washington: Smithsonian Contributions to Knowledge, no. 673, 1889.
Irvine, W. *Apes, Angels, and Victorians: A Joint Biography of Darwin and Huxley.* Reprint. Cleveland: Meridian, 1959.
Isaac, Glynn Ll. "Aspects of Human Evolution." In *Evolution from Molecules to Men,* ed. D. S. Bendall, 509–43. Cambridge: Cambridge University Press, 1983.

Johanson, Donald C., and Edey, Maitland A. *Lucy: The Beginnings of Humankind.* New York: Simon & Schuster, 1981.

Jones, Frederic Wood. *Aboreal Man.* 2d impression. London: Edward Arnold, 1918.

———. "The Origin of Man." In *Animal Life and Human Progress,* ed. Arthur Dendy, 99–131. London: Constable, 1919.

———. *Man's Place among the Mammals.* London: Edward Arnold, 1929.

———. *Design and Purpose.* London: Kegan Paul, Trench, Trubner & Co., 1942.

———. *Habit and Heritage.* London: Kegan Paul, Trench, Trubner & Co., 1943.

———. *Hallmarks of Mankind.* London: Baillière, Tindall & Cox, 1948.

———. *Trends of Life.* London: Edward Arnold, 1953.

Jones, Greta. *Social Darwinism and English Thought: The Interaction between Biological and Social Theory.* London: Harvester, 1980.

Kammerer, Paul. *The Inheritance of Acquired Characteristics,* trans. A. Paul Maerker-Brandon. New York: Boni & Liveright, 1924.

Keith, Arthur. "The Extent to Which the Posterior Segments of the Body Have Been Transmuted and Suppressed in the Evolution of Man and Allied Primates." *J. Anat. and Physiol.* 37 (1903):18–40.

———. "A New Theory of the Descent of Man." *Nature* 85 (1910):206.

———. *Ancient Types of Man.* New York: Harper, 1911.

———. "Klaatsch's Theory of the Descent of Man." *Nature* 85 (1911):509–10.

———. "The Early History of the Gibraltar Cranium." *Nature* 87 (1911):313–14.

———. "Modern Problems Relating to the Antiquity of Man." *Report of the British Association for the Advancement of Science,* 1912:753–59.

———. "President's Address: The Reconstruction of Fossil Human Skulls." *J. Roy. Anthrop. Inst.* 44 (1912):12–31.

———. "The Piltdown Skull and Brain Cast." *Nature* 92 (1913):197–99, 292, 345–46.

———. *The Antiquity of Man.* London: Williams & Norgate, 1915; 2d ed. 2 vols. London: Williams & Norgate, 1925.

———. "President's Address: On Certain Factors Concerned in the Evolution of Human Races." *J. Roy. Anthrop, Inst.* 46 (1916):10–34.

———. "War as a Factor in the Evolution of Races." *L'Anthropologie* 27 (1916): 172–77.

———. "The Adaptational Machinery Concerned in the Evolution of Man's Body." *Nature* 112 (1923):257–68.

———. "Man's Posture: Its Evolution and Disorders." *Brit. Medical J.* 1 (1923):451–54, 499–502, 545–48, 587–90, 624–26, 669–72.

———. *The Human Body.* London: Williams & Norgate, 1925.

———. "The Fossil Anthropoid Ape from Taungs." *Nature* 115 (1925):234–35.

———. "The Taungs Skull." *Nature* 116 (1925):462–63.

———. *Concerning Man's Origins.* London: Watts, 1927.

———. *Darwinism and What it Implies.* London: Watts, 1928.

———. *New Discoveries Relating to the Antiquity of Man.* London: Williams & Norgate, 1931.

———. "The Evolution of Human Races: Past and Present." In *Early Man: His Origin, Development and Culture,* by Grafton Elliot Smith et al., 47–64. London: Ernest Benn, 1931.

———. *The Construction of Man's Family Tree.* London: Watts, 1934.

———. *Darwinism and Its Critics.* London: Watts, 1935.

———. "Origins of Modern Races of Mankind." *Nature* 138 (1936):194.

———. "A Resurvey of the Anatomical Features of the Piltdown Skull, with some Observations on the Recently Discovered Swanscombe Skull." *J. Anat. and Physiol.* 73 (1938–39):155–85, 234–54.

———. "Fifty Years Ago." *Am. J. Phys. Anthrop.* 26 (1940):251–67.

———. *Essays on Human Evolution.* London: Scientific Book Club, 1946.

———. *A New Theory of Human Evolution.* London: Watts, 1948.

———. *An Autobiography.* London: Watts, 1950.

———, and McCown, Theodor. "Mount Carmel Man: His Bearing on the Ancestry of Modern Races." In *Early Man,* ed. G. G. MacCurdy, 41–52. Philadelphia and New York: Lippincott, 1937.

Keller, Ferdinand. *The Lake Dwellings of Switzerland and Other Parts of Europe,* trans. J. E. Lee. 2d ed. 2 vols. London: Longmans Green, 1872.

Kellogg, Vernon L. *Darwinism Today.* New York: Henry Holt, London: George Bell, 1908.

Kelvin, Sir William Thomson. Baron. *Popular Lectures and Addresses.* 3 vols. London: Macmillan, 1891–94.

Klaatsch, Hermann. "Entstehung und Entwickelung des Menschengeschlechtes." In *Weltall und Menscheit: Geschichte der Erforschung der Natur und der Verwertung der Naturkräfte im Dienste der Völker,* ed. Hans Kramer, vol. 2, 1–338. 5 vols. Berlin: Bong & Co., 1902.

———. "Anthropologische und paläontologische Ergebnisse eine Studienreise durch Deutschland, Belgien, und Frankreich." *Zeitschrift für Ethnologie* 35 (1903):92–132.

———. "Die neuersten Ergebnisse der Paläontologie des Menschen und ihre Bedeutung für das Abstammungsproblem." *Zeitschrift für Ethnologie* 41 (1909): 537–80.

———. "Die Aurignac-Rasse und ihre Stellung im Stambaum der Menschheit." *Zeitschrift für Ethnologie* 42 (1910):513–77.

———. "Menschenrassen und Menschenaffen." *Korrespondenzblatt Deutsche Geschellschaft fur Anthropologie* 41 (1910):91–100.

———. *The Evolution and Progress of Mankind,* ed. Adolf Heilborn; trans. Joseph McCabe. London: T. Fisher Unwin, 1923.

———, and Hauser, O. "Homo Mousteriensis Hauseri. Ein altdiluvier Skelettfund im Department Dordogne und seine Zugehörigkeit zum Neanderthaltypus." *Archiv für Anthropologie* n.s. 7 (1909):287–97.

———. "Homo Aurignacensis Hauseri: Ein palëolithischer Skeletfund aus dem unteren Aurignacien der Station Combe-Capelle bei Montferrand (Périgord)." *Praehistorisches Zeitschrift* 1 (1910):273–338.

Knox, Robert. *The Races of Man: A Philosophical Enquiry into the Influence of Race over the Destiny of Nations.* 2d ed. London: Henry Renshaw, 1862.

Koestler, Arthur. *The Case of the Midwife Toad.* London: Hutchinson, 1971.

Kollmann, J. "Neue Gedanken über das alte Problem von der Abstammung des Menschen." *Korrespondenzblatt Deutsche Geschellschaft für Anthropologie* 36 (1905):9–20.

Kottler, Malcolm J. "Alfred Russel Wallace, the Origin of Man, and Spiritualism." *Isis* 65 (1974):145–92.

Kropotkin, Peter. *Mutual Aid: A Factor in Evolution.* 1914. Reprint, intro. Ashley Montagu. Boston: Extending Horizon Books, n.d.

Kuhn, Thomas S. *The Structure of Scientific Revolutions.* Chicago: University of Chicago Press, 1959.

Laing, S. *Human Origins.* London: Chapman & Hall, 1894.

Lamarck, J.B.P.A. de Monet, *Philosophie zoologique: Ou exposition des considérations relatives à l'histoire naturelle des animaux.* New ed. 2 vols. Paris: Savy, 1873.

———. *Zoological Philosophy: An Exposition with Regard to the Natural History of Animals,* trans. Hugh Elliot. London: Macmillan, 1914. Reprint. New York: Hafner, 1963.

Laming-Emperaire, A. *Origines de l'archéologie préhistorique en France.* Paris: A. & J. Picard, 1964.

Landau, Misia. "The Anthropogenic: Paleoanthropological Writing as a Genre of Literature." Ph.D. diss., Yale University, 1981.

———. "Human Evolution as Narrative." *American Scientist* 72 (1984):262–68.

———. "Paradise Lost: The Theme of Terrestriality in Human Evolution." Paper read to University of Iowa Symposium on the Humanities, March 1984.

Langham, Ian. *The Building of British Social Anthropology: W.H.R. Rivers and His Cambridge Disciples in the Development of Kinship Studies, 1898–1931.* Dordrecht: D. Reidel, 1981.

———. "Sherlock Holmes, Circumstantial Evidence, and Piltdown Man." *Physical Anthropology News* 3 (1984):1–4.

Lanham, Url. *The Bone Hunters.* New York: Columbia University Press, 1975.

Lankester, Sir E. Ray. *Science from an Easy Chair.* London: Methuen, 1910.

———. *The Kingdom of Man.* London: Watts, 1911.

———. "On the Discovery of a Novel Type of Flint Implements below the Base of the Red Crag of Suffolk, Proving the Existence of Skilled Workers of Flint in the Pliocene Age." *Phil. Trans. Roy. Soc. Lond.* 102 B (1912):283–336.

———. *Diversions of a Naturalist.* London: Methuen, 1915.

Lartet, Edouard. "Note sur un grand singe fossile qui se rattache au groupe des singes supérieures." *Comtes rendus de l'Académie des Sciences* 43 (1856):219–23.

———. "Nouvelles recherches sur la coexistence de l'homme et des grands mammifères fossiles." *Annales des Sciences Naturelles* 4th ser. 15 (1861):177–253.

———, and Christy, Henry. *Reliquiae Aquitanicae: Being Contributions to the Archaeology and Palaeontology of Périgord and the Adjoining Provinces of Southern France,* ed. Thomas Rupert Jones. London: Williams & Norgate, 1875.

Leakey, Louis S. B. *Adam's Ancestors: An Up-to-date Outline of What Is Known about the Origin of Man.* 3d ed. London: Methuen, 1934; 4th ed., 1953.

———. *The Stone Age Races of Kenya.* London: Oxford University Press, 1935.

———. *Stone Age Africa: An Outline of Prehistory in Africa.* London: Oxford University Press, 1936.

———. *White African.* London: Hodder & Stoughton, 1937.

———. *By the Evidence: Memoirs, 1932–1951.* New York: Harcourt Brace Jovanovich, 1974.

Le Conte, Joseph. *Evolution: Its Nature, Its Evidences, and Its Relation to Religious Thought.* 2d ed. New York: Appleton, 1899.

Lewin, Roger. *Human Evolution: An Illustrated Introduction.* Oxford: Blackwell Scientific Publications, 1984.

Lubbock, Sir John (Lord Avebury). *Prehistoric Times: As Illustrated by Ancient Remains and the Manners and Customs of Modern Savages.* London: Williams & Norgate, 1865. 7th ed., 1913.

———. *The Origin of Civilisation and the Primitive Condition of Man.* London: Longmans Green, 1870.

Lull, Richard Swan. *Organic Evolution: A Textbook.* New York: Macmillan, 1917.

———. "The Antiquity of Man." In *The Evolution of Man,* ed. G. A. Baitsell, 1–38. New Haven: Yale University Press, 1922.

Lurie, Edward. *Louis Agassiz: A Life in Science.* Chicago: University of Chicago Press, 1960.

Lydekker, R. "Review of Dubois, *Pithecanthropus erectus.*" *Nature* 51 (1894–95):291.

Lyell, Sir Charles. *Geological Evidences of the Antiquity of Man: With Remarks on Theories of the Origin of Species by Variation.* London: Murray, 1863.

———. *Principles of Geology: Or the Modern Changes of the Earth and Its Inhabitants Considered as Illustrative of Geology.* 11th ed. 2 vols. London: Murray, 1872.

Lyon, John. "The Search for Fossil Man: Cinq Personnages à la Recherche du Temps Perdu." *Isis* 61 (1970):68–84.

MacAlister, R.A.S. *A Textbook of Europena Archaeology.* Vol. 1. *The Palaeolithic Period.* Cambridge: Cambridge University Press, 1921.

MacBride, E. W. *An Introduction to the Study of Heredity.* London: Williams & Norgate, 1924.

McCabe, Joseph. *Prehistoric Man.* Manchester: Milner, n.d.

———. *The Evolution of Mind.* 2d ed. London: Watts, 1921.

McCown, Theodor D., and Keith, Sir Arthur. *The Stone Age of Mount Carmel.* Vol. 2. *The Fossil Human Remains from the Levalloiso-Mousterian.* Oxford: Clarendon Press, 1939.

McCown, Theodore D., and Kennedy, Kenneth A. R. *Climbing Man's Family Tree: A Collection of Major Writings on Human Phylogeny, 1699–1971.* Englewood Cliffs, N.J.: Prentice-Hall, 1972.

MacCurdy, George Grant. "Recent Discoveries Bearing on the Antiquity of Man." *Annual Report, Smithsonian Institution,* 1909, pp. 531–83.

———. *Human Origins: A Manual of Prehistory.* New York: Appleton, 1924.

———, ed. *Early Man: As Depicted by Leading Authorities at the International Symposium, the Academy of Natural Sciences, Philadelphia.* Philadelphia and New York: Lippincott, 1937.

McDougall, William. *An Introduction to Social Psychology.* 15th ed. London: Methuen, 1920.

McGee, W. J. "The Trend of Human Progress." *Am. Anthropologist* 1 (1899):401–47.

McLennan, John Ferguson. *Studies in Ancient History: Comprising a Reprint of Primitive Marriage, an Inquiry into the Origin of the Form of Capture in Marriage Ceremonies.* London: Bernard Quaritch, 1876.

Maine, Sir Henry. *Lectures on the Early History of Institutions.* London: Murray, 1875.

———. *Village-Communities in the East and West.* 3d ed. London: Murray, 1876.

———. *Dissertations on Early Law and Custom: Chiefly Selected from Lectures Delivered at Oxford.* London: Murray, 1883.

———. *Ancient Law.* Reprint. London: Everyman, 1917.

Malinowski, Bronislaw. *Sex and Repression in Savage Society.* London: Routledge & Kegan Paul, 1927. 4th impression, 1953.

Mandelbaum, Maurice. *History, Man, and Reason: A Study in Nineteenth-Century Thought.* Baltimore: Johns Hopkins University Press, 1971.

Manouvrier, L. "On *Pithecanthropus erectus.*" *Am. J. Science* n.s. 4 (1897):213–34.

Marchant, James, ed. *Alfred Russel Wallace: Letters and Reminiscences.* New York: Harper, 1916.

Marsh, Othniel C. "Recent Discoveries of Extinct Mammals." *Am. J. Science* 12 (1876):59–61.
———. "On the *Pithecanthropus erectus* from the Tertiary of Java." *Am. J. Science* n.s. 1 (1896):475–82.
Marston, Alvan T. "The Swanscombe Skull." *J. Roy. Anthrop. Inst.* 67 (1937):339–406.
Matthew, William D. "The Arboreal Ancestry of the Mammalia." *Am. Naturalist* 38 (1904):811–18.
———. "Climate and Evolution." *Ann. New York Acad. Sci.* 24 (1914):171–318.
———. "The Evolution of the Mammals in the Eocene." *Proc. Zool. Soc. Lond.* 1927: 947–85.
———. *Climate and Evolution.* 2d ed., revised, ed. E. H. Colbert. New York: New York Academy of Sciences, 1939.
Mayr, Ernst. *Systematics and the Origin of Species.* New York: Columbia University Press, 1942.
———. "Taxonomic Categories in Fossil Hominids." *Cold Spring Harbor Symposia on Quantitative Biology* 15 (1950):109–17.
———. *Evolution and the Diversity of Life: Selected Essays.* Cambridge, Mass.: Harvard University Press, 1976.
———. "Reflections on Human Paleontology." In *A History of American Physical Anthropology,* ed. Frank Spencer, 231–37. New York: Academic Press, 1982.
———, and Provine, W. B., eds. *The Evolutionary Synthesis: Perspectives on the Unification of Biology.* Cambridge, Mass.: Harvard University Press, 1980.
Millar, Ronald. *The Piltdown Men: A Case of Archaeological Fraud.* London: Victor Gollancz, 1972.
Miller, Gerrit S., Jr. "The Jaw of Piltdown Man." *Smithsonian Institution Miscellaneous Collections* 65, no. 12 (1916): 31 pp.
———. "The Piltdown Jaw." *Am. J. Phys. Anthrop.* 1 (1918):25–52.
———. "Conflicting Views on the Problem of Man's Ancestry." *Am. J. Phys. Anthrop.* 3 (1920):213–45.
———. "The Controversy over Human 'Missing Links.' " *Smithsonian Institution, Annual Report,* 1928, pp. 413–65.
Millhauser, Milton. *Just before Darwin: Robert Chambers and Vestiges.* Middletown, Conn.: Wesleyan University Press, 1959.
Mivart, St. George Jackson. *On the Genesis of Species.* 2d ed. London, Macmillan, 1871.
———. *Man and Apes: An Exposition of Structural Resemblances Bearing upon Questions of Affinity and Origin.* London: Robert Hardwicke, 1873.
———. *Essays and Criticism.* 2 vols. London: James R. Osgood, McIlvane & Co., 1892.
Moir, J. Reid. *Pre-Palaeolithic Man.* Ipswich: W. E. Harrison, 1919.

———. *The Earliest Men. Huxley Memorial Lecture, Imperial College, 1939.* London: Macmillan, n.d.

Montagu, M. F. Ashley. *Man's Most Dangerous Myth: The Fallacy of Race.* 5th ed. New York: Oxford University Press, 1974.

Montandon, George. "L'ologénisme, ou ologenèse humaine." *L'Anthropologie* 39 (1929):103–22.

Moore, James R. *The Post-Darwinian Controversies: A Study of the Protestant Struggle to Come to Terms with Darwin in Britain and America, 1870–1900.* Cambridge: Cambridge University Press, 1979.

Moore, Ruth. *Man, Time, and Fossils: The Story of Evolution.* London: Jonathan Cape, 1954.

Morgan, Lewis H. *Ancient Society: Or Researches in the Lines of Human Progress from Savagery through Barbarism to Civilization.* 1877. Reprint, ed. Leslie A. White. Cambridge, Mass.: Harvard University Press, 1964.

Morgan, Thomas Hunt. *Evolution and Genetics.* Princeton: Princeton University Press, 1925.

Morris, Charles. "The Making of Man." *Am. Naturalist* 20 (1886):493–505.

———. "From Brute to Man." *Am. Naturalist* 24 (1890): 341–500.

Morton, Dudley J. "Evolution of the Human Foot." *Am. J. Phys. Anthrop.* 5 (1922):305–36.

———. "Evolution of Man's Erect Posture." *J. Morphology* 43 (1926):147–79.

———. "Human Origin: Correlation of Previous Studies of Primate Feet and Posture with Other Morphological Evidence." *Am. J. Phys. Anthrop.* 10 (1927):173–203.

Morton, Samuel G. *Crania Americana: Or a Comparative View of the Skulls of Various Aboriginal Nations of North and South America.* Philadelphia: Dobson, 1839.

———. *Crania Aegyptica: Or Observations on Egyptian Ethnography Derived from Anatomy, History, and the Monuments.* Philadelphia: Pennington, 1844.

Müller, Max. *Lectures on the Science of Language Delivered at the Royal Institution of Great Britain in April, May, and June, 1861.* London: Longmans Green, Longman & Roberts, 1861.

Munro, Robert. "President's Address, Anthropology Section." *Report of the British Association for the Advancement of Science,* 1893, pp. 885–95.

———. *Prehistoric Problems: Being a Selection of Essays on the Evolution of Man and Other Controverted Problems in Anthropology and Archaeology.* Edinburgh and London: William Blackwood, 1897.

Myres, John L. "Primitive Man in Geological Time" and "Neolithic and Bronze Age Cultures." In *The Cambridge Ancient History,* ed. J. B. Bury, S. A. Cook, and F. E. Adcock, vol. 1, chaps 1, 2. 2d ed. Cambridge: Cambridge University Press, 1924.

Nott, Josiah Clark, and Giddon, George R. *Indigenous Races of the Earth: Or New Chapters of Ethnological Inquiry.* London: Trubner, 1857.

Nuttall, George H. F. *Blood Immunity and Blood Relationships: A Demonstration of Certain Blood Relationships amongst Animals by Means of the Precipitin Test for Blood.* Cambridge: Cambridge University Press, 1904.

Oakley, Kenneth P., and Hoskins, C. Randall. "New Evidence on the Antiquity of Piltdown Man." *Nature* 165 (1950): 379–82.

Obermaier, Hugo. *Fossil Man in Spain,* introd. Henry Fairfield Osborn. New Haven: Yale University Press, for the Hispanic Society of America, 1924.

Odom, Herbert H. "Generalizations on Race in Nineteenth-Century Physical Anthropology." *Isis* 58 (1967):5–18.

Oldroyd, David, and Langham, Ian, eds. *The Wider Domain of Evolutionary Thought.* Dordrecht: D. Reidel, 1983.

Oppenoorth, W.F.F. "The Place of *Homo soloensis* among Fossil Men." In *Early Man,* ed. G. G. MacCurdy, 349–60. Philadelphia and New York: Lippincott, 1937.

Osborn, Henry Fairfield. "The Geological and Faunal Relations of Europe and America during the Tertiary Period, and the Theory of the Successive Invasions of an African Fauna." *Science* n.s. 11 (1900):561–74.

———. *Men of the Old Stone Age: Their Environment, Life, and Art.* London: George Bell, 1916. 3d ed., 1927.

———. *The Origin and Evolution of Life on the Theory of Action, Reaction, and Interaction of Energy.* New York: Charles Scribner's Sons, 1917.

———. "Hesperopithecus, the Anthropoid Primate from Western Nebraska." *Nature* 110 (1922):281–83.

———. *Evolution and Religion in Education: Polemics of the Fundamentalist Controversy of 1922 to 1926.* New York: Charles Scribner's Sons, 1926.

———. "Why Central Asia?" *Natural History* 26 (1926):263–69.

———. "Recent Discoveries Relating to the Origin and Antiquity of Man." *Proc. Am. Philosophical Soc.* 66 (1927):373–89.

———. *Man Rises to Parnassus: Critical Epochs in the Prehistory of Man.* 2d ed. Princeton: Princeton University Press, 1928.

———. "The Plateau Habitat of Pro–Dawn Man." *Science* 67 (1928):570–71.

———. *The Titanotheres of Ancient Wyoming, Dakota, and Nebraska.* 2 vols. Washington: U.S. Geological Survey Monograph no. 55, 1929.

———. "Is the Ape-man a Myth?" *Human Biology* 1 (1929):4–9.

———. "Aristogenesis, the Creative Principle in the Origin of Species." *Am. Naturalist* 68 (1934):193–235.

Owen, Richard. *On the Anatomy of the Vertebrates.* 3 vols. London: Longmans Green, 1866–68.

Owen, Rev. Richard, ed. *The Life of Richard Owen.* 2 vols. London: Murray, 1894.

Oxnard, Charles. *The Order of Man: A Biomathematical Anatomy of the Primates.* New Haven: Yale University Press, 1984.

Pearson, Karl. *The Grammar of Science.* 2d ed. London: A. & C. Black, 1900.

Peel, J. D. Y. *Herbert Spencer: The Evolution of a Sociologist.* London: Heinemann, 1971.

Penniman, T. K. *A Hundred Years of Anthropology.* 2d ed. London: Duckworth, 1952.

Perry, W. J. *The Children of the Sun: A Study of the Early History of Civilization.* London: Methuen, 1923.

Pilbeam, David. "The Descent of Hominoids and Hominids." *Scientific American* 250 (1984):60–69.

Pilgrim, Guy E. "New Siwalik Primates and Their Bearing on the Question of the Evolution of Man and the Anthropoidea." *Records of the Geological Society of India* 45 (1915):1–72.

Plate, Robert. *The Dinosaur Hunters: Othniel C. Marsh and Edward D. Cope.* New York: D. McKay, 1964.

Poliakov, Léon. *The Aryan Myth: A History of Racist and Nationalist Ideas in Europe,* trans. Edmund Howard. London: Chatto & Windus, for Sussex University Press, 1971.

Pouchet, Georges. *The Plurality of the Human Race,* trans. Hugh J. C. Beavan. London: Longmans Green, Longman & Roberts, for the Anthropological Society, 1864.

Prichard, James Cowles. *Researches into the Physical History of Mankind.* 5 vols. London: Sherwood, Gilbert & Piper, 1841–47.

Provine, William B. *The Origins of Theoretical Population Genetics.* Chicago: University of Chicago Press, 1971.

Pyecraft, William Plane et al. *Rhodesian Man and Associated Remains.* London: British Museum, 1928.

Ranke, J. "Ueber die individuellen Variationen im Schädelbau des Menschen." *Anthropologisches Korrespondenzblatt* 1897:139–46.

Ratzel, Friedrich. *The History of Mankind,* trans. A. J. Butler, introd. E. B. Tylor. 3 vols. London: Macmillan, 1896.

Read, Carveth. *The Origin of Man and His Superstitions.* Cambridge: Cambridge University Press, 1920.

Reader, John. *Missing Links: The Hunt for Earliest Man.* London: Collins, 1981.

Ripley, William Z. *The Races of Europe: A Sociological Study.* London: Kegan Paul, Trench, Trubner & Co., 1900.

Rivers, W.H.R. "President's Address, Anthropology Section." *Report of the British Association for the Advancement of Science,* 1911, pp. 490–99.

Romanes, George John. *Mental Evolution in Animals: With a Posthumous Essay on Instinct by Charles Darwin.* London: Kegan Paul, Trench, & Co., 1883.

———. *Mental Evolution in Man: Origin of Human Faculty.* London: Kegan Paul, Trench, Trubner & Co., 1888.

———. *Essays.* ed. C. Lloyd Morgan. New ed. London: Longmans Green, 1897.

Rudwick, M.J.S. "The Strategy of Lyell's *Principles of Geology.*" *Isis* 61 (1970):5–33.

———. "Uniformity and Progression: Reflections on the Structure of Geological Theory in the Age of Lyell." In *Perspectives in the History of Science and Technology,* ed. Duane H. D. Roller, 209–27. Norman: University of Oklahoma Press, 1971.

———. *The Meaning of Fossils: Episodes in the History of Palaeontology.* 2d ed. New York: Science History Publications, 1976.

Ruse, Michael. *The Darwinian Revolution: Science Red in Tooth and Claw.* Chicago: University of Chicago Press, 1979.

Sanford, William F., Jr. "Dana and Darwinism." *J. Hist. Ideas* 26 (1965):531–46.

Schaafhausen, D. "On the Crania of the Most Ancient Races of Man," trans George Busk. *Natural History Review* 8 (1861):155–72.

Schmidt, R. R. *The Dawn of the Human Mind,* trans. R.A.S. Macalister. London: Sidgwick & Jackson, 1936.

Schultz, Adolphe H. "Fetal Growth of Man and Other Primates." *Quart. Rev. Biology* 1 (1926):465–521.

Schwalbe, Gustav. "Studien über *Pithecanthropus erectus,* Dubois." *Zeitschrift für Morphologie und Anthropologie* 1 (1899):16–240.

———. *Die Vorgeschichte des Menschen.* Braunschweig: Friedrich Viewig, 1904.

———. *Studien zur Vorgeschichte des Menschen.* Stuttgart: E. Schweizerbartsche, 1906.

———. "Uber das Gehirnrelief der Schläfengegend des menschlichen Schädels." *Zeitschrift für Morphologie und Anthropologie* 10 (1907):1–93.

———. "The Descent of Man." In *Darwin and Modern Science,* ed. A. C. Seward, 112–36. Cambridge: Cambridge University Press, 1909.

———. "Kritische Besprechung von Boule's Werk: 'L'homme fossile de La Chapelle-aux-Saints.'" *Zeitschrift für Morphologie und Anthropologie* 16 (1914):527–610.

Schwartz, Joel S. "Darwin, Wallace, and the *Descent of Man.*" *J. Hist. Biology* 17 (1984):271–89.

Sergi, Giuseppe. *The Mediterranean Race: A Study of the Origin of European Peoples.* London: Walter Scott, 1901.

———. *Le origine umane: Ricerche paleontologiche.* Torino: Fratelli Bocca, 1913.

Shand, Alexander F. *The Foundations of Character: Being a Study of the Tendencies of the Emotions and Sentiments.* London: Macmillan, 1914.

Shapin, Steven. "History of Science and Its Sociological Reconstruction." *History of Science* 20 (1982):157–211.

Shellshear, Joseph L., and Smith, Grafton Elliot. "A Comprehensive Study of the Endocranial Cast of *Sinanthropus.*" *Phil. Trans. Roy. Soc. Lond.* 123 B (1934):469–87.

Shor, Elizabeth N. *The Fossil Feud between E. D. Cope and O. C. Marsh.* New York: Exposition Press, 1974.

Simpson, George Gaylord. "Mesozoic Mammalia. III. Preliminary Comparison of Jurassic Mammals except Multituberculates." *Am. J. Sci.* 5th ser. 10 (1926):558–69.

———. "Mesozoic Mammalia. IV. The Multituberculates." *Am. J. Sci.* 5th ser. 11 (1926):228–50.

———. *A Catalogue of the Mesozoic Mammalia in the Geological Department of the British Museum.* London: British Museum, 1928.

———. *Tempo and Mode in Evolution.* New York: Columbia University Press, 1944.

———. "Some Principles of Historical Biology Bearing on Human Origins." *Cold Spring Harbor Symposia on Quantitative Biology* 15 (1950): 55–66.

———. *Concession to the Improbable: An Unconventional Autobiography.* New Haven: Yale University Press, 1978.

Slobodin, Richard. *W.H.R. Rivers.* New York: Columbia University Press, 1978.

Smith, Fred H., and Spencer, Frank, eds. *The Origins of Modern Humans: A World Survey of the Fossil Evidence.* New York: Alan R. Liss, 1984.

Smith, Sir Grafton Elliot. "On the Relationship of Lemurs and Apes." *Nature* 76 (1907):7–8.

———. "Discussion on the Origin of Mammals." *Report of the British Association for the Advancement of Science,* 1911, pp. 424–28.

———. "President's Address, Anthropology Section." *Report of the British Association for the Advancement of Science,* 1912, pp. 575–98.

———. "The Piltdown Skull and Brain Case." *Nature* 92 (1913):267–68, 318–19.

———. "Anthropology." In *Evolution in the Light of Modern Knowledge: A Collected Work,* by James H. Jeans et al., 287–320. London: Blackie, 1923.

———. *The Evolution of Man: Essays.* London: Humphrey Milford and Oxford: Oxford University Press, 1924. 2d ed., 1927.

———. "The Fossil Anthropoid Ape from Taungs." *Nature* 115 (1925):235.

———. *The Search for Man's Ancestors.* London: Watts, 1931.

———. "The Discovery of Primitive Man in China." *Smithsonian Institution, Annual Report,* 1931, pp. 531–47.

———. "The Evolution of Man." In *Early man,* by Grafton Elliot Smith et al., 13–46. London: Ernest Benn, 1931.

———. *The Diffusion of Culture.* London: Watts, 1933.

———. *Human History.* New ed. London: Jonathan Cape, 1934.

Smith, Sir Grafton Elliot et al. *Early Man: His Origin, Development, and Culture.* London: Ernest Benn, 1931.

Sollas, William Johnson. "On the Cranial and Facial Characters of the Neanderthal Race." *Phil. Trans. Roy. Soc. Lond.* 199 (1908):281–339.

———. "Presidential Address." *Quart. J. Geol. Soc. Lond.* 65 (1909):i–cxxii.

———. "Presidential Address." *Quart. J. Geol. Soc. Lond.* 66 (1910):xlviii–lxxxviii.

———. *Ancient Hunters and Their Modern Representatives.* London: Macmillan, 1911. 3d ed., 1924.

———. "The Taungs Skull." *Nature* 115 (1925):908–9.

———. "On a Sagittal Section of the Skull of *Australopithecus africanus.*" *Quart. J. Geol. Soc. Lond.* 82 (1926):1–10.

———. "The Chancellade Skull." *J. Roy. Anthrop. Inst.* 56 (1927):89–122.

Spencer, Frank, ed. *A History of American Physical Anthropology.* New York: Academic Press, 1982.

Spencer, Herbert. *Principles of Biology.* 2 vols. London: Williams & Norgate, 1864.

———. *Principles of Psychology.* 2d ed. 2 vols. London: Williams & Norgate, 1870–72.

———. *The Principles of Sociology.* 3 vols. London: Williams & Norgate, 1876–96.

———. *Essays, Scientific, Political, and Speculative.* 3 vols. London: Williams & Norgate, 1883.

———. *An Autobiography.* 2 vols. London: Williams & Norgate, 1904.

Spurrell, Herbert G. F. *Modern Man and His Forerunners: A Short Study of the Human Species, Living and Extinct.* 2d ed. London: George Bell, 1918.

Stanton, William. *The Leopard's Spots: Scientific Attitudes toward Race in America, 1815–1859.* Chicago: Phoenix Books, 1960.

Stebbins, Robert E. "France." In *The Comparative Reception of Darwinism,* ed. Thomas F. Glick, 117–67. Austin: University of Texas Press, 1972.

Stepan, Nancy. *The Idea of Race in Science: Great Britain, 1800–1960.* London: Macmillan, 1982.

Stephens, Lester G. "Joseph Le Conte's Evolutionary Idealism: A Lamarckian View of Cultural History." *J. Hist. Ideas* 39 (1978):465–80.

———. *Joseph Le Conte: Gentle Prophet of Evolution.* Baton Rouge: Louisiana State University Press, 1982.

Stocking, George W., Jr. *Race, Culture, and Evolution: Essays in the History of Anthropology.* New York: Free Press, London: Collier-Macmillan, 1968.

———. "What's in a Name? The Origins of the Royal Anthropological Institute (1837–71)." *Man* 6 (1971):369–90.

Straus, William L., Jr. "The Riddle of Man's Ancestry." *Quart. Rev. Biology* 24 (1949):200–223.

———, and Cave, A.J.E. "Pathology and the Posture of Neanderthal Man." *Quart. Rev. Biology* 32 (1957):348–63.

Taylor, Griffith. *Environment and Race: A Study of the Evolution, Migration, Settle-*

ment, and Status of the Races of Man. London: Oxford University Press/Humphrey Milford, 1927.

Taylor, Isaac. *The Origin of the Aryans: An Account of the Prehistoric Ethnology and Civilization of Europe.* London: Walter Scott, 1889.

Teilhard de Chardin, Pierre. *The Phenomenon of Man.* London: Collins, 1959.

Thomson, J. Arthur. *What Is Man?* London: Methuen, 1923.

Thompson, M. W. *General Pitt-Rivers: Evolution and Archaeology in the Nineteenth Century.* Bradford on Avon: Moonraker Press, 1977.

Topinard, Paul. *Anthropology,* trans. Robert T. H. Bartley. London: Chapman & Hall, 1878.

———. *Eléments d'anthropologie générale.* Paris: Delahaye & Lecrosnier, 1885.

Turner, Frank Miller. *Between Science and Religion: The Reaction to Scientific Naturalism in Late Victorian England.* New Haven: Yale University Press, 1974.

Tylor, Edward B. *Researches into the Early History of Mankind and the Development of Civilization.* 2d ed. London: Murray, 1870.

———. "On the Survival of Palaeolithic Conditions in Tasmania and Australia." *J. Roy. Anthrop. Inst.* 28 (1898): 199.

———. "On Stone Implements from Tasmania." *J. Roy. Anthrop. Inst.* 30 (1900):257–59.

———. *Anthropology: An Introduction to the Study of Man and Civilization.* London: Macmillan, 1913.

Urry, James. "Englishmen, Celts, and Iberians: The Ethnographic Survey of the United Kingdom, 1892–1899." In *Functionalism Historicized,* ed. George W. Stocking, 83–105. Madison: University of Wisconsin Press, 1984.

Vacher de Lapouge, G. *L'Aryen: Son rôle social.* Paris: Albert Fontemoing, 1899.

Vallois, Henri-V. "Les preuves anatomiques de l'origine monophyletique de l'homme." *L'Anthropologie,* 39 (1929):77–101.

———. "Neanderthals and Praesapiens." *J. Roy. Anthrop. Inst.* 84 (1954): 111–30.

———. "La grotte de Fontéchevade: Anthropologie." *Archives de l'Institut de Paléontologie Humaine* 29 (1957):1–164.

Verneau, R. "La race de Néanderthal et la race de Grimaldi: Leur rôle dans l'humanité." *J. Roy. Anthrop. Inst.* 54 (1924):211–30.

Villeneuve, L. de, ed. *Les Grottes de Grimaldi.* 4 vols. Monaco: Imprimerie de Monaco, 1906–19.

Vogt, Carl. *Lectures on Man: His Place in Creation, and in the History of the Earth,* ed. James Hunt. London: Longmans Green, Longman & Roberts, for the Anthropological Society, 1864.

———. "The Primitive Period of the Human Species." *Anthropological Review* 5 (1867):204–21, 334–50.

Von Bonin, Gerhart. "Klaatsch's Theory of the Descent of Man." *Nature* 85 (1911):508–9.

Von Buttel-Reepen, H. *Man and His Forerunners,* trans. A. G. Thacker. London: Longmans Green, 1913.

Von Koenigswald, G.H.R. "A Review of the Stratigraphy of Java and Its Relations to Early Man." In *Early Man,* ed. G. G. MacCurdy, 23–32. Philadelphia and New York: Lippincott, 1937.

———, and Weidenreich, F. "Discovery of an Additional *Pithecanthropus* Skull." *Nature* 142 (1938):715.

———. "The Relationship between *Pithecanthropus* and *Sinanthropus.*" *Nature* 144 (1939):926–29.

Vorzimmer, Peter J. *Charles Darwin: The Years of Controversy. The "Origin of Species" and Its Critics, 1859–1882.* Philadelphia: Temple University Press, 1970.

Wallace, Alfred Russel. *Darwinism: An Exposition of the Theory of Natural Selection with Some of Its Applications.* London: Macmillan, 1889.

———. *Natural Selection and Tropical Nature.* New ed. London: Macmillan, 1895.

———. *My Life: A Record of Events and Opinions.* 2 vols. New York: Dodd, Mead & Co., 1905.

Ward, Lester Frank. "The Relation of Sociology to Anthropology." *Am. Anthropologist* 8 (1895): 241–56.

Washburn, S. L. "The Analysis of Primate Evolution with Paticular Reference to the Origin of Man." *Cold Spring Harbor Symposia on Quantitative Biology* 15 (1950):67–77.

Waterston, David. "The Piltdown Mandible." *Nature* 92 (1913):319.

Watson, D.M.S. *Palaeontology and the Evolution of Man.* Oxford: Clarendon Press, 1928.

Wegner, Richard N. "A New Theory of the Descent of Man." *Nature* 85 (1910):119–21.

Weidenreich, Franz. "Entwicklungs- und Wassentypen des *Homo primigenius.*" *Natur und Museum* 58 (1928):1–13, 51–62.

———. "Lamarckismus." *Natur und Museum* 62 (1932):298–300.

———. "Some Problems Dealing with Ancient Man." *Am. Anthropologist* 42 (1940):375–83.

———. "The 'Neanderthal Man' and the Ancestors of *Homo sapiens.*" *Am. Anthropologist* 45 (1943): 39–48.

———. "The Skull of *Sinanthropus pekinensis:* A Comparative Study of a Primitive Hominid Skull." *Palaeontologica Sinica* n.s. D, no. 10. Peking: Geological Survey of China, 1943.

———. "Giant Early Man from Java and South China." *Anthropological Papers of the American Museum of Natural History* 40, pt. 1 (1945).

———. *Apes, Giants, and Man.* Chicago: University of Chicago Press, 1946.

Weiner, J. S. *The Piltdown Forgery.* London: Oxford University Press, 1955.

Weinert, Hans. *Der Schädel des eiszeitlichen Menschen von le Moustier im neuer Zussammensetzung.* Berlin: Springer, 1925.

———. *"Pithecanthropus erectus." Zeitschrift für Anatomie und Entwickelungsgeschichte* 87 (1928):429–547.

———. *Ursprung der Menschheit: Uber den engeren Anschluss des Menschengeschlechts an die Menschenaffen.* Stuttgart: Ferdinand Enke Verlag, 1932.

Wells, Herbert George. *The Outline of History: Being a Plain History of Life and Mankind.* Revised ed. London: Cassell, 1925.

———. *The Short Stories of H. G. Wells.* London: Benn, 1927.

———; Huxley, Julian; and Wells, G. P. *The Science of Life.* London: Cassell, 1931.

Wendt, Herbert. *In Search of Adam: The Story of Man's Quest for Truth about His Earliest Ancestors,* trans. James Cleugh. Boston: Houghton Mifflin, 1956.

———. *From Ape to Adam: The Search for the Ancestry of Man.* London: Thames & Hudson, 1971.

Westermark, Edward. *The History of Human Marriage.* 5th ed. 3 vols. London: Macmillan, 1921.

Wiedersheim, Robert. *Der Bau des Menschen als Zeugniss für seine Vergangenheit.* 2d ed. Freiburg and Leipzig: J.C.B. Mohr, 1893.

———. *The Structure of Man; An Index to His Past History,* trans. H. and M. Bernard, introd. G. B. Howes. London: Macmillan, 1895.

Wilder, Harris Hawthorne. *The Pedigree of the Human Race.* New York: Henry Holt, 1926.

Willey, Arthur. *Convergence in Evolution.* London: Murray, 1911.

Willey, Basil. *Darwin and Butler: Two Versions of Evolution.*London: Chatto & Windus, 1940.

Williams-Ellis, Amabel. *Darwin's Moon: A Biography of Alfred Russel Wallace.* London and Glasgow: Blackie, 1966.

Wilson, Daniel. *Prehistoric Man: Researches into the Origin of Civilisation in the Old and New Worlds.* 2 vols. Cambridge and London: Macmillan, 1862.

Winslow, John, and Meyer, Alfred. "The Perpetrator at Piltdown." *Science 83* 4 (1983):32–43.

Woodward, Sir Arthur Smith. "President's Address, Geological Section." *Report of the British Association for the Advancement of Science,* 1909, pp. 462–71.

———. "Missing Links among Extinct Animals." *Report of the British Association for the Advancement of Science,* 1913, pp. 783–87.

———. "On the Lower Jaw of an Anthropoid Ape (Dryopithecus) from the Upper Miocene of Lérida (Spain)." *Quart. J. Geol. Soc. Lond.* 70 (1914):316–20.

———. "Fourth Note on the Piltdown Gravel, with Evidence of a Second Skull of *Eoanthropus dawsoni.*" *Quart. J. Geol. Soc. Lond.* 73 (1917):1–7.

———. *A Guide to the Fossil Remains of Man in the Department of Geology and*

Palaeontology in the British Museum (Natural History). 3rd ed. London: British Museum, 1922.
———. "The Fossil Anthropoid Ape from Taungs." *Nature* 115 (1925):235–36.
———. "Recent Progress in the Study of Early Man." *Report of the British Association for the Advancement of Science,* 1935:128–42.
———. *The Earliest Englishman.* London: Watts, 1948.
———. et al. "Discussion on the Zoological Position and Affinities of Tarsius." *Proc. Zool. Soc. Lond.* 1919:465–98.
Worsaae, J. J. A. *The Primeval Antiquities of Denmark,* trans. William J. Thomas. London: J. H. Parker, 1849.
Wright, G. Frederick. *The Origin and Antiquity of Man.* London: Murray, 1913.
Young, Robert M. "The Historiographical and Ideological Context of the Nineteenth-Century Debate on Man's Place in Nature." In *Changing Perspectives in the History of Science,* ed. M. Teich and R. M. Young, 344–438. London: Heinemann, 1973.
Zuckerman, Sir Solly. "Sinanthropus and Other Fossil Men." *Eugenics Review* 24 (1933):273–84.
———. "Human Genera and Species." *Nature* 145 (1940):510–11.
———. "Sir Grafton Elliot Smith, 1871–1937." In "The Concepts of Human Evolution," ed. S. Zuckerman. *Symposia of the Zoological Society of London* 33, (1973):3–21.
———. *From Apes to Warlords: The Autobiography (1904–1946).* London: Hamish Hamilton, 1978.
———. *The Social Life of Monkeys and Apes: Reissue of 1932 Edition Together with a Postscript.* London: Routledge & Kegan Paul, 1981.
———., ed. "The Concepts of Human Evolution." *Symposia of the Zoological Society of London* 33 (1973).

Index